ACS SYMPOSIUM SERIES 771

Pesticides and Wildlife

John J. Johnston, Editor
USDA National Wildlife Research Center

American Chemical Society, Washington, DC

Library of Congress Cataloging-in-Publication Data

Pesticides and wildlife / John J. Johnston, editor
p. cm.—(ACS symposium series ; 771)
Includes bibliographical references.
ISBN 0-8412-3709-3
1. Pesticides and wildlife—Congresses. I. Johnston, John J., 1958- II. American Chemical Society. Meeting (218th : 1999 : New Orleans, La.) III. Series.

QH545.P4 P48135 2000
599.17—dc21 00-056578

The paper used in this publication meets the minimum requirements of American National Standard for Information Sciences—Permanence of Paper for Printed Library Materials, ANSI Z39.48-1984.

Distributed by Oxford University Press

PRINTED IN THE UNITED STATES OF AMERICA

Foreword

THE ACS SYMPOSIUM SERIES was first published in 1974 to provide a mechanism for publishing symposia quickly in book form. The purpose of the series is to publish timely, comprehensive books developed from ACS sponsored symposia based on current scientific research. Occasionally, books are developed from symposia sponsored by other organizations when the topic is of keen interest to the chemistry audience.

Before agreeing to publish a book, the proposed table of contents is reviewed for appropriate and comprehensive coverage and for interest to the audience. Some papers may be excluded in order to better focus the book; others may be added to provide comprehensiveness. When appropriate, overview or introductory chapters are added. Drafts of chapters are peer-reviewed prior to final acceptance or rejection, and manuscripts are prepared in camera-ready format.

As a rule, only original research papers and original review papers are included in the volumes. Verbatim reproductions of previously published papers are not accepted.

ACS BOOKS DEPARTMENT

Contents

Assessment of Risk to Non-Target Wildlife

Analysis of Pesticide Residues in Wildllife

Reducing Wildlife Exposure to Pesticides

Preface

Pesticides provide humankind with a plethora of benefits including increased agricultural productivity, decreased sickness and mortality from pest-associated diseases, increased longevity of physical structures, and improved horticultural opportunities. However, the use of pesticides may be accompanied by undesirable effects to a variety of non-target species including wildlife. Weighing the potential benefits versus risks of pesticide use is less than straightforward. It is not surprising that entities with differing agendas can arrive at differing conclusions when comparing these benefits and risks. As a chemist working with pesticides for the U.S. Department of Agriculture's National Wildlife Research Center, I interact with parties harboring a wide variety of viewpoints regarding the use of pesticides. In dealing with such a diverse client base as represented by society, it has been my experience that the best approach is to rely on science to elucidate and support objective conclusions.

This symposium series volume brings together a wide variety of scientists working in the area of pesticides and wildlife. It is my intent that the chapters presented here offer a variety of scientifically valid approaches for evaluating the impact of pesticides on both target and non-target species. It is further intended that the approaches offered here will be used and improved upon by other scientists working in the area of pesticides and wildlife. Finally, I sincerely hope that these offerings will permit a variety of entities to begin a dialog, to find some common ground, and to arrive at procedures for the safe use of pesticides. In this way, it is hoped that this publication will help us to continue to reap the benefits of pesticides while being good stewards of the environment.

JOHN J. JOHNSTON
APHIS/WS/National Wildlife Research Center
U.S. Department of Agriculture
4101 La Porte Avenue
Fort Collins, CO 80521–2154

Chapter 1

Introduction to Pesticides and Wildlife

John J. Johnston

APHIS/WS/National Wildlife Research Center, U.S. Department of Agriculture, 4101 La Porte Avenue, Fort Collins, CO 80521–2154

A pesticide is a substance intended for destroying, repelling or mitigating any animal, microorganism or plant pest. While pesticides are commonly chemical agents, biological or physical agents can also be pesticides. Many people equate the term pesticide with insecticide. However, there are a plethora of pesticides for which insects are not the target pest. Examples include herbicides, fungicides, rodenticides, acaricides, larvacides, etc. Additionally, pesticides are not limited to toxicants, but also include repellents, attractants and growth regulators (*1*).

Pesticides are nearly ubiquitous in today's world. In agriculture, insecticides, nematicides and herbicides are applied to plants and soil to improve the yield and/or the appearance of the harvested commodities (*2*). Such uses are common knowledge among the populace. A variety of other uses of pesticides are less obvious to the average citizen, but are widespread nonetheless. For example, insecticides and rodenticides are released into the environment in an effort to decrease populations of pathogen carrying animals such as insects and rodents. Pesticides such as mildicides and insecticides are routinely incorporated into building materials and/or applied to structures to prevent pest infestation. Pesticides are routinely applied to boat bottoms to prevent the attachment and growth of a mollusks. The list goes on and on.

The ubiquitous nature of pesticides suggests that exposure to wildlife is inevitable. Primary exposure results from wildlife directly consuming the pesticide formulation. Primary exposure scenarios include birds feeding on granular insecticide formulations, squirrels consuming rodenticide baits intended for rats, or wildlife being exposed to pesticide drift resulting from the spray application of an insecticide in an agricultural situation. Secondary exposure results from the consumption of animals containing pesticide residues. Secondary exposure scenarios include scavenging mammals such as ferrets, feeding on rodent carcasses containing rodenticide residues. Tertiary exposure results from another animal such as a raptor feeding on the secondarily exposed ferret in the previous example. Unfortunately, the possible routes of non-target exposure of wildlife to pesticides are varied and many.

Pesticides are bioactive compounds. As such, pesticides exert biological effects on living organisms (*3*). Fortunately, most pesticides are somewhat selective. The intended effect of the pesticide is usually more pronounced towards the target species than non-targets. Selectivity can also be enhanced by adopting application practices to minimize non-target exposure. However, even in the best of situations, selectivity is less than absolute. Thus, when non-target wildlife are exposed to pesticides, unintended biological effects often follow.

Historical Perspective

Insecticides

Even after Rachael Carson's *Silent Spring* (*4*) focused international attention on the deleterious effects of pesticides on wildlife, the scientific community was shocked to learn that insecticide residues had been discovered in penguins and seals in the Antarctic. The fact that these animals were located thousands of miles from the nearest pesticide application site illustrated to many that pesticide transport throughout the environment was widespread. In response to such findings, during the later half of the 1960's, the Canadian government collected and analyzed a variety of wildlife including mammals, fish, migratory and non-migratory birds, and invertebrates for pesticide residues. The results indicated that nearly every sample of Canadian wildlife contained detectable levels of pesticides. Furthermore, many of these pesticides were linked to declines in bird and fish populations, illustrating that unintended effects of pesticides on wildlife may be quite significant (*5*). As a result of these and additional studies, many of the popular pesticides of the era, such as DDT, were banned from use in Canada and the U.S.(*6*). Nearly 30 years after being banned, studies presented in this book such as "Bioaccumulation of Pesticides in Bats"(Chpt. 6) indicate that organochlorine pesticides are still present in U.S. wildlife. Additionally, "Tsetse and Locust Control in Africa"(Chpt. 7) points out that organochlorine pesticides are still being used in third world countries.

In the 1970's, the use of organophosphate and carbamate insecticides increased significantly. These insecticides were thought to be environmentally safer because they are less stable and less prone to widespread transport and bioaccumulation in the food chain. However, as pointed out in the chapters "Carbofuran Spill Site on a Remote Island of the Hawaiian National Wildlife Refuge"(Chpt. 3), "Forensic Investigations of Impacts from Three Groups of Pesticides on Raptors"(Chpt. 4) and "Chemical and Biochemical Evaluation of Swainson's Hawk Mortalities in Argentina" (Chpt. 21), organophosphate and carbamate insecticide exposure to non-target wildlife is not uncommon and can result in acute toxicity to significant quantities of wildlife.

In the 1980's, the next class of major insecticides, pyrethroids, was introduced. Even though pyrethroids have become widely popular, reports of non-target poisonings have been minimal. It appears that for pyrethroids, non-target wildlife poisonings are less of a concern than with previously the cited classes of insecticides.

However, pyrethrum, the biologically active chrysanthemum extract which pyrethroids are based upon, is toxic to reptiles as illustrated in the chapter on the development of toxicants for brown tree snake control on Guam (Chpt. 16).

Herbicides

Another major class of pesticides addressed in this book is herbicides. Potential wildlife exposure is significant because herbicides represent the most rapidly growing class of pesticides in the last two decades. Further contributing to the potential exposure of wildlife to herbicides is that these pesticides have been developed for application at multiple stages of agricultural production. For example, preplanting herbicides are applied to the soil prior to seeding, preemergent herbicides are applied prior to the appearance of the pest plants and postemergent herbicides are applied after the crop or weeds are visible. Due to the high level of research and development in this area, herbicides are composed of a wide variety of chemical classes. In general, herbicides tend to be more water soluble than most other types of pesticides. Potential wildlife exposure via herbicide contaminated surface water is a concern. Such concerns are addressed in the chapter "History and Risk Assessment of Triazine Herbicides in the Lower Mississippi River"(Chpt. 17).

Rodenticides

Rodenticides are used to reduce pest rodent populations in many countries. Though rodenticides are less widely used than insecticides or herbicides, non-target wildlife exposure to rodenticides is a significant concern due to similarities in the physiology and behavior of the target species and non-target wildlife. With rodenticides, development of safe application practices is extremely important to reduce non-target exposure. Several chapters in this text (Chpts. 11-14)address the development of safe rodenticides and the assessment of rodenticide related risks to non-target wildlife.

Avicides

Avicides are used to reduce pest bird populations. That such pesticides exist is a surprise to many people. Similar to rodenticides, a challenge for the development and safe use of avicides is to minimize potential exposure and toxic effects to non-target birds and other animals. In addition to the development of selective chemicals, as is the case with rodenticides, careful application is required to reduce exposure of non-target species. Such approaches are presented in the chapter on bird preferences for multiple diets (Chpt. 3). Newly developed risk assessment approaches are applied to the use of an avicide in the chapters "Primary and Secondary Hazards associated with the use of Avicides"(Chpt. 15) and "Use of a Poisson Distribution to Estimate Blackbird Mortality from Pesticide Application"(Chpt. 8).

Analytical Chemistry

Chemical analyses are generally required to unequivocally determine the magnitude of pesticide residues in the environment. Without such techniques, determining pesticide exposure to wildlife is speculative at best. While approaches and analytical techniques to quantify pesticide residues in wildlife are contained in many chapters, these topics are covered in detail in "Ion Trap GC-MS Analysis of Tissue Samples for Chlorinated Pesticides", "Recent Advances in Analytical Techniques to Investigate Pesticide Poisoning in Wildlife", "Environmental Specimen Banking – a Useful Tool for Retrospective and Prospective Environmental Monitoring" and "Chorioallantoic Membranes for Non-lethal Assessments of Pesticide Exposure and Effects"(Chpts. 18-20). Similarly, new approaches for assessing the affects of pesticides on wildlife populations are contained in many chapters, but is the main topic of "Large Scale Monitoring of Non-target Pesticide Effects on Farmland Arthropods in England"(Chpt 5).

Protecting Non-target Wildlife

Finally, this book concludes with what may be the most important chapters, ways to protect wildlife from unintentional pesticide exposure. "Comparisons of Primary and Secondary Repellents for Aversive Conditioning of European Starlings"(Chpt. 22) describes studies to develop bird repellents. These repellents could potentially be used in lieu of avicides to protect agricultural commodities or could be added to pesticide formulations to minimize consumption by non-target birds. Regulatory aspects of protecting wildlife from pesticides are presented in "Hurdles in Compiling Pesticide Registration Incidents" and "Regulatory Aspects of Protecting Endangered Species with County Pesticide Bulletins"(Chpts. 24,25).

References

1. *Toxicology: The Basic Science of Poisons*; Amdur, M.O.; Doull, J; Klaassen, C.D., Eds.; McGraw Hill; New York, 1993.
2. *Pesticide Effects on Terrestrial Wildlife*; Somerville, L.; Walker, C.H., Eds; Taylor & Francis Publ.; London, 1990.
3. *Merriam Webster's Collegiate Dictionary*; Mish, F.C.; Morse, J.M., Eds; Merriam-Webster Publ.; Springfield, MA, 1993.

4. Carson, R. *Silent Spring*; Houghton-Mifflin Publ., Boston, MA, 1962.
5. *Pesticides & Wildlife.* Canadian Wildlife Service, Ministries of Fisheries and Forestry; Ottawa, Canada, 1971.
6. Johnston, J.J. Toxicity of the Organophosphate Insecticide Fenitrothion and Its Metabolism by Blue Crabs, *Callinectes sapidus.* Ph.D. dissertation. Univeristy of Florida; Gainesville, FL, 1986.

Routes of Pesticide Exposure to Wildlife

Chapter 2

Bioaccumulation of Pesticides in Bats from Missouri

Angela Schmidt[1], Virgil Brack, Jr.[1], Russ Rommé[1], Karen Tyrell[2], and Alan Gehrt[3]

[1]BHE Environmental, Inc., 11733 Chesterdale Road, Cincinnati, OH 45246
[2]BHE Environmental, Inc., 7039 Maynardville Highway, Suite 7, Knoxville, TN 37918
[3]Kansas City District, U.S. Army Corps of Engineers, Kansas City, MO 64106

Bats may be exposed to pesticides through direct exposure (e.g., ingestion of contaminated food or water, inhalation of aerosols, dermal absorption) and through indirect exposure (e.g., ingestion of contaminated prey by adults and contaminated milk by young). This study assesses the potential for bioaccumulation of pesticides by federally endangered Indiana bats (*Myotis sodalis*) and gray bats (*M. grisescens*). Both species occur on Fort Leonard Wood, Missouri where the pesticides chlorpyrifos and malathion have been used since the mid-1980's. This study began in 1997 and will continue until 2003. Surrogate species for the endangered bats were captured during spring and fall from Fort Leonard Wood and nearby Mark Twain National Forest. Insects were collected to determine pesticide concentration of the bats' food. Bat and insect samples also were screened for organochlorine and organophosphorous pesticides. Surrogate bats and insects collected in 1997 and 1998 had detectable amounts of DDE, heptachlor epoxide, and dieldrin.

Introduction

Two federally endangered species, the Indiana bat (*Myotis sodalis*) and gray bat (*M. grisescens*) occur on Fort Leonard Wood (FLW), a 63,000 acre U.S. Army facility in Missouri. As required by the Endangered Species Act, the National Environmental Policy Act (NEPA), and applicable Army regulations, FLW assesses effects of actions on the Installation that have the potential to affect endangered species. This study is being conducted to evaluate the effects of pesticides used at FLW on Indiana bats and gray bats.

After reviewing results of a Biological Assessment completed to evaluate effects of pesticide use and other activities at FLW, the U.S. Fish and Wildlife Service (FWS) issued a Biological Opinion (BO) and Take Statement containing Reasonable and Prudent Measures (RPMs), designed to monitor, reduce, or eliminate harmful effects to endangered bats from certain activities and training. One RPM required FLW to collect and analyze surrogate bats that reside on the installation and have a similar potential for exposure to pesticides as Indiana bats and gray bats.

FLW is conducting other biomonitoring studies to assess secondary effects from pesticides and other training materials to Indiana bat and gray bat insect prey populations and habitat quality for the bats and prey. These studies are not addressed here.

The Installation uses the pesticides chlorpyrifos (O,O-diethyl O-3,5,6 trichloro-2-pyridyl phosphorothioate or Dursban) and malathion (S-{1,2-bis (ethoxy carbonyl) ethyl} O,O-dimethyl phosphorodithioate). Chlorpyrifos is used to control webworms, ticks, and chiggers; malathion is used to control mosquitoes. Both chlorpyrifos and malathion are organophosphorous pesticides and latent cholinesterase inhibitors. Fort Leonard Wood has minimized use of these pesticides, however, they have been applied in large areas and in locations at or near where Indiana bats and gray bats occur. Chlorpyrifos is generally applied by a handsprayer and malathion is applied as a fog from a pressurized aerosol generator.

The Indiana bat hibernates in caves (hibernacula) on FLW that provide suitable conditions for efficient hibernation. Indiana bats occur in significant numbers on FLW between August and April. However a few Indiana bats have been captured on FLW during the summer. Prior to hibernation, swarming (autumn active foraging period) begins in August and lasts several weeks. They roost in trees during the spring, summer, and fall. Typically, maternity roosts of up to 100 adult females are formed under slabs of exfoliating bark; no such roosts have been found on FLW. Indiana bats may form nursery roosts a few miles from hibernacula, or travel hundreds of miles to summer habitat. In autumn and spring, these bats may frequent hibernacula but roost in trees and forage outside the caves.

The Indiana bat appears to be a selective opportunist, foraging around tree crowns in both riparian and non-riparian habitat (*1, 2, 3*). Their diet varies by location, often containing insects with a terrestrial-based life cycle as well as insects that are aquatic-based (*3, 4, 5*). Terrestrial-based insect prey include moths, leafhoppers, and beetles. Aquatic-based prey include mayflies, stone flies, and caddis flies (*1, 2, 6, 7*), and comprise a large portion of the bats' diet.

Gray bats hibernate in caves in winter and form maternity roosts in caves during summer. No gray bats are known to hibernate on FLW. There are several gray bat maternity roosts on FLW. This species frequently forages over streams and other water bodies, and a large portion of the diet consists of insects with an aquatic larval stage, such as mayflies, stone flies, and caddis flies (*1, 2, 6, 7*).

Purpose

This study is being conducted to determine if Indiana bats or gray bats bioaccumulate chlorpyrifos or malathion used on FLW. The primary objective of this multi-year study is to assess whole-body burdens of the pesticides in species of bats (surrogates) with diets and habitat preferences similar to those of Indiana bats and gray bats. Additionally, insect samples are being evaluated to characterize the presence of pesticides in prey.

Bioaccumulation of pesticides in bats is affected by how and when the exposure to the pesticide occurs. Pesticide contamination/exposure has been suspected as a contributing factor in population declines of Indiana bats and gray bats, but only limited data exist upon which to base this premise. An Exposure Pathway Analysis was completed to determine when Indiana bats and gray bats occur on FLW, what activities (e.g., forage, hibernate, swarm) they perform while on the installation, and other relevant life history information (e.g., body size). The Exposure Pathway Analysis indicated both species may inhale, ingest, or dermally absorb pesticides.

To assess the bioaccumulation of chlorpyrifos and malathion in bat tissue and insects, sample locations were selected on FLW (exposure sites) and on Mark Twain National Forest (reference sites located approximately 75 miles from FLW). A comparison of pesticide residues in bats and insects at reference sites to those in bats and insects inhabiting FLW will help establish the source of pesticides detected in samples.

In addition to concerns for chlorpyrifos and malathion, FLW is assessing bioaccumulation of other pesticides in bats and insects. Surrogate bats and insect samples were screened for organochlorine and other organophosphorous pesticides used in Missouri during the past 30 years.

Methods

Surrogate Species Selection

Fifteen species of bats (Order Chiroptera) are known from Missouri (*8*). Four species having similar habitat and diet preferences to Indiana bats and gray bats were selected as surrogate species. The most appropriate surrogate species that occur on FLW are the eastern pipistrelle bat (*Pipistrellus subflavus*), red bat (*Lasiurus borealis*), and little brown bat (*Myotis lucifugus)*. Eastern pipistrelles and red bats, as well as big brown bats (*Eptesicus fuscus*), and little brown bats are insectivorous, and as established from extensive mist net surveys on FLW, occur in the same areas as Indiana bats and gray bats. Little brown bats and big brown bats are relatively uncommon on FLW and subsequently only a few were included in this study. The eastern pipistrelle and red bats are the most common species collected on the installation.

The eastern pipistrelle and red bat have similar lifespans (7 to 10 years), similar body weights (5 to 12 grams), and similar low fecundity rates (1 to 2 offspring per year) as Indiana bats and gray bats.

The eastern pipistrelle hibernates in caves during winter and forms maternity roosts in trees during summer. The species is frequently captured in riparian habitat. The eastern pipistrelle often forages over or near streams (*2*), and aquatic-based insects may be common in the diet (*6*).

The red bat roosts in woodlands during summer. Although the species may be migratory, the wintering location of a majority of the population is unknown. Red bats forage over trees and in open areas (*1,2*) and fly in corridors along or over streams. The diet consists largely of terrestrial-based insects (*6, 9*).

Field Collection Techniques

Eleven exposure sites were selected on FLW. Two of the sites are located in areas where malathion and chlorpyrifos are used regularly (during spring, summer, and fall); Cantonment Area and golf course. Both malathion and chlorpyrifos have been applied on at least 20 acres in these two areas, four times a year. The other nine exposure areas selected for this study have variable pesticide use. Pesticides are applied only during military operations. All exposure sites are within 20 miles (approximately) of each other and occur throughout the entire installation. FLW applies the pesticides on an as-needed basis and follows the installation's Pest Management Plan. When possible, exposure sites were sampled immediately after application of malathion or chlorpyrifos.

Surrogate bats were collected with mist nets between late April and early June, and between August and September in 1997 and 1998. The mist nets used in this study were made of monofilament nylon with a mesh size of approximately 3.8 cm, and varied in length and height to cover the bats' flight corridor at each site. The nets were placed across roads or streams likely to be used by bats as travel corridors or for foraging. Specific mist net sites were selected to maximize canopy closure above the nets whenever possible. One or two nets were deployed at 11 exposure and 3 reference sites for at least one calendar night in 1997 and 1998. Biologists raised nets at dusk and monitored them approximately every 20 minutes until sufficient surrogate bats were captured for analysis, or until 0400 hours. A minimum of two surrogate bats were collected from each exposure site and at least five surrogate bats from each reference site.

The species, sex, maturity, and reproductive condition of captured bats were identified. Body weight and right forearm length of each captured bat were measured. The height at which the bat was captured and time of capture were recorded. Indiana bats and gray bats were banded with a numbered, colored, celluloid, split-ring forearm band, and released.

Insects were collected during fall using black (ultraviolet) and white (fluorescent) light traps. Insects were sampled in fall to ensure adequate representation of insects available for consumption by Indiana bats and gray bats (all insects were adults). Additionally, based on surrogate bats using the installation and their feeding habits, it is believed their diets may differ slightly from that of gray bats. In order to adequately represent gray bats' dietary exposure to pesticides, we collected insects during the time when gray bats would have increased energy demands. Traps consisted of a light source surrounded by Plexiglas® dividers over a funnel with a mesh bag to retain insects. Traps were suspended from trees at heights of at least 2-m above ground to capture flying insects. One black light trap and one white light trap were deployed in separate trees at dusk (approximately between 1900-2130 hours) at exposure and reference sites. Insects captured in black light traps and white light traps were composited into a single sample per site. Insect samples were weighed and information including date, sample location, collection time, sample number, and sample weight was recorded.

Laboratory Techniques

A 5-g to 10-g sample of composited insects or an entire bat was weighed into a Teflon extraction jar and spiked with the appropriate surrogate internal standards (these contain organochlorine and phosphorous pesticides). Samples were combined with methylene chloride (100-mL of 75%/25%) and approximately 50 g of sodium sulfate and macerated with a Tekmar Tissumizer equipped with a stainless steel probe and centrifuged at 2000 RPM. An additional 5-g aliquot of tissue was removed for percent moisture determination. The extract from the Teflon extraction jar was decanted into an Erlenmeyer flask. After each maceration (total of two 50 mL solvent additions) the centrifuged solvent extracts were combined in the flask. An additional extraction (the third) was performed using shaking techniques, the sample centrifuged a third time, and the extracts combined with the others. A 10-mL aliquot of the combined extracts was removed from the Erlenmeyer flask and air-dried for lipid weight determination (methylene chloride extractables). The remainder of the extract was dried over sodium sulfate, processed through a gravity fed alumina cleanup column (20 g, 2% deactivated), and concentrated to 900 μL in a Kuderna-Danish apparatus under nitrogen. The concentrated extract was further cleaned using size-exclusion High Performance Liquid Chromatography (HPLC). This procedure removed common contaminants which could interfere with instrumental analysis, including lipid and elemental sulfur. The post-HPLC extract was concentrated to approximately 500 μL under nitrogen and the recovery internal standards added to quantify extraction efficiency. The extract was solvent-exchanged with hexane for chlorinated hydrocarbon analysis by GC/ECD (gas chromatography/ electron capture detector).

Sample extracts were analyzed for pesticides by using a GC equipped with electron capture detectors (ECD) following procedures defined in and modified from EPA's SW846 Method #8081. A five-level calibration was analyzed before any sample extracts, with the lowest level just above the detection limit, and the range of the calibration encompassing the expected concentration range of the samples. The instruments were equipped with two ECD detectors and two 60-m capillary columns of different polarities (DB-5 and DB-1701). The 60-m DB-5 column, electronic pressure controlled inlet, hydrogen carrier gas, and optimized long temperature program, provide an extra measure of accuracy above the standard 30-m column (i.e., reduction in coelution, more stable baseline) to provide high quality GC/ECD data in a single data set. A portion of the samples was also injected into a Hewlett Packard 5890 gas chromatograph second with a nitrogen phosphorous detector (NPD). The NPD detector was used to detect the presence of nitrogen or phosphorous containing compounds (e.g., organophosphorous pesticides). Confirmation analysis was performed using the dual column GC and ECD.

Results

Exposure Pathway Analyses

An Exposure Pathway Analysis was completed for Indiana bats and gray bats on FLW to determine potential direct and indirect exposure pathways and routes (Figure 1). This study addresses only primary (direct) exposure pathways. Review of the life histories of Indiana bats and gray bats indicated individuals of both species may be exposed to pesticides through inhalation, ingestion, and dermal absorption on FLW (Figure 2).

Sample Locations and Bat Capture Results

Four or more bats were collected from each of 11 exposure sites during 1997 and 1998. Reference sites were selected in Mark Twain National Forest where we believe chlorpyrifos, malathion, or other pesticides have not been used. Eighteen bats were collected during both 1997 and 1998 from each of three reference sites.

One hundred-twenty five bats were collected in spring and fall of 1997; 38 were analyzed for pesticides. Some bats were released due to their status (e.g., state-listed) and many of the bats collected were analyzed to assess the presence of polycyclic aromatic hydrocarbons (PAHs), metals, PAH metabolites, and Cytochrome P450 activity. Eighty percent of those analyzed were eastern pipistrelles (n = 15) or red bats (n = 20). Other species sampled and analyzed for pesticides included big brown bat (n = 2), and little brown bat (n = 1). Approximately half of the bats collected for laboratory analysis were female.

Bat capture results in 1998 were similar to those of 1997. One hundred-twenty nine bats were collected and 36 were analyzed for pesticides. Eighty percent of those

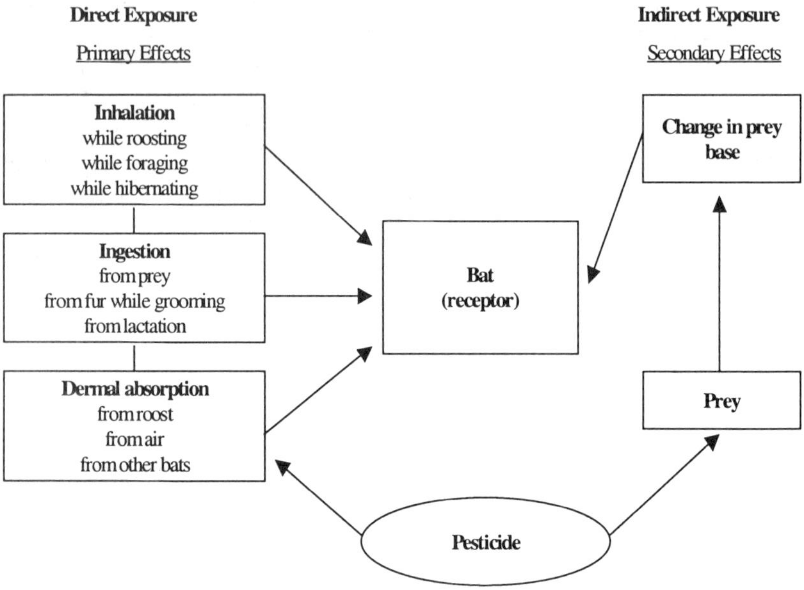

Figure 1. Pesticide Exposure Pathways for Indiana bats and gray bats on Fort Leonard Wood.

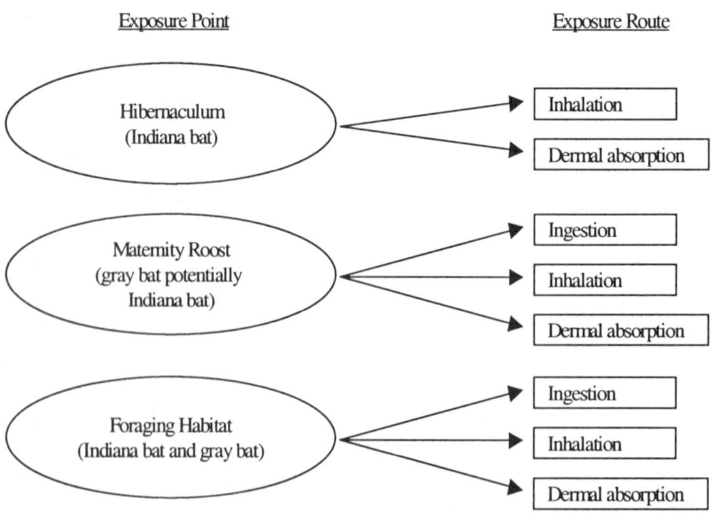

Figure 2. Points and routes of pesticide exposure for Indiana bats and gray bats on Fort Leonard Wood.

analyzed were eastern pipistrelles (n = 11) or red bats (n = 24). The only other species collected and analyzed for pesticides was big brown bat (n = 1).

Insect samples were collected from each of 11 exposure sites and each of 3 reference sites in both 1997 and 1998. Each of the 28 samples were analyzed for pesticides.

Pesticide Analysis and Screening

Neither chlorpyrifos or malathion were detected (method detection limit of 0.001µg/g) in any bat or insect sample collected on FLW or at reference sites during 1997 or 1998.

Every bat sample (whole body sample) (38 bats from 1997 and 36 from 1998) contained detectable concentrations of DDE and heptachlor epoxide. Dieldrin was detected in 22 of the 36 bats tested in 1998. Because the sample size is small, the usefulness of statistical analysis is limited. An preliminary review of pesticide residue content in bats yielded no identifiable or correlative relationships with sex, species, body size, or body weight (e.g., larger and heavier bats did not have the greatest concentration of pesticide in their tissues). A thorough statistical analysis of the data will be conducted at the end of the study (year 2003).

The mean concentration of DDE in bats collected in 1997 and 1998 was 0.652 µg/g (sd ± 1.314) and ranged from < 0.001 to 8.9 µg/g (Table 1). The mean concentration of heptachlor epoxide in bats collected in 1997 and 1998 was 0.020 µg/g (sd ± 0.063) and values ranged from < 0.001 to 4.8 µg/g (Table 1).

Dieldrin was only detected in 22 bat samples in 1998. The mean concentration was 0.060 µg/g (sd ± 0.102), and concentrations ranged from < 0.001 to 0.38 µg/g (Table 1).

Fourteen of the 28 insect samples contained low levels of DDE, with the mean ranging from < 0.001 to 0.005µg/g; 8 insect samples had detectable concentrations of heptachlor epoxide, with the mean ranging from < 0.001 to 0.005 µg/g.

Table 1. Mean concentration, standard deviation, and range of heptachlor epoxide and DDE in bat samples in spring and fall of 1997 and 1998, and dieldrin in spring and fall of 1998 at 11 exposure sites on Fort Leonard Wood and three reference sites in Missouri.

Pesticide	*Sample Size*	*Mean Concentration**	*SD*	*Range **
Heptachlor epoxide	74	0.020	0.063	<0.001 to 4.8
DDE	74	0.652	1.314	<0.001 to 8.9
Dieldrin	22	0.060	0.102	<0.001 to 0.38

*Units are *(µg/g)

Discussion

Indiana Bat Exposure

Indiana bats are most likely to be exposed to pesticides while foraging (exposure via ingestion) in autumn and spring, during their time on FLW. Based upon surveys completed on the Installation, the number of swarming bats around Indiana bat hibernaclum is greatest from mid-August through early October. Following winter hibernation on FLW, numbers of swarming bats increase substantially in mid-April, with peak swarming activity in late April and May. In the spring, Indiana bats may migrate up to several hundred miles to summer habitat, or may remain near their hibernaculum.

Indiana bats are more susceptible to low doses of pesticides (or any toxicant) during spring staging and fall swarming, when energy demands are high. The metabolic processes of these insectivores are rapid, and bats must consume large quantities (i.e., over 20% of their body weight in two hours of foraging) of food to survive. In addition, fat stores must be accumulated in spring prior to migration and reproduction, and in autumn preceding hibernation. Spring and autumn are spent on FLW. Light-tagging studies of many species of bats have shown that individuals sometimes concentrate foraging in small patches of habitats (less than 20 acres) where insect prey are abundant. If foraging is concentrated in an area where there was a recent pesticide application, the bat may receive an increased dose of pesticide, regardless of energy demands. Should this occur during periods of high energy demands, the dose to which the bat is subjected would be even greater. If a pesticide is stored in fatty tissue, a bat may be exposed chronically (over several months) during hibernation and migration when fat deposits are depleted.

In addition to ingesting pesticides, Indiana bats may inhale or absorb pesticides through their dermis. The Indiana bat may inhale pesticides while foraging (bats often begin foraging at dusk, near the time malathion is used on FLW) or they could get the pesticide on their fur and ingest it while grooming or absorb it through the skin.

Gray Bat Exposure

Gray bat exposure to pesticides on FLW differs from that of Indiana bats because they use FLW at different times of their annual cycle. Gray bats arrive on FLW in April and most leave by late October. They consume insect prey throughout the summer and fall until they depart for fall migration.

Gray bats utilize several caves in and around FLW as maternity roosts where females give birth. During gestation, parturition, and lactation, female gray bats are more likely to be affected (i.e., depressed immune system) if exposed to pesticides or other toxicants. For many mammalian species, pregnant females are more likely to get ill than non-pregnant ones. The bat expends energy to produce a single offspring that grows to adult size in 4 to 6 weeks. While the female replenishes energy supplies (by consuming copious quantities of insects), she has additional energy requirements to produce milk.

Gray bat pups may nurse for up to 60 days, however, they begin supplemental foraging upon becoming volant (when approximately 30 days old). Volant pups could get increased doses of a pesticide if they fed in an area where a pesticide had recently been applied and received milk containing pesticide from the mother. As with Indiana bats, gray bats deplete their fat supply at least twice (e.g., during preparation for spring migration and fall migration) while they are on FLW. They must consume large quantities of insects to meet energy requirements and build fat supplies for future energy needs. Because gray bats are insectivores and have a fluctuating energy demand, they may be more susceptible to toxicants/disease when energy demands are great. Anything that depresses their immune system would increase their vulnerability to the toxic effects (increase their sensitivity) of pesticides or to disease.

Organophosphorous Pesticides – Chlorpyrifos and Malathion

As expected with organophosphorous pesticides, chlorpyrifos and malathion were not detected in bat tissue or insect samples collected on FLW and at reference sites. Organophosphorous pesticides are generally hydrolyzed (hydrophilic) and are not known to bioaccumulate in mammals. These pesticides' mode of action is the inhibition of cholinesterase. They do not bioconcentrate in food chains (*10*). Chlorpyrifos and malathion must undergo an oxidative desulfuration step for maximum antichloinesterase activity. Mixed function oxidase metabolism (MFO) occurs in the liver of vertebrates, and in fat bodies, Malpighian tubules, and digestive tracts of invertebrates. MFO metabolism is responsible for the toxification (oxidative desulfuration) process of organophosphorous pesticides. MFO activity varies between taxon. We were unable to find MFO activity information for bats. The major concern with toxicity of organophosphorous pesticides is acute toxicity. No toxicological data that indicates a dose/response relationship has been developed for Indiana bats or gray bats for organophosphorous pesticides.

The ecological consequences of organophosphorous pesticides are typically short-term and acute. Besides being acutely toxic to vertebrates, organophosphorous pesticides may affect predator-prey relationships by disrupting population dynamics of either population *(10)*. Daily activity patterns, energy requirements, and behaviors of vertebrates could be affected. These types of pesticides have been shown to interfere with thermoregulation in birds and mammals. Repeated applications of pesticides may cause cumulative physiological effects, although they are not accumulating in the organism (*10*). To assess effects not related to bioaccumulation of pesticides, FLW is conducting long-term monitoring studies of the insect populations preyed upon by the bats, assessing foraging habitat through water quality, soil quality, and vegetation surveys; conducting histopathological examination of lungs of bats captured for analysis; and other studies. These studies are intended to detect certain sublethal effects before they irreversibly affect endangered bats.

Organochlorine Pesticides

The most unexpected finding of this study was that surrogate bats contained measurable concentrations of DDE, heptachlor epoxide, and dieldrin. These organochlorine pesticides have not been used in the United States since the early 70's. The environmental residence time for chlorinated hydrocarbons can be over 30 years, however, new sources of the pesticides may be contributing to their continued presence in the environment.

Although organochlorine pesticides are no longer used in the United States, they may continue to pose a threat to wildlife here. Organochlorine pesticides and their degradation byproducts, may persist in the environment for decades following use. DDT may be transformed into DDD and DDE. The transformation from DDT to DDD is a chemical process and not a result of microbial action (*11*). DDE is the byproduct of an incomplete microbial metabolic pathway and does not further degrade from biological activity (*11, 12*). While DDT may not persist long in the environment, the two degradation byproducts are persistent, and both DDD and DDE retain toxic properties (*13*). Heptachlor was used to control pests in soil, and insects in feed, and is biotransformed and metabolized into the more toxic epoxide (heptachlor epoxide). Dieldrin and aldrin are similar compounds. Aldrin does not persist in the environment like dieldrin; after environmental release, aldrin is quickly degraded into dieldrin through an incomplete microbial metabolic pathway (*12,13*).

Organochlorine pesticides bioaccumulate and have been banned for all uses within the United States. Heptachlor, dieldrin, and aldrin were banned for most uses in the United States by the U.S. EPA in early 1970's (*13*). These compounds were approved for limited use after 1974. Most uses of heptachlor were withdrawn in the United States by 1983 and were banned in 1987 (*13*). DDT was banned in the United States in 1972 (*13*). However, banning these pesticides in the United States did not stop their production. These organochlorine pesticides and many others continue to be produced and used in countries outside the United States. Areas where these chemicals were used within the United States continue to contain detectable levels of these pesticides or their degradation residues 25 or more years after use was discontinued (*13*). In addition, current use of organochlorine pesticides outside the United States and mobilization from sites with contaminated soils within the United States may result in atmospheric transport of these chemicals (*14*).

DDT, heptachlor, and dieldrin are readily transported through the atmosphere for great distances from their use points. Modeling indicates atmospheric concentrations of DDT will remain above detectable limits (0.1 pgm^{-3}) until the year 2010. Atmospheric concentrations of DDE will remain above detectable concentrations until the year 2060. Atmospheric concentrations of dieldrin will remain at detectable levels until sometime between these two extremes (*14*). Dieldrin and DDT also concentrate in aquatic soils and may continue to enter the food web through the actions of bottom-feeding fishes (*14*). Organochlorine insecticides are poorly soluble in water, but may be found in measurable quantities in aquatic systems near contaminated soil (*11*).

Heptachlor epoxide, a metabolite of heptachlor, was detected in all bat samples analyzed. Heptachlor is rapidly metabolized to heptachlor epoxide when it enters an organism. Heptachlor was used in Missouri as a replacement for aldrin to control corn pests in 1978 and 1979.

Concentrations of heptachlor epoxide in bats collected in 1997 and 1998 were below values reported as lethal levels for bats (concentrations found in dead gray bats analyzed in 1983) (*15*). In 1983, dead gray bats were found in Missouri with whole body concentrations of heptachlor epoxide ranging from 0.6 to 2.6 μg/g (*15*). Red bats collected from Missouri were reported to have heptachlor epoxide concentrations ranging up to 1.7 μg/g. (*15*). Bats collected and analyzed in 1998 appear to have less heptachlor epoxide than bats caught in Missouri in 1983. This decrease may indicate concentrations of heptachlor epoxide bioavailable to bats are decreasing. Sample sizes in 1997 and 1998 were very small, therefore, it is impossible to assess differences in concentrations of heptachlor epoxide in bats collected on and off FLW.

Most of the surrogate bats submitted for pesticide analyses were either red bats or eastern pipistrelles. Heptachlor epoxide concentrations in red bats were greater than those in eastern pipistrelles. The reason for this interspecies difference may be related to food preferences and foraging habitats of the two species. Red bats utilize terrestrial foraging habitats and consume more terrestrial insects than eastern pipistrelles. Eastern pipistrelles forage near aquatic habitats and consume aquatic-based insects. Because no heptachlor was detected in sediment samples from FLW in 1998 and 1997, it is likely heptachlor does not remain in aquatic environments and is not as available to eastern pipistrelles.

DDE (dichlorodiphenyldichloroethane), a component and metabolite of DDT, was detected in all surrogate bats analyzed in 1997 and 1998. Two dead gray bats found in Missouri in 1991 had DDE concentrations of 0.6 mg/kg and 3.5 mg/kg (*16*). DDE (as well as heptachlor epoxide and chlordane) killed these gray bats (*16*). Concentrations measured in bats collected on or near FLW exceeded those measured in 1991 in bats from Missouri. DDE is persistent in the environment and appears to be bioavailable and accumulating in bats, although DDT has not been used for more than 20 years.

Dieldrin is the generic name for an insecticide that contains no less than 85% 1,2,3,4,10,10-hexachloro-6-7-epoxy-1,4,4a,5,6,7,8,8a-octahydro-1,4-endo,exo-5,8-dimethano-naphthalene. Dieldrin was detected in some bats tested in 1998. Dieldrin is a breakdown product of Aldrin and is found in organisms within days after application of Aldrin. Mortality of several colonies of gray bats in Franklin and Boone counties Missouri was associated with Dieldrin (*17*). Dieldrin has been shown to cause an increase in metabolic requirements to maintain thermal homeostasis in mammals (*10*). Animals that hibernate or enter torpor to cope with decreasing food and water supplies will be additionally stressed if exposed to dieldrin. An affected animal would use substantial energy reserves to feed an inefficient metabolic system during a time of year when food is scarce and foraging is metabolically expensive.

Because of bioaccumulation, persistence in soil, and continued use around the world, organochlorine pesticides continue to pose potential threats to wildlife. Bats may encounter these compounds through contaminated prey, atmospheric contact, consumption of contaminated water, or through any of the exposure pathways described previously for organophosphorous pesticides. The risk associated with these compounds should decrease through time if production and use of these chemicals around the world declines and contaminated sites within the United States are remediated.

Fort Leonard Wood plans to continue monitoring bioaccumulation of pesticides in surrogate bat populations for several years. These data may help identify the role of exposure to organochlorine pesticides in the range-wide decline of the Indiana bat and gray bat. It is possible that continuous exposure to low doses of organochlorine pesticides may have sublethal effects on Indiana bats and gray bats, and these sublethal effects may go undetected.

Conclusions

Based upon our preliminary findings, the two pesticides chlorpyrifos and malathion used by FLW do not appear to bioaccumulate in bat tissues or in insects preyed upon by Indiana bats and gray bats.

Four species of bats analyzed in this study contained detectable concentrations of DDE and heptachlor epoxide. Several bats collected in 1998 (both spring and fall) had detectable concentrations of dieldrin. A thorough statistical analysis will be completed in the future, at the end of this study.

Insects preyed upon by Indiana bats and gray bats bioaccumulate DDE and heptachlor epoxide. These insects serve as a reservoir for DDE and heptachlor, increasing the amount of pesticides bioavailable to bats.

Acknowledgments

The authors would like to thank the U.S. Army Corps of Engineers, Kansas City District and Fort Leonard Wood, Missouri for funding this study and allowing publication of the results. We also would like to thank Dr. Rich Reaves for his editorial contributions.

Literature Cited

1. LaVal, R. K.; Laval, M. L. Ecological studies and management of Missouri bats with emphasis on cave-dwelling species. Missouri Dept. of Conserv. Terrest. Series. 1980, Vol. 8, pp 1-53.

2. LaVal, R. K.; Clawson, R.L.; LaVal, M. L.; Caire, W. Foraging behavior and nocturnal activity patterns of Missouri bats, with emphasis on the endangered species *Myotis grisescens* and *Myotis sodalis*. J. Mamm. 1977, Vol. 58, pp 592-599.
3. Brack, V., Jr. Ph.D. dissertation, Purdue University, West Lafayette, Indiana, 1983.
4. Brack, V., Jr.; LaVal, R. K. Food habits of the Indiana bat in Missouri. J. Mamm. 1985, Vol. 66, pp 308-315.
5. Kurta, A.; Whitaker, J. O., Jr. Diet of endangered Indiana bat (Myotis sodalis) on the northern edge of its range. American Midl. Nat. 1998, Vol. 140, pp 280-286.
6. Brack, V., Jr. The foraging ecology of some bats in Indiana. Indiana Acad. Sci. 1985, Vol. 94, pp 231-237.
7. Lacki, M. J.; Burford, L.; Whitaker, J. O. Food habits of gray bats in Kentucky. J. Mamm. 1995, Vol. 76, Issue 4, p 1256.
8. Schwartz, C. W.; Schwartz, E. R. The wild mammals of Missouri. University of Missouri Press, Columbia, Missouri, 1981; pp 356.
9. Whitaker, J. O., Jr. Food habits of bats from Indiana. Canadian J. Zool. 1972, Vol. 50, pp 877-883.
10. Handbook of Ecotoxicology: Hoffman, D. J.; Rattner, B. A.; Burton, G. A., Jr.; Cairns, J., Jr.; Lewis Publishers: Boca Raton, FL, 1995; pp 755.
11. Chapelle, F. H. Ground-water Microbiology and Geochemistry. John Wiley & Sons: New York, NY, 1993: pp 424.
12. Manahan, S. E. Environmental Chemistry: Lewis Publishers: Chelsea, MI, 1991: pp 583.
13. Agency for Toxic Substances and Disease Registry (ATSDR). 1993. Toxicological profile for aldrin/dieldrin. Atlanta. U.S. Department of Health and Human Services, Public Health Service Publ. TP-93/05.
14. U.S. Environmental Protection Agency. Great Lakes Binational Toxics Strategy Quarterly Update, URL http://www.epa.gov/glnpo/bns/pest/April99update.html.
15. Clawson, R. L.; Clark, D. R., Jr. Pesticide contamination of endangered gray bats and their food base in Boone County, Missouri, 1982. Bulletin Environmental Contamination and Toxicology. 1989, Vol. 42, pp 431-437.
16. Clawson, R. L. Pesticide contamination of endangered gray bats and their prey on Boone, Franklin, and Camden counties, Missouri. Transactions of Missouri Academy of Science. 1991, pp 7.
17. Clark, D. R., Jr.; Clawson, R. L.; Stafford, C. J. Gray bats killed by Dieldrin at two additional Missouri caves: aquatic macroinvertebrates found dead. Bulletin Environmental Contamination and Toxicology. 1983, Vol. 30, pp 214-218.

Chapter 3

Characterization of a Carbofuran-Contaminated Site in the Hawaiian Islands National Wildlife Refuge

Michael D. David[1,3], Sonia Campbell[1], LeeAnn Woodward[2], and Qing X. Li[1]

[1]Department of Environmental Biochemistry, University of Hawaii, Honolulu, HI 96822
[2]U.S. Fish and Wildlife Service, Hawaiian Islands NWR, Honolulu, HI 96850

An area of shore on Laysan Island in the Hawaiian Islands National Wildlife Refuge has been observed over the last 10 years to cause mortality to several species of birds, crabs, and insects which come into contact with the sand. Previous analysis of two composite samples from 1993 and 1997 indicated contamination of carbofuran in the sand at levels in the low µg/g range. In November of 1997, and July and October of 1998, 129 additional samples were collected in order to define the extent of the carbofuran contamination. Sand samples were extracted with acetone in a pressurized fluid extractor and analyzed by GC-MS. Localized contamination was present at 1,639 µg/g of carbofuran and over 8,832 µg/g of its primary breakdown product, carbofuran-7-phenol. In addition to carbofuran-7-phenol, four other known transformation products were observed in sample extracts, including 3-hydroxy carbofuran, 3-keto carbofuran, 3-keto carbofuran phenol, and carbofuran diol. Two additional compounds which have not been previously reported as carbofuran breakdown products were also found. Spectra are presented and structures are proposed for these two compounds. The structure of one compound, carbofuran acetate, was confirmed by synthesis and GC/MS. To assess relative toxicity, Microtox bioassays were conducted for standards of carbofuran, 3-hydroxy carbofuran, and carbofuran-7-phenol, as well as for water extracts of native contaminated and clean sand. These assays supported the conclusion that the toxicity at the site can be explained by the presence of carbofuran and its products.

[3]Current address: American Cyanamid, P.O. Box 400, Princeton, NJ 08534.

Introduction

The Hawaiian Islands National Wildlife Refuge extends about 800 miles from the main Hawaiian Islands toward Midway island (Figure 1). Because of its remote location and near isolation from anthroprogenic activities, this area is an important breeding area for many marine species, including the endangered Hawaiian Monk seal, the threatened Hawaiian Green turtle, and four species of endangered seabirds (1,2). The largest island in the refuge is Laysan Island, encompassing approximately 1700 acres.

In May of 1988, researchers stationed on Laysan Island observed an area near the northern shore which contained an unusually large number of dead seabirds. Dead scavengers such as ghost crabs and insects were also observed within close proximity of the bird carcasses within this area. The area was marked and referred to as the "dead zone" (DZ). The first samples analyzed to determine the cause of the observed mortality were taken in August of 1993. The site was initially partitioned in a 30 x 30 ft grid for reference of sampling locations. Site investigations included observations of background radiation, an electromagnetic survey, metals analysis, and sampling of air and sand. Two sand samples were analyzed for presence of organic contaminants and tested positive for carbofuran at 11 and 22 μg/g. These two samples were composites consisting of small samples taken at several locations over the surface of the site, and therefore did not pinpoint the exact location of the contamination.

A second investigation was done in June of 1997 in which the original sampling grid was extended to 60 x 60 ft. Sampling included carbofuran immunoassays of 22 samples, which revealed contamination in the 10-20 ng/g range, and five samples analyzed for carbofuran by commercial analytical laboratories, which were reported at similar contamination levels (1). The initial reports of carbofuran contamination at this site led to speculation that there may have been an underground source of carbofuran, such as a buried drum. The analytical results from the first two site surveys failed to definitively establish carbofuran as the source of the observed toxicity, and led directly to the samples collected, analyzed, and reported in this chapter.

The insecticide carbofuran is a potent acetylcholinesterase inhibitor (3,4). It has been demonstrated to be toxic to non-target species such as fish and mammals (5) and has been shown to alter avian neurochemical function (6). Physical transport of carbofuran in the environment is dictated by its water solubility: 0.32 g/L ($25^{\circ}C$), vapor pressure: 2.7 mPa ($33^{\circ}C$) (4), and other physical and chemical properties. Carbofuran has been estimated to be leachable through soil to groundwater (7), and has been observed in agricultural field runoff (8).

The primary degradation pathways of carbofuran are hydrolysis, photolysis, and biodegradation (3,9-12). Hydrolysis has been demonstrated to be the dominant breakdown mechanism under environmental conditions (11). Half-life values of carbofuran degradation have been reported in the range of 107 days (13), 46-117 days (14), 14-73 days at three depths in soil (15), and 20-40 days for biodegradation (12).

The primary breakdown product observed in degradation studies was carbofuran-7-phenol, which is produced by the of hydrolysis of the carbamate (5,12,16). Other degradation and oxidation products which have been reported include 3-keto

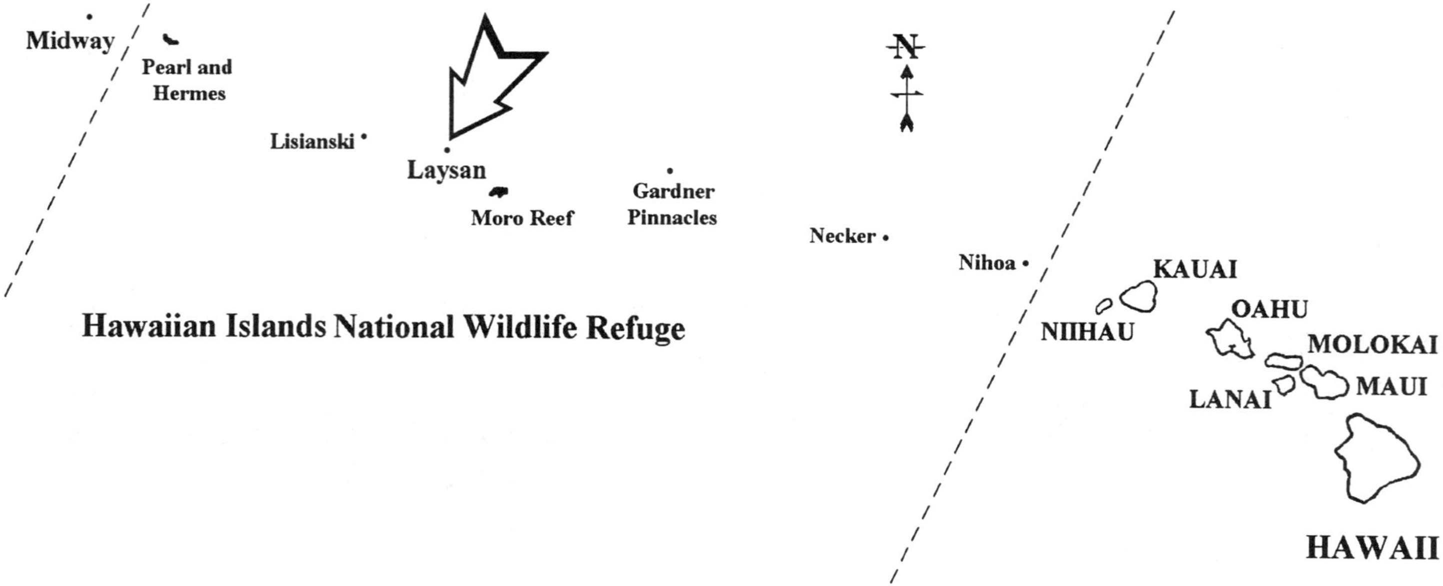

Figure 1. Location of Laysan Island in the Hawaiian Islands National Wildlife Refuge

carbofuran, 3-hydroxy carbofuran, 3-keto carbofuran phenol, and carbofuran 3,7-diol (5,14-20).

Although the degradation and environmental dynamics of carbofuran have been extensively studied in soil and freshwater environments, little is known about its behavior in a marine environment. In fact, degradation of insecticides in marine environments has received little attention (21).

This paper reports the results from analysis of sand samples taken from October 1997 to September 1998 on Laysan Island. It presents unique observations of pesticide breakdown in a marine setting. The purpose of this study is threefold: 1) to confirm the existence of carbofuran at the dead zone site and delineate the extent of the contamination, 2) to observe the degradation of carbofuran in this unique coastal / marine environment, with particular emphasis on identification of transformation products, and 3) to estimate the ecological impact of the contamination present by assessing the relative toxicity. Gas chromatography / mass spectrometry (GC/MS) was used to identify carbofuran and its breakdown products. The toxicity of these levels of carbofuran was assessed by the use of the Microtox bioassay on standards of carbofuran, carbofuran-7-phenol, and 3-hydroxy carbofuran as well as aqueous extracts of site samples.

Experimental

Chemicals

Carbofuran (2,3-dihydro-2,2-dimethyl-7-benzofuranyl methyl carbamate) standard was an EPA reference standard (lot # N708, 99.5%). Standards of the breakdown products carbofuran-7-phenol (2,3-dihydro-2,2-dimethylbenzofuran-7-ol), carbofuran diol (2,3-dihydro-2,2-dimethylbenzofuran-3,7-diol), and 3-hydroxy carbofuran were obtained from Chem Service (West Chester, PA). Solvents used in sample extractions, including acetone and methylene chloride, were Optima grade from Fisher Scientific (Hampton, NH). Microtox reagents, including active bacterial culture (Lot ACV016-3), resuspension solution, osmotic adjusting solution, and diluent, were purchased from their manufacturer, Azur Environmental (Carlsbad, CA).

Carbofuran acetate was synthesized from carbofuran-7-phenol and acetyl chloride (98%, Acros, Fisher Scientific) in anhydrous ethyl ether (Fisher Scientific) in the presence of triethylamine (99%, Aldrich, Milwaukee, WI). The reaction mixture was stirred constantly at room temperature for 30 min, filtered, diluted in ethyl ether, and analyzed by GC/MS.

Sampling and Extraction

The sampling site was on the north shore of Laysan Island, located at 25°46' north latitude, 171°44' west longitude. Seven samples were collected the week of November 7, 1997, an additional 27 were collected the week of July 27, 1998, and 95 more were taken over October 21-27, 1998. During these three sampling events, the southwest corner of the 60 x 60 ft grid established in June of 1997 was arbitrarily assigned the coordinates 0 N (for the north-south axis) and 0 E (for the east-west axis). The sampling area was extended to include an east-west dimension of 160 ft (from -60 to 100 ft E based on the previous grid) with a north-south dimension of 130 ft (-50 to 80 ft N). An east-west transect of the site was completed with surface samples collected at 15 ft intervals at the 15 N line. A north-south transect was done with surface samples collected every 15 ft at the 30 E line. Additional samples were taken near each of the four corners of the grid and near carcasses of birds observed in the DZ area. Samples DZ95 and DZ105 consisted of loose sand collected from the exterior of carcasses of a bird and a crab, respectively. The two carcasses were collected from unspecified locations within the sampling grid during the pre-sampling removal of debris from the DZ area.

Surface samples of coastal sand were collected with plastic spoons which were disposed after a single collection. Depth samples were collected by hand with a metal trowel after excavating a broad pit to reach the lower samples while minimizing cross-contamination with the more superficial sand. The metal trowel was cleaned with acetone between each sample collection. Samples were collected, transported, and stored in previously unused glass or plastic sampling containers. Sample transport from Laysan Island to Oahu, Hawaii for analysis took five days after each sampling trip. While in transit, samples were kept in refrigerated storage. Upon arrival at the laboratory all samples were frozen at -20° C until the day they were extracted. Average moisture content of the dead zone sand as delivered was 3.1%; the observed pH of an aqueous mixture of sand (2:1) was 6.5.

Sand samples were extracted using an automated pressurized fluid extraction device, the ASE200 from Dionex (Salt Lake City, UT) using a method similar to the EPA method 3545 (SW-846 CD-ROM, v. 2, 1997). Extractions used 6-8 g of sand with a solvent of acetone at extraction conditions of 1500 psi, 5 m cell preheating and 5 m static extraction times. Approximately 20-25 mL solvent was collected from each extraction and was reduced under nitrogen and adjusted to 2.0-3.0 mL in acetone. Samples were analyzed directly by GC/MS. Mean recovery for five subsamples of a batch of carbofuran-spiked sand was 106%, with relative standard deviation of 14%.

Analysis

A Hewlett-Packard 5890II GC interfaced with a 5989A MS was employed for all analyses. A DB-5ms (J&W Scientific, Folsom, CA) GC column with dimensions of 30 m x 0.25 mm i.d. with a 0.25 mm film thickness was utilized. The injection port was kept at 170°C while the oven temperature was programmed from 70°C to 180°C at 5°C min and transfer line from the GC to the MS was 280°C. A freshly silanized injection port liner was installed in the GC before every set of analyses to prevent breakdown of carbofuran in the injector. The mass spectrometer was operated in electron impact mode at 70 *eV*. The quantification limit was 50 μg/g of carbofuran in the sand. The presence of carbofuran was confirmed by comparison to its standard and by mass spectral library matching. Breakdown products were identified by comparison to their standards, mass spectral library matching, and spectral interpretation.

Toxicity Evaluation

The Microtox toxicity test is a bioassay used to rank relative toxicity of environmental pollutants, typically in an aqueous phase. It uses the bacterium *Photobacterium phosphoreum*, a light-producing organism, as an indicator species by exposing it to several levels of the chemical of interest and correlating the reduction in light with toxicity. The Microtox method reports toxicity as the effective concentration of the test chemical which results in 50% mortality of the reagent organism (EC_{50}), after 5 or 15 min exposure. This test has been applied to many pesticides and their breakdown products, including carbofuran and carbofuran-7-phenol (22,23).

A Microtox model 500 analyzer was used to determine EC_{50}s of an aqueous blank, solutions of pure carbofuran standard and two known breakdown products, and aqueous extracts of four samples from the site. The instrument and Microtox reagent were validated with phenol standard at 100 mg/L, for which a 5 min EC_{50} was 16.2 +/- 5.3 mg/L (the supplier's reported phenol toxicity of 20.4 mg/L for this lot of Microtox reagent was within our range). Aqueous solutions of carbofuran, carbofuran-7-phenol, and 3-hydroxy carbofuran were prepared in distilled water at known concentrations.

Four DZ samples were analyzed using the Microtox system: two contaminated samples (DZ103 and 104), and two uncontaminated samples used as controls (DZ61 and 63). Sand from these four samples (10-20 g) was extracted with 150 mL deionized water for 18-20 h. Sand was removed from the water by filtration and 10 mL was used in the assay (assay was completed within 4 h of filtration). The remaining filtered water was extracted in a separatory funnel with 3x15 mL methylene chloride. The three methylene chloride extracts were combined, reduced to 2.5 mL under nitrogen, and analyzed directly by GC/MS using the previously reported conditions in order to quantify aqueous-phase carbofuran. All EC_{50} values were calculated by Microtox Omni software as concentration percentages of the tested solution, and were converted to mg/L by adjusting for known concentration of toxicant in the aqueous solution. Comparison of means was done by a t-test and is reported at the 95% confidence level ($\alpha < 0.05$).

Results and Discussion

Site Contamination

The total number of samples extracted and analyzed from three site collections was 129. Residues of carbofuran and/or carbofuran-7-phenol were detected in 34 samples. Quantifications of these samples are listed in Table 1, ranging from 0.1 to 1639 μg/g for carbofuran, and 0.4 to 8832 μg/g for carbofuran-7-phenol. This table includes grid coordinates describing their location as well as depth and sampling date. Figure 2 shows the sampling grid of the site with the locations of 32 samples in which quantifiable amounts of carbofuran or carbofuran-7-phenol were detected. The two

Table I. Dead Zone Samples Containing Carbofuran and Carbofuran Phenol

DZ site no.	*Sampling date*	*Depth ft*	*Grid coordinates* E	N	*Carbofuran* µg/g	*Carbofuran-7-phenol* µg/g
4	10/22/98	0	-6	15	1639	122
95	10/22/98	0	bird		1019	2036
129	11/7/97	0	33	30	892.6	3361
104	10/20/98	0	-6	15	457	8832
103	10/20/98	0	-6	15	189	3563
40	10/23/98	1	-6	15	37.4	27.3
31	10/23/98	2	-6	15	2.6	1.8
102	10/23/98	3	-6	15	3.5	9.7
101	10/23/98	4	-6	15	0.1	0.8
39	10/23/98	5	-6	15	0.2	bql
35	10/21/98	0	-6	21	11.2	9.6
15	10/23/98	0	-6	21	8.1	1.6
87	10/21/98	0	-6	15	1.0	10.2
52	10/22/98	0	-6	15	0.6	1.8
14	10/23/98	0	-6	9	0.2	-
32	10/1/98	0	-6	9	0.2	bql
16	10/23/98	0	-12	21	0.2	-
72	10/22/98	0	75	75	0.2	-
77	10/21/98	0	3	15	0.2	-
6	10/23/98	0	0	9	0.2	-
81	10/21/98	0	63	15	0.1	-
82	10/21/98	0	78	15	0.1	-
84	10/21/98	0	48	15	0.1	-
91	10/21/98	0	30	45	0.1	bql
96	10/21/98	0	30	-45	0.1	0.8
98	10/22/98	0	-40	-30	0.1	bql
99	10/21/98	0	-18	15	0.1	0.4
100	10/21/98	0	-12	15	0.1	0.5
120	7/27/98	0	-8	14	bql	113
105	10/20/98	0	crab		-	1.5
107	7/27/98	2	45	30	-	0.7
110	7/27/98	0	33	30	bql	1.0
121	7/27/98	2	-8	14	-	16.9
122	7/27/98	2	-8	14	-	1.6

bql = mass ions detected below quantification limit

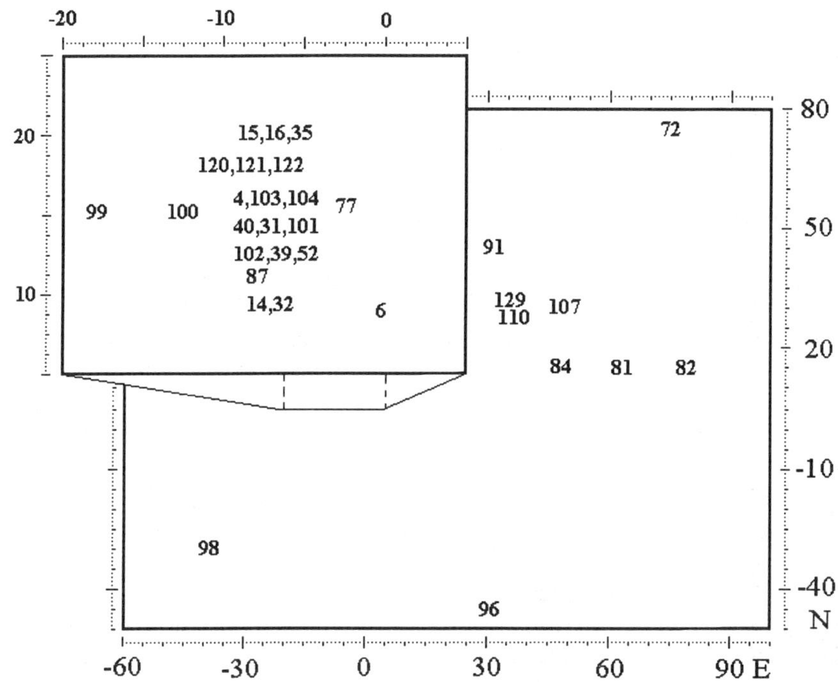

Figure 2. Dead Zone sampling grid with locations of carbofuran-contaminated samples listed in Table 1.

samples from animal carcasses with unspecified coordinates are noted in Table 1 but not on Figure 2.

The carbofuran at the DZ site is primarily near-surface contamination. Of the 28 samples in which carbofuran was detected, 23 were surface samples. Carbofuran concentration dissipates rapidly below the surface, as demonstrated in samples 103, 40, 31, 102, 101, and 39. The concentration at the surface is fivefold greater than that at 1 ft. At a depth of 5 ft, there is less than 1% of the contamination existing at 1 ft. Because of the high permeability and low organic carbon content of sand, surface carbofuran would be expected to leach readily during any significant rain event. If rapid leaching does occur, the reduced concentration of carbofuran at 1 to 5 ft depths implies the stability of carbofuran is lower below the surface than at the surface. Because of the effects of wind and sun, the surface sand has a lower moisture content than that below 1 ft. This higher moisture content at depth encourages hydrolysis and may support a unique coastal-marine microbial consortium effective at degrading carbofuran. The lack of significant contamination below the surface also suggests that there is not a buried source continuously releasing carbofuran to the site as has been previously speculated (1,2).

Minor Transformation Products

In addition to carbofuran and carbofuran-7-phenol, six other related compounds were also detected at lower concentrations. Four of these compounds were previously reported carbofuran breakdown products, including 3-hydroxy carbofuran, carbofuran diol, 3-keto carbofuran, and 3-keto carbofuran-7-phenol. They were not quantified but their presence is noted in Table 2. Figure 3 presents the structures of carbofuran and the five known breakdown products observed in these extracts and the transformation pathways. Their presence in DZ samples was confirmed by comparison of mass spectra and GC retention times to standards and by mass spectral library matching.

Beyond the previously reported transformation products of carbofuran, two additional related compounds were observed in site extracts. Proposed structures for these previously unreported carbofuran transformation products, along with possible transformation pathways, are presented in Figure 3. A compound referred to here as carbofuran acetate was observed in five DZ extracts, noted in Table 2. The carbofuran acetate structure was confirmed by comparison of mass spectra and GC retention time to the synthesized product. Mass spectra of this compound from an extract of a site sample is compared to that of the synthesized chemical in Figure 4. A transformation pathway which produces carbofuran acetate directly from carbofuran is not apparent. It may be produced in a secondary reaction from carbofuran-7-phenol, possibly with a natural chemical with an acetyl moiety. It may also be an impurity of the carbofuran formulation originally released at this site. It is correlated with large concentrations of carbofuran-7-phenol in the samples, which supports it being a transformation product.

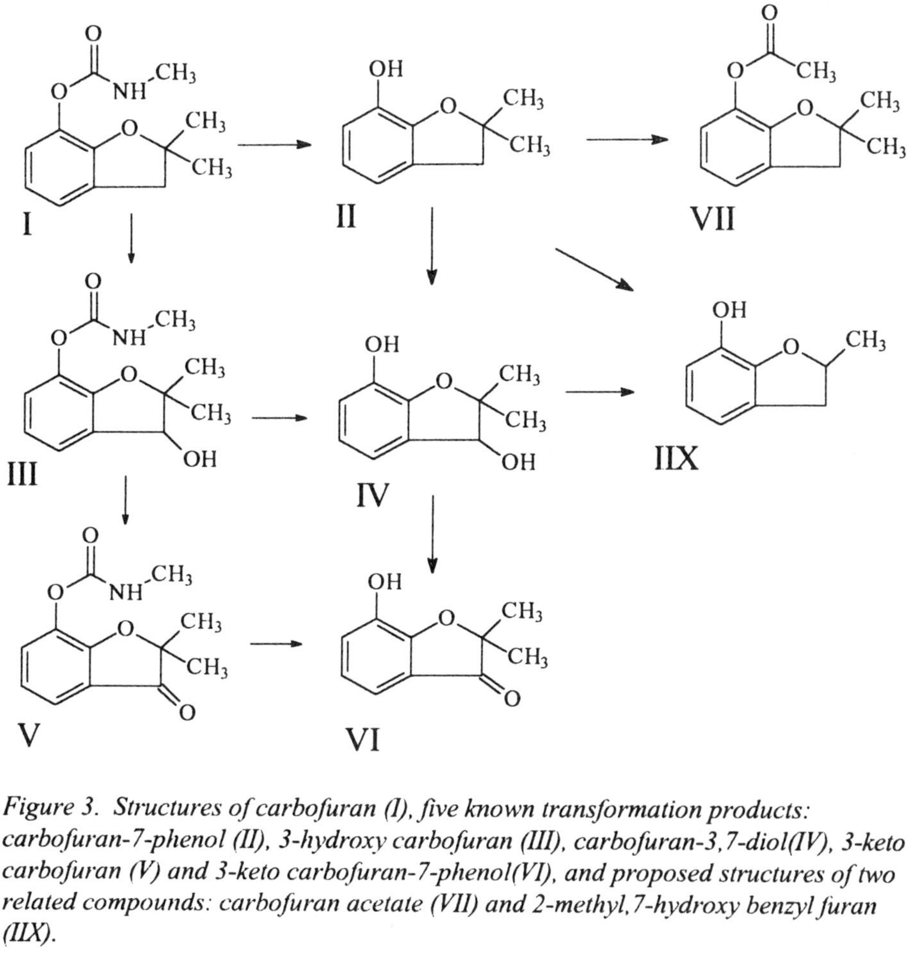

Figure 3. Structures of carbofuran (I), five known transformation products: carbofuran-7-phenol (II), 3-hydroxy carbofuran (III), carbofuran-3,7-diol(IV), 3-keto carbofuran (V) and 3-keto carbofuran-7-phenol(VI), and proposed structures of two related compounds: carbofuran acetate (VII) and 2-methyl,7-hydroxy benzyl furan (IIX).

A

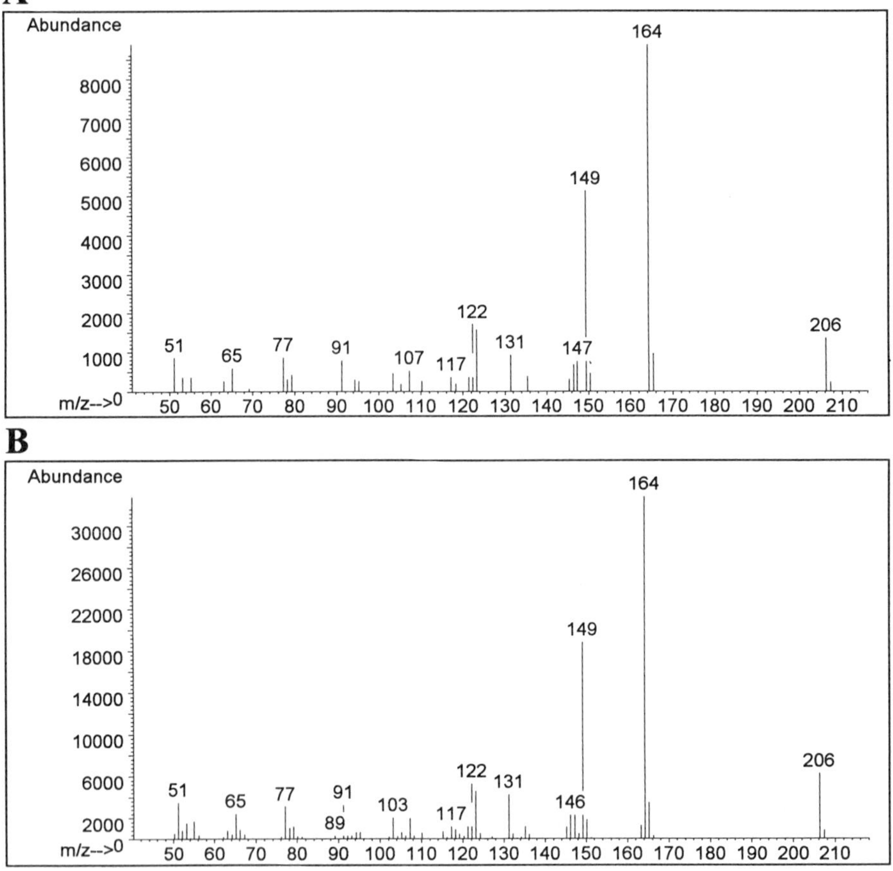

Figures 4 A,B. Mass spectra of carbofuran acetate from extract of Dead Zone sample 120 (A) and synthesized product (B).

Table II. Minor Carbofuran Breakdown Products

DZ Site	*OH-C* *III* [a]	*C-D* *IV*	*K-C* *V*	*K-CP* *VI*	*C-A* *VII*	*MHBF* *IIX*
4				+		+
95		+		+		+
103	+	+	+	+	+	+
104		+	+	+	+	+
129						+
87					+	
120					+	
105					+	

OH-C= 3-hydroxy carbofuran, C-D= carbofuran diol, K-C= 3-keto carbofuran, K-CP= 3-keto carbofuran phenol, C-A= carbofuran acetate, MHBF= 2-methyl 7-hydroxy benzyl furan
[a] Roman numerals refer to structures in Figure 3.

The mass spectra of the second related compound is presented in Figure 5. The proposed structure, based on spectral interpretation, is depicted in Figure 3 as 2-methyl 7-hydroxy benzyl furan. As with carbofuran acetate, it could have been present in the original contamination as an impurity of carbofuran, or it could be a transformation product. The assumption that it is a transformation product is supported by its presence in the 5 samples most contaminated with carbofuran-7-phenol, as noted in Table 2.

Toxicity Evaluation

Two sets of control samples were analyzed using the Microtox assay and are reported in Table 3. The first control was a single replicate done using deionized water, from the same source as the water used to prepare the carbofuran standards and extract the site samples. This sample exhibited no toxicity, with the percent effect observed being unquantifyable (essentially zero). The second set of control tests were triplicate analyses of aqueous extracts of two carbofuran-free sand samples from the site. Because the carbofuran concentration was zero in these samples, a mg/L EC_{50} value could not be calculated, and the data is reported only as the percentage of the solution required for 50% reduction in light production at the given time. The Microtox bacteria were affected by the two DZ control samples. The difference between the zero effect of the deionized water and the positive effect of the aqueous extracts of the control sand can be explained by unquantified factors present in the natural sand. These undetermined factors could include the presence of other organic compounds or natural products, the natural salinity of the sand, and the endemic microbial fauna of the site.

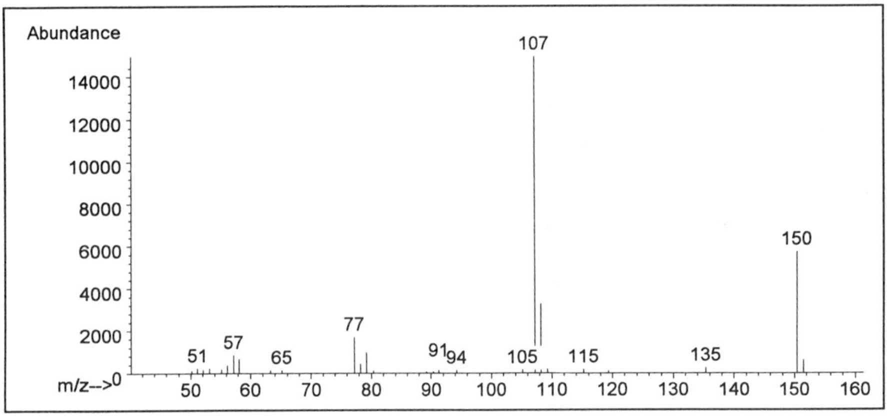

Figure 5. Mass spectra of 2-methyl,7-hydroxy benzyl furan (proposed structure) from extract of Dead Zone sample 95.

Table III. Microtox Assay Results

Microtox Sample	aqueous conc. mg/L	n	EC_{50} 5 min % effect range	mean mg/L +/-σ	15 min % effect range	mean mg/L +/-σ
DI water [a]	0	1	0	-	0	-
DZ61	0	3	63.3 -83.9	[b]	59.4 -72.3 [c]	[b]
DZ63	0	3	50.5 -60.6	[b]	46.5 - 55.9	[b]
carbofuran	50, 100	5	7.7 -20.9 [d]	8.9 +/- 1.4	5.4 -14.5 [d]	6.3 +/-1.0
carbofuran phenol	65	3	53.7 -89.9	48.3 +/- 12	63.7 -70.0	43.8 +/-2.1
3-hydroxy carbofuran	50	2	22.6 -26.8	12.4	10.5 -11.0	5.4
DZ 103 carbofuran	66	3	3.3 -3.9	2.36 +/- 0.2	2.1 -2.6	1.59 +/-0.2
DZ104 carbofuran	179	3	2.7 -3.6	5.57 +/- 0.8	1.7 -2.7	3.71 +/-1.0

[a] deionized water control
[b] not calculated because of zero carbofuran
[c] only 2 replicates calculated
[d] using two different concentrations

Toxicity data from the Microtox assays of aqueous standards and DZ site extracts are also presented in Table 3. Microtox assays of chemical standards indicate that the toxicity of carbofuran is similar to that of 3-hydroxy carbofuran, and that of carbofuran-7-phenol is significantly lower. The mean EC_{50}s at both 5 and 15 min levels for extracts of DZ samples 103 and 104 are significantly different from the respective means of carbofuran. The mg/L EC_{50} values reported for these samples are based on the carbofuran level in the aqueous extracts, and do not allow for the presence of carbofuran-7-phenol and other transformation products. The difference between the EC_{50}s of the carbofuran standards and the contaminated DZ samples can be accounted for by presence of carbofuran breakdown products in the site extracts or the unquantified factors of the natural sand, as seen in the control samples. Synergistic effects, which exist when a combination of two or more chemicals proves to be more toxic than would be predicted by measuring the toxicity of the individual components separately, was not calculated with this simple bioassay, but synergism cannot be ruled out.

Comparison of the toxicity of the two site extract control samples with the two carbofuran-contaminated samples reveals that the Microtox bioassay measures the toxicity of the carbofuran samples as significantly greater than the carbofuran-free samples. This supports the conclusion that the observed toxicity in this area is caused by the presence of carbofuran and its breakdown products.

Conclusions

The analysis of sand samples taken from the Dead Zone area of Laysan Island during 1997-1998 has established carbofuran and up to seven of its transformation products as contaminants of this remote site. Two of these compounds, carbofuran acetate and 2-methyl 7-hydroxy benzyl furan, have not been reported in previous carbofuran degradation studies. The marine environment and the extended time over which the carbofuran has degraded may be responsible for these unique products. From the results of Microtox bioassays conducted with extracts of site samples, it can be concluded that carbofuran and its transformation products present a significant toxicity and exposure risk toward the endemic fauna of this area. Proposals for the remediation of this site are being considered at the time of this publication.

Acknowledgments

This work was supported by a contract agreement from the U.S. Fish and Wildlife Service. The authors would like to thank Bunnie Yoneyama for providing the Microtox assay hardware and technical assistance.

References

1. Foster Wheeler Environmental Corporation. Laysan Island Dead Zone Supplemental site investigation summary report. October, 1997, USFWS Contract No. 14-48-0010-93-004
2. Enserch Environmental Corporation. Preliminary Assessment/Investigation final summary report; "Dead Zone" site, Laysan Island. February, 1994, USFWS Contract No. 14-48-0010-93-004
3. Bachman, J.B.; Patterson, H.H. *Environ. Sci. Technol.* **1999,** *33*, 874-881.
4. Worthing, C.W., ed. The Pesticide Manual, ninth edition, 1991, British Crop Protection Council, London.
5. Lalah, J.O.; Wandiga, S.O.; Dauterman, W.C. *Bull Environ. Contam. Toxicol.* **1996,** *56*, 37-41
6. Mortenson, S.R.; Johnson, K.A.; Weisskopf, C.P.; Hooper, M.J.; Lacher, T.E.; Kendall, R.J. *Bull. Environ. Contam. Toxicol.* **1998**, *60*, 562-568.
7. Li, Z.C., Yost, R.S.; Green, R.E. *Journal of Contaminant Hydrology* **1998**, *29*, 285-299.

8. Nicosa, S.; Carr, N.; Gonzales, D.A.; Orr, M.K. *J. Environ. Qual.* **1991**, *20*, 532-539.
9. Falah , I.P.; Hammers, W.E. *Toxicological and Environmental Chemistry* **1994**, *42*, 9-18.
10. Barcelo, D; Chiron, S.; Fernandez-Alba, A. *Organohalogen Compd.* **1995**, *24*, 517-522.
11. Seiber, J.N.; Catahan, M.P.; Barril *J. Environ. Sci. Health B* **1978**, *13*, 131-148.
12. Venkateswarlu, K.; Gowda, T.K.S.; Sethunathan, N *J. Ag. Food Chem* **1977**, *25*, 533-536.
13. Oshiro, W.; Miles, C.J.; Green, R.E. Physical-chemical properties of selected pesticides for environmental fate modeling. HITAR Res. Ser. 069, University of Hawaii, Honolulu, **1993**.
14. Caro, J.H.; Freeman, H.P.; Glotfelty. D.E.; Turner, B.C;. Edwards, W.M. *J. Ag. Food Chem.* **1973**, *21*, 1010-1015.
15. Johnson, W.G.; Lavy, T.L. *J. Env. Qual.* **1994**, *23*, 556-562.
16. Archer, T.E.; Stokes, J.D.; Bringhurst, R.S. *J. Ag. Food Chem.* **1977**, *25*, 536-541.
17. Deuel, L.E.; Price, J.D.; Turner, F.T.; Brown, K.W. *J. Environ. Qual.* **1979**, *8*, 23-26.
18. Ou, L.T.; Gancarz, D.H.; Wheeler, W.B.; Rao, P.S.C.; Davidson, J.M. *J. Environ. Qual.* **1982**, *11*, 293-298.
19. Racke, K.R.; Coats, J.R. *J. Agric. Food Chem.* **1988**, *36*, 1067-1072.
20. Rajagopal, B.S.; Brahmaprakash, G.P.; Reddy, B.R.; Singh, U.D.; Sethunathan, N. *Res. Rev.* **1984**, *93*, 1-199.
21. Cotham, W.E.; Bidleman, T.F. *J. Ag. Food Chem.* **1989**, *37*, 824-828.
22. Somasundaram, L.; Coats, J.R.; Racke, K.D.; Stahr, H.M. *Bull. Environ. Contam. Toxicol.* **1990**, *44*, 254-259.
23. Ruiz, M.J.; Lopez-Jaramillo, L.; Redondo, M.J. ;Font, G. *Bull. Environ. Contam. Toxicol.* **1997**, *59*, 619-625.

Chapter 4

Forensic Investigative Techniques to Identify Impacts (Primary and Secondary) from Three Groups of Pesticides on Raptors in California

Robert C. Hosea[1], Brian J. Finlayson[1], and E. E. Littrell[2]

[1]Pesticide Investigations Unit, California Department of Fish and Game, 1701 Nimbus Road Suite F, Rancho Cordova, CA 95670
[2]California Department of Fish and Game, 1701 Nimbus Road, Suite A, Rancho Cordova, CA 95670

When investigating losses of raptors and identifying or eliminating pesticide exposure as a causative or contributing factor it is important to establish and follow a standardized investigative protocol. Investigators need to gather information on the circumstances surrounding the loss such as land use patterns in the area, time of year and observed behavior of the bird prior to death in conjunction with conducting necropsies of dead birds. The collection of specific tissue samples for specific chemical analyses needs to be directed towards the suspected pesticide(s) and the possible route(s) of exposure (dermal absorption, ingestion, or inhalation). Samples can also be collected from incapacitated birds and the analytical results used to assist in determining a course of treatment.

Between 1992 and 1999, 63 raptors were documented as exposed to organophosphate or carbamate insecticides or anticoagulant rodenticides in California. The species involved have included Red-tailed Hawks, Red-shouldered Hawks, Northern Harriers, Golden Eagles, Bald Eagles, Sharp-shinned Hawks, American Kestrels, Barn Owls, and Great Horned Owls. These birds have been recovered from various locations in the state associated with both agricultural and semi-urban settings. Tissue samples collected for chemical analyses during necropsies included contour feathers, footskin, liver, crop and stomach contents and brains. Other tissues were collected for chemical analyses during the course of necropsies if observed signs or available information indicated additional analyses were necessary. Analytical methods employed included identification of organophosphate and carbamate insecticide residues by gas chromatography thermionic specific detector (GC-TSD), and identification of anticoagulant rodenticide residues by HPLC with post column fluorescence. Brain cholinesterase activity and reactivation levels were determined colorimetrically using modifications to the Ellman method. Detection of one or more organophosphate or carbamate insecticides or anticoagulant rodenticides in specific tissues was coupled with other information to determine the degree that exposure to pesticides may have been responsible for the loss. Three case histories, one each, involving exposure to organophosphate and carbamate insecticides and anticoagulant rodenticides are presented.

Pesticides and Raptors

Three main classes of insecticides have been reported to exhibit adverse impacts on raptors (1). These include: A) Organochlorines such as DDT, dieldrin, and chlordane (2-4); B) Organophosphates including diazinon, methidathion, methyl and ethyl parathion, and fenthion (5-9), and C) Carbamates such as carbofuran, aldicarb, and methomyl (5,10-13). In addition to the three classes of insecticides, adverse impacts on raptors have been reported for anticoagulant rodenticides, primarily brodifacoum (1,14-19), and for other rodenticides such as strychnine (20).

The impacts of organochlorine insecticides on raptors and diagnostic methods used have been extensively discussed in the scientific literature (1,21,22). With the cancellation of many of the registered uses of these compounds in the 1970's the frequency and magnitude of adverse impacts has greatly decreased (1). Organophosphate and carbamate insecticides are acutely toxic, cholinesterase inhibiting compounds that have replaced many of the chronically toxic, bioaccumulative organochlorine insecticides (5). New, more toxic, "second generation" anticoagulant rodenticides were developed and introduced in response to an increasing occurrence of resistance in commensal rodents to "first generation" anticoagulant compounds like warfarin (23).

Organophosphate and carbamate insecticides are formulated (granular, powder, and liquid) for use depending on the crop treated and the pest controlled. Many of the use patterns for these insecticides pose unique risks of primary and secondary exposure for raptors. With the exception of two tracking powders, all rodenticides used in California are currently formulated as grain based baits (pelletized, wax block, or coated grain). The only likely route of exposure for raptors to rodenticides is through consumption of exposed prey.

The California Department of Fish and Game (DFG) through the Pesticide Investigations Unit (PIU) investigates reported fish and wildlife losses that may be due to exposure to pesticides. The PIU is notified of fish or wildlife losses by DFG enforcement personnel (wardens), department biologists, and staff of other State or local (county) agencies. Cooperating wildlife rehabilitation groups and private citizens also notify DFG of suspicious fish or wildlife losses (24). Investigation of a wildlife loss involves gathering all available information concerning the loss, conducting necropsies on any carcasses that may be recovered from the site of the loss, collecting appropriate tissue samples for chemical residue analyses, and interpreting the results of the analyses in conjunction with all previously gathered information.

In order to fully investigate fish and wildlife losses it is necessary to establish and follow a uniform protocol. ASTM standard E 1849-96 (25) outlines one such basic investigative protocol. In California this protocol is followed by the DFG's PIU when investigating fish or wildlife losses that may be due to exposure to pesticides.

In California, the primary use of organophosphates resulting in impacts on raptors is as a dormant spray in fruit or nut orchards. Applications of these dormant sprays typically take place from late October until early February prior to the opening of flower buds on the trees (bud break). The primary insecticides used have been ethyl parathion, diazinon, methidathion, and chlorpyrifos. These applications have affected raptors (5, 7, 26, 27).

Carbamate insecticides are used on rice, grapes and alfalfa. These applications have affected raptors (5, 28, 29). The greatest number of carbamate related incidents have involved applications of the granular form of carbofuran. Applications of granular carbofuran to rice are made at planting in April or early May. Losses of raptors have occurred both in the spring at the time of application and in the fall after harvest (October). Applications of granular carbofuran to grapes occur between October (immediately post harvest) through January. Raptor losses in grapes typically occur within a short time period after application.

Urban uses of anticoagulant rodenticides occur throughout the year; exposed raptors may be recovered at any time. Agricultural and non-structural applications usually occur between June and October. It is often very hard to pinpoint the source or time of exposure because of the time delay between ingestion and necropsy and analysis. Raptor losses due to secondary exposure to strychnine have not been reported in California for over seven years. Use of strychnine is now limited to subterranean applications only, and the use of Compound 1080 for rodent or predator control is currently prohibited, minimizing any risk of exposure.

Forensic Investigative Techniques

The three exposure pathways for pesticides are inhalation, dermal absorption, and ingestion. All three pathways can serve as primary exposure pathways. Ingestion of exposed prey by raptors is the classic route for secondary exposure.

Standard Protocol

As noted in standard E 1849-96 (25) it is important to gather specific information to assist in a loss investigation. This information includes:

1) Location (Where was/were the animal(s) found.);

2) Surrounding land use (What crops were being grown, or non-agricultural activities were occurring in the area.);
3) Time of year (In cases where pesticide exposure is suspected it is important to note that some pesticides are only applied at specific times during crop development, others have unique application methods or specific formulations that are dependent on the time of year);
4) History of the animals involved in the loss (Observations of an animal's behavior prior to death can provide important information on possible causes for the loss), and
5) A thorough physical examination and necropsy of any animals involved in the loss (This includes collection of tissue samples for exposure testing, performance of specific diagnostic tests such as fluoroscopy, and collection of samples for histopathology as observed signs may warrant).

Sample Collection

Losses involving wildlife either involve individuals that have been incapacitated (with supportive care can recover and be released), or moribund (dying) and dead individuals. Typically, if these losses involve raptors, there will be one or, very rarely, multiple individuals found. Some aspects of an investigation, including examination and sample collection, will vary depending on whether the raptor is dead or could potentially recover.

Incapacitated Birds

Certain samples should be collected immediately, prior to administration of supportive care. This helps to minimize the risk of accidental sample contamination or invalidation of the results of any analytical procedures. These samples would include blood, a footwash, and twelve to fifteen contour feathers. Initial castings and faeces should be collected, and a fluoroscopic examination or x-ray be performed. If the crop is and the contents are removed they should also be saved for later analysis. Blood samples collected for cholinesterase activity measurement should be immediately spun down and the plasma decanted. The cholinesterase activity measurement should then immediately be run or the sample frozen and kept in an ultra-cold freezer until the analysis can be performed. All other tissue samples collected for pesticide residues should be placed in chemically clean glass jars and frozen until analysis.

Blood samples can be used for measurement of plasma cholinesterase activity levels. These should be depressed if exposure to organophosphate or carbamate insecticides has occurred. Recently developed cholinesterase reactivation techniques (23, 30) allow documentation of depression in individuals, even if measured cholinesterase activity levels fall within the normal range for the species. Spontaneous reactivation of depressed cholinesterase activity has been documented following exposure to carbamate insecticides (10, 28). Measurement of lead concentrations may also provide valuable information that can assist in treatment for incapacitated birds. Quite frequently hawks and eagles scavenge animal carcasses or gut piles in the field. If these animals have been shot, subsequent ingestion of any lead shot or bullet fragments can result in lead toxicosis. Recent exposure to anticoagulant rodenticides can also be diagnosed, in part by chemical analysis of blood plasma (31). Impacts to clotting times can also be measured (32). The volume of blood needed, collection tube(s) used and sample preservation is dependent on the analyses planned.

A footwash can be performed using up to 50 milliliters of a mild solvent such as methanol or acetone on the non-feathered portion of the bird's feet. The rinsate can be analyzed for residues of organophosphate or carbamate insecticides by means of GC-TSD. Residues present as a result of dermal (perching) exposure within the previous 72 hours can usually be recovered (32). After 72 hours the compounds appear to be too tightly bound to the dermal surface (32).

Twelve to fifteen contour feathers can be clipped from random locations on the bird's body for analysis to identify surface residues of organophosphate or carbamate insecticides that may have been applied in an area as an aerosol (aerial or airblast applications). Care should be taken not to collect the feathers from a single area to avoid compromising the bird's thermoregulatory ability.

Regurgitated pellets of undigested fur, feathers and bones (castings) from consumed prey, or crop contents that have been removed, can be analyzed for residues of organophosphate and carbamate insecticides or anticoagulant rodenticides (1, 32). These analyses are important in documenting occurrences of secondary poisoning.

Excreta may contain residues of anticoagulant rodenticides that have been recently consumed. Excreta can also carry residues of metabolites of insecticides that will help in identification of exposed individuals (27).

Moribund or Dead Birds

A thorough necropsy should be performed if the raptor is dying (moribund) and subsequently euthanized or is found dead. A basic necropsy protocol can be found in many veterinary medical texts. This should include a fluoroscopic examination, collection of the samples previously discussed for a bird being placed on supportive care, and additional tissue samples for residue analyses and histopathology. Tissue samples should be collected from proventriculus contents, liver, kidney, and skin from the plantar area of the feet for residue analyses, and the brain for measurement of cholinesterase activity levels. During the course of the necropsy it is important to document the presence of any subcutaneous haematomas, or haematomas in the area of the thigh or the shoulder. Note any evidence of haemorrhage on the surface of the heart muscle, presence of blood in the pericardial sac or free blood or chocolate brown to brick red color in the gastrointestinal tract. The presence of haematomas or evidence of internal haemorrhage are important indicators of possible exposure to anticoagulant rodenticides. Remove the skin from the back of the cranium. Be sure to note if small or pinpoint (petechial) haemorrhages are detected on the surface of the cranium. These are often present if the bird has struck an object such as a window, car, etc. Even if there are observable evidence of significant trauma, broken leg or wing, samples should still be collected and analyzed for the presence of pesticide residues. Exposure to cholinesterase-inhibiting pesticides (organophosphates or carbamates) may result in some degree of impairment (B. Wilson pers. comm., 7, 33, 34) that could result in fatal injuries from collisions with vehicles, structures or power lines.

The contents of the proventriculus and glandular stomach can be analyzed for pesticide residues. Carbamate and organophosphate insecticides are characteristically acutely toxic and residues should still be present (8,10,13). Residues of anticoagulant rodenticides in the stomach indicate a very recent exposure, probably within the previous four days (32).

The principle binding sites for anticoagulant rodenticides are plasma proteins. High concentrations of anticoagulants can be found in the liver, spleen, and kidney (35). Retention times in the liver can vary significantly depending on the compound(s) (36, 37). Newton et al. (32) reported that residues of flocoumafen and other rodenticides may be non-uniformly distributed in liver tissue. To yield the most accurate representation of residue concentrations, the entire liver, less necessary histopathology samples, should be excised, homogenized, and analyzed. To avoid contamination of the sample by residues contained in the bile, the gall bladder should be removed, intact from the liver, prior to homogenizing the organ. A sample of the homogenate can be analyzed for residues. In areas with rangeland, or which support activities such as hunting, a sample of the liver homogenate should be analyzed to rule out lead toxicosis as the cause of death. Kidney tissue is also acceptable for analysis of lead levels.

One of the few areas on a bird where the skin is exposed and dermal absorption of pesticides can occur is the feet. When not in flight, birds are in direct contact with a surface through the plantar (bottom) region of their feet. Organophosphate or carbamate insecticides present on perching surfaces can be absorbed through the skin of the raptors resulting in exposure and possible death (7, 27, 38). Remove the skin from the plantar portion of the raptor's feet until the tendons that control flexing of the talons are exposed. This skin sample can be analyzed for the presence of organophosphate and carbamate insecticide residues. An incident involving dermal exposure to ethyl parathion was reported for a Griffon Vulture (39). The bird apparently came into direct contact with the pesticide probably while scavenging in a disposal site or a dumpster. The point of contact in this case however, was not the feet but the exposed skin on the neck of the bird.

Cholinesterase activity levels in the brain can be measured using the Ellman method (40) or modifications (29, 41). Cholinesterase reactivation techniques used for plasma cholinesterase measurements (23, 30) can be used to document depression of activity that still falls within the normal range of values given by Hill (42). Depression of brain cholinesterase activity levels below 50% of normal is considered clinical evidence that death was due to exposure to cholinesterase–inhibiting agent (43, 44). As with plasma cholinesterase activity, brain cholinesterase can spontaneously reactivate if the bird was exposed to a carbamate insecticide (10, 28). The reactivation process can be slowed significantly if the brain is frozen to at least -80^{0}C immediately after it is removed from the bird (B. Wilson pers. comm.). Brain cholinesterase activity and reactivation measurements have been successfully performed up to eight days post mortem on carcasses of laboratory reared Northern Bobwhite Quail *Colinus virginianus* killed by oral gavage exposure to carbofuran. The carcasses were held in the laboratory for up to eight days under different temperature and humidity regimes intended to simulate field conditions (29).

Analytical Methods

Feather, footwash, footskin, and crop/stomach content samples collected by the PIU during a loss investigation are analyzed by the Department of Fish and Game Water Pollution Control Laboratory (WPCL) for residues of organophosphate and carbamate insecticides. Feather samples are extracted in methylene chloride, the solvent refiltered and evaporated to 10 ml. Footskin and crop/stomach content samples are dried and macerated with anhydrous granular sodium sulfate until the samples are free flowing. The samples are then extracted in methylene chloride, and the solvent is filtered and evaporated to 10 ml. Solvents used to collect footwash samples were evaporated to 10 ml. All samples extracts are then resuspended in petroleum ether and evaporated to 10 ml. The extracts are placed on a florisil column and sequentially eluted with 200 ml of 6% (v:v) ethyl ether/petroleum ether, 200 ml 15% (v:v) ethyl ether/petroleum ether, 50% (v:v) ethyl ether/petroleum ether, and 35% (v:v) ethyl acetate/petroleum ether. Each of the resulting fractions is collected separately. One ml of iso-octane is added to each fraction and the solution evaporated to 1 ml. The resulting solutions are then transferred to 10 ml concentrator tubes and diluted to 10 ml with iso-octane. All extracts are analyzed using a Varian 3600 GC with a TSD and a DB-1 megabore capillary column. Confirmation of any detected residues is done using the same instrument with a DB-17 megabore capillary column. The ethyl ether/petroleum ether fractions are analyzed for residues of organophosphate insecticides and the ethyl acetate/petroleum ether fraction is analyzed for carbamate insecticides. Detection limits for samples analyzed for carbofuran were 0.1 ppm. Detection limits for samples analyzed for organophosphate insecticide residues varied by sample type and sample size. The detection limit ranged between 0.1 ppb and 200 ppb depending on the compound and the size of the individual samples. Red-tailed Hawks provided a larger footskin sample than an American Kestrel. Large numbers of contour feathers could be collected from a hawk carcass compared to the 12 to 15 collected from a living bird.

Liver and kidney tissue submitted to the WPCL for analysis of lead concentration were processed by open tube nitric acid (HNO_3) reflux digestion using a heating block. Digestion was allowed to continue until all NO_x fumes had dissipated. The remaining HNO_3 was evaporated to 0.5 ml without boiling. The residue was brought to a constant volume of 25 ml using 16-18 megaOhm water. Analysis of the extract is by graphite furnace atomic absorption spectrophotometry. The detection limit was 0.1 ppm.

Measurement of brain cholinesterase activity for raptors suspected of exposure to cholinesterase inhibiting pesticides was made at the University of California at Davis either in the Department of Avian Sciences or at the California

Veterinary Diagnostic Laboratory System, Toxicology Laboratory (CVDLS). All measurements were done on a plate reader using modifications of the Ellman method (29, 41). Cholinesterase reactivation activity was measured using methods developed by Hooper (23, 30). All results are compared to published normal activity levels for representative species and to internal method normal values to determine level of depression.

Liver tissue from raptors suspected of exposure to anticoagulant rodenticides is submitted to CVDLS for analysis. The samples are analyzed for the presence of nine different anticoagulant compounds (coumafuryl, difethialone, brodifacoum, bromadiolone, chlorophacinone, diphacinone, warfarin, pindone and coumachlor) using HPLC with both photodiode array and fluorescence detection (31). Detection limits for samples ranged from 0.01ppm to 0.25 ppm depending on the compound.

Case Histories

Nine species of raptors have been identified as exposed to organophosphate and carbamate insecticides or anticoagulant rodenticides in California (Table 1). Three case histories investigated by the PIU are presented here to demonstrate the forensic techniques utilized.

Case #1: Organophosphate Insecticide (Methidathion)

The carcass of an adult female Northern Harrier *Circus cyaneus* was recovered from a fruit orchard on December 19, 1997 in Stanislaus County, located in the San Joaquin Valley of California (DFG case accession # P-1934). The surrounding land use was primarily fruit and almond orchards. Orchards adjacent to the site where the bird was found had been sprayed with organophosphate dormant sprays within the previous few days. A fluoroscopic examination of the bird was made with negative results. The extremities were palpated for evidence of fractures. A non-perforating fracture of the right femur was identified. Petechial haemorrhages were identified on the cranium following removal of the skin. Both of these indicate some form of physical trauma, possibly resulting from the bird striking one or more tree limbs in the orchard where it was found. The crop was palpably full when presented. During the course of the necropsy the bird was judged to have good fat reserves with no gross evidence of disease. The crop contents, consisting of tissue, mammal hair and bird feathers, were removed. Skin from the plantar region of both feet was collected as were contour feathers from several areas of the body. These samples were submitted for analysis to determine the presence of organophosphate insecticides. The brain was removed and submitted for measurement of cholinesterase activity.

Table 1. Raptor Species Exposed[1], by Pesticide Group, in California (1992-1999)			
Species	Organophosphates[2]	Carbamates[3]	Anticoagulants[4]
Red-tailed Hawk *Buteo jamaicensis*	20	14	1
Red-shouldered Hawk *Buteo lineatus*	0	0	2
Northern Harrier *Circus cyaneus*	2	3	0
American Kestrel *Falco sparverius*	1	1	1
Bald Eagle *Haliaeetus leucocephalus*	0	1	0
Golden Eagle *Aquila chrysaetos*	0	0	8
Barn Owl *Tyto alba*	1	0	4
Great Horned Owl *Bubo virginianus*	0	0	3
Sharp-shinned Hawk *Accipiter striatus*	0	1	0

1. Documented as Exposed (DFG Records)
2. Organophosphate Insecticides: Diazinon, Methidathion, Chlorpyrifos, Fenthion, Phorate
3. Carbamate Insecticides: Carbofuran
4. Anticoagulant Rodenticides: Brodifacoum, Diphacinone, Bromadiolone

Analysis of the samples by GC-TSD found residues of the organophosphate methidathion in the crop contents (15 ppm) and in the footskin (12 ppm). The feather sample was negative for pesticide residues. These results indicate that the bird fed on an animal that had been exposed to methidathion. . The presence of the insecticide on the footskin indicates that the bird came into direct contact with a surface that had been exposed. This could either be the body of its prey or a perch. Because the feather sample was negative for the presence of organophosphate insecticides it is possible to eliminate preening as the source of the ingested pesticide. The cholinesterase activity level in the brain was measured at 4.7 μmol/gram/minute. Hill (42) lists the normal range for brain cholinesterase activity various species of raptors as 14 to 24 μmol/gram/minute. The measured activity level was depressed by greater than 50%. The calculated exposure concentration of methidathion was below the published acute oral LD_{50} values for avian species (45). However, the concentration of methidathion in the crop/stomach contents only represents that portion of the exposure that had no been metabolized by the hawk. For example, dermal toxicity levels have not been developed for raptors thus, the degree that dermal exposure contributed to the loss can not be easily determined. The concentration of methidathion in the footskin sample was approximately twelve times higher than any previously identified organophosphate insecticide residues. The presence of methidathion in the crop and on the feet, in conjunction with the significantly depressed brain cholinesterase activity strongly suggests that this loss was due to methidathion poisoning. The physical injuries to the bird, the fracture and the haemorrhages on the cranium, may have occurred as the result of an impact that occurred in the orchard while the hawk was impaired.

Case #2: Carbamate Insecticide (Carbofuran)

The carcasses of several Red-Tailed Hawks were recovered from a vineyard near Yountville in Napa County by DFG personnel in January 1991 (DFG case accession # P-1347). Carcasses of European Starlings and American Robins were also recovered at the site. The birds were found in, and immediately adjacent to, the vineyard. The vineyard had been treated with an application of granular carbofuran approximately one month prior to the loss. A survey of the vineyard revealed the presence of carbofuran granules on the surface of the soil. Physical examination of the bird carcasses by PIU staff did not reveal any evidence of trauma. The crops of four of the hawks were palpably full. The contents were characterized as feathers and starling body parts, including in one case, the intact head of a starling. The brain excised from this starling head was analyzed for cholinesterase activity levels. Samples of the hawks' crop contents were submitted for organophosphate and carbamate analysis. The brains of the hawks were also submitted for measurement of cholinesterase activity levels. Contour feather and footskin samples were not collected because a dermal route of exposure was unlikely.

The crop contents of several of the starlings had residues of carbofuran (2.0-110 ppm). Initial measurement of depressed brain cholinesterase activity levels in the starlings and subsequent spontaneous *in vitro* reactivation (23) of cholinesterase activity supports the finding that the starlings had died as a result of exposure to a carbamate (M. Hooper pers. comm., 10, 28).

Carbofuran was also detected in the crop contents from the hawks (0.1-7.6 ppm). Cholinesterase activity levels for the recovered starling head were measured and found to be depressed by greater than 50% from published values of 22 μmol/gram/minute (42). This indicated that the consumed starling had probably died as a result of exposure to a cholinesterase-inhibiting compound. The brain cholinesterase activity levels for the hawks were also depressed and exhibited spontaneous *in vitro* reactivation (M. Hooper pers. comm.). The presence of carbofuran residues in the crop contents of the hawks, in conjunction with the depressed brain cholinesterase activities strongly suggests that the hawks may have died as a result of exposure to carbofuran. The presence of the Starling with depressed brain cholinesterase activity suggests the exposure to carbofuran was secondary. This resulted from the raptors preying on dying European Starlings, or scavenging carcasses of starlings which had consumed carbofuran.

Case #3: Anticoagulant Rodenticide (Brodifacoum)

The carcass of an adult male Golden Eagle was recovered from its breeding territory in Contra Costa County on March 11, 1999 (DFG case accession # P-2060A). The bird had been part of a long term radio telemetry study of eagles in the area. Based on telemetry data the breeding territory consisted mainly of open rangeland and random outbuildings with some areas of urban development.

The bird was not recovered in the vicinity of power lines and the feathers did not have the "singed" odor characteristic of accidental electrocution. The necropsy indicated no other evidence of physical trauma. The animal was skinned to determine the presence of puncture wounds from conflicts with other eagles or from a gunshot. The pericardial sac contained serum and blood. Approximately 65% of the surface of the heart muscle was haemorrhagic. The major vessels associated with the heart contained unclotted blood. The lung tissue was haemorrhagic, bleeding from a cut surface. The cerebro-spinal fluid was blood stained, indicating cranial haemorrhage. These clinical signs were consistent with previously published symptoms of anticoagulant toxicosis in raptors (14, 15, 17, 46). Liver tissue was analyzed for residues of anticoagulant rodenticides. Kidney tissue was also analyzed for lead concentrations. Kidney tissue had a lead concentration of 1.1 ppm, well below the level that would indicate acute toxicosis (47). Liver tissue had a brodifacoum concentration of 0.04 ppm. The presence of the rodenticide in liver tissue alone does not support a diagnosis of anticoagulant toxicosis. However, if considered in conjunction with the observed clinical signs consistent with anticoagulant toxicosis, a diagnosis of anticoagulant toxicosis is supported.

Brodifacoum is only registered in California for use in and adjacent to structures to control commensal rodents (rats and mice). It is unlikely that rodents will remain solely within the confines of a particular structure but will move in and out so that they might be captured and consumed by a raptor. The anticoagulant rodenticide baits are pelletized and a raptor would not consume them directly unless they were placed in some type of meat bait. The evidence supports the conclusion that the death of this eagle was a result of secondary exposure to brodifacoum.

Conclusions

These three case histories illustrate the importance of gathering all available information during a loss investigation. This information includes the physical and biological conditions of the bird, use practices on adjacent lands, and tissue residues of chemicals used on adjacent lands. Physical symptoms by themselves may be misleading (the trauma to the Northern Harrier) or they can be insufficient by themselves to diagnose a cause for the loss (the haemorrhaging in the eagle). Information on time of year and surrounding land use are also important in determining the scope of residue analyses. The Northern Harrier was found in an orchard during the time of year when applications of organophosphate dormant sprays would be expected. It was important to look for residues of rodenticides registered for use against commensal and agricultural because the home range of the eagle included urban developments and rangeland. Analysis of tissues can also help to eliminate possible sources of exposure or causes for a loss. The Northern Harrier did not ingest organophosphate insecticides by preening because the analysis of the feather sample was negative for residues. The Golden Eagle had been exposed to lead, but the concentration in the kidney tissue wasn't high enough to support lead toxicosis as a cause for the loss. By following a standard protocol during a loss investigation it is unlikely that vital information will be overlooked.

Literature Cited

1. Porter, S.L. in *Raptor Biomedicine*. Redig, P.T., Cooper, J.E., Remple, J.D., and Hunter, D.B. (eds) 1993, pp. 239-245.
2. Stone, W.B. *Kingbird* 1981, 31:79-81.
3. Hickey, J.J., Anderson, D.W. *Science* 1968, 162:271-273.
4. Littrell, E.E. *Calif. Fish and Game.* 1986, 72(3):182-185.
5. Mineau, P., Fletcher, M.R., Glaser, L.C., Thomas, N.J., Brassard, C., Wilson, L.K., Elliot, J.E., Lyon, L.A., Henny, C.J., Bolinger, T., Porter, S.L. *J. Raptor Research* 1999, 33(1):1-37.
6. Henny, C.J., Blus, L.T., Kolbe, E.J., and Fitzner, R.E. *J. Wildlife Management* 1985, 49:648-658.
7. Hosea, R.C. California Department of Fish and Game, *Environmental Services Division Administrative Report* 96-5. 1996, 22pp.
8. Stone, W.B., Overmann, S.R., Okoniewski, J.C. *The Condor* 1984, 86:333-336.

9. Henny, C.J., Kolbe, E.J., Hill, E.F., Blus, L.J. *J. Wildlife Diseases,* 1987, 23(2):292-295.
10. Littrell, E.E. *Calif. Fish and Game* 1988, 74(4):226-231.
11. Allen, G.T., Veatch, J.K., Stroud, R.K., Vendel, C.G., Proppenga, R.H., Thompson, L., Shafer, J.A., Braselton, W.E. *J. Wildlife Diseases,* 1996, 32(2):385-389.
12. Balcomb, R. *J. Wildlife Mgmt.* 1983, 47:1129-1132.
13. Elliott, J.E., Langelier, K.M., Mineau, P., Wilson, L.K. *J. Wildlife Diseases* 1996, 32(3):486-491.
14. Hegdal, P.L., Colvin, B.A. *Environ. Toxicol. and Chem.* 1988, 7:245-260
15. Mendenhall, V.M., Pank, L.F. *Wildl. Soc. Bull.* 1980, 8(4):311-315
16. Merson, M.H., Byers, R.E., Kaukeinen, D.E. *J. Wildl. Manage.* 1984, 48(1):212-216.
17. Newton, I., Wyllie, I., Freestone, P. *Environmental Pollution* 1990, 68:101-117.
18. Shawyer, C.R. *The British Barn Owl.* The Hawk Trust publ. 1987, 113pp
19. Stone, W.B., Okoniewski, J.C., Stedelin, J.R. *J. Wildlife Diseases* 1999, 35(2):187-193.
20. Littrell, E.E. In *Proceedings of the 7th Annual Wildlife Conference* (San Francisco). Venizelos, N., Grijalva, C. (eds). 1983, pp. 106-110.
21. Stickel, L.F. In *Environmental Pollution By Pesticides.* Edwards, C.A. (ed). 1973, pp. 254-313.
22. Wiemeyer, S.N., Frenzel, R.W., Anthony, R.G., McClelland, B.R., Knight, R.L. *J. Raptor Research.* 1989, 23:140-146.
23. Hunt, K.A., Hooper, M.J. *Analytical Biochemistry* 1993, 212:335-343.
24. Littrell, E.E. in Proceedings *of 14th Vertebrate Pest Conference.* L.R. Davis and R.E. Marsh (eds) university of California, CA, 1990, pp. 59-61.
25. Standard Guide For Fish and Wildlife Incident Monitoring And Reporting (E 1849-96) in *Annual Book of ASTM Standards,* 1997, vol. 11.05.
26. Wilson, B.W., Hooper, M.J., Littrell, E.E., Detrich, P.J., Hansen, M.E., Weisskopf, C.P., Seiber, J.N. *Bull. Environ Contam. Toxicol.* 1991, 47:717-724.
27. Hooper, M.J., Detrich, P.J., Weisskopf, C.P., Wilson, B.W. *Bull. Environ. Contam. Toxicol.* 1989, 42:651-659
28. Hill, E.F., Fleming, W.J. *Environmental Toxicol. And Chem.* 1982, 1:27-38.
29. Hunt, K.A., Hooper, M.J., Littrell, E.E. *J. Wildlife Diseases* 1995, 31(2):186-192.
30. Wilson, B.W., Hooper, M.J., Hansen, M.E., Nieberg, P.S. In *Organophosphates: Chemistry, Fate and Effects.* Chambers, J.E., Levi, P.E. (eds) 1992, pp. 108-132.
31. Palazoglu, M.G., Tor, E.R., Holstege, D.M., Galey, F.D. *J. Agric. Food Chem.* 1998, 46:4260-4266.
32. Newton, I., Wyllie, I., Gray, A., Eadsforth, C.V. *Pesticide Science,* 1994, 41:187-193.
33. Yamamoto, J.T. PhD. Thesis, University of California, Davis, California, 1994
34. Wilson, B.W., Seiber, J.N., Fry, D.M. *Report to the Almond Board of California.* 1993
35. Buck, W.B., Osweiler, G.D. in *Clinical and Diagnostic Veterinary Toxicology 2nd ed.* Van Gelder, G.A. (ed) 1976, pp. 253-256

36. Eason, C.T., Wright, G.R., Batcheler, D. *New Zealand Journal of Agricultural Research.* 1996, 39:397-400.
37. Laas, F.J., Forss, D.A., Godfrey, M.E.R. *New Zealand Journal of Agricultural Research* 1985, 28:357-359.
38. Henderson, J.D., Yamato, J.T., Fry, D.M., Seiber, J.N., Wilson, B.W. *Bull. Environ. Contam. Toxicol.* 1994, 52:636-640.
39. Shlosberg, A., Lublin, A., Baht, O., Hermann, G.D., Hadji, V., Bellyache, M., Mechanic, S., Nyssa, A. *Vet. Human Toxicol.* 1994, 36(3):224-226.
40. Ellman, G.L., Courtney, K.D., Andres, V., Featherstone, R.M. *Biochem. Pharm.* 1961, 7:88-95.
41. Tor, E.R., Holstege, D.M., Galey, F.D. *J. of AOAC International* 1994, 77(5): 1308-1313.
42. Hill, E.F. *J. of Wildlife Diseases.* 1988, 24:51-61.
43. Grue, C.E., Fleming, W.J., Busby, D.G., Hill, E.F. *Trans. 48th North American Wildlife and Natural Resources Conf.* 1983.
44. Burn, J.D., Light, F.A. *J. Wildlife Diseases.* 1996, 32(2):216-224.
45. Smith, G.J. *Pesticide Use And Toxicology In Relation To Wildlife: Organophosphorus And Carbamate Compounds.* U.S. Fish and Wildlife Resource Publication 170. 1987.
46. Radvanyi, A., Weaver, P., Massari, C., Bird, D., Broughton, E. *Bull. Environ. Contam. Toxicol.* 1988, 41:441-448.
47. *The Merck Veterinary Manual* 8th edition; Aiello, S.E. Ed.; Merck and Co. Whitehouse Station, NJ, 1998; pp 2072-2073.

Monitoring Wildlife Exposure to Pesticides

Chapter 5

Large-Scale Monitoring of Non-Target Pesticide Effects on Farmland Arthropods in England: The Compromise between Replication and Realism of Scale

Geoffrey K. Frampton

Biodiversity and Ecology Division, School of Biological Sciences, University of Southampton, Bassett Crescent East, Southampton, Hampshire SO16 7PX, United Kingdom

Replication of experimental treatments is necessary for the unambiguous interpretation of pesticide effects on wildlife using statistical hypothesis testing but is incompatible with the large temporal and spatial scales required for a fully realistic field study. In this paper I consider the pros and cons of some large-scale experimental approaches used during the last two decades in England to investigate long-term effects of pesticide use on arthropods and other farmland wildlife. I present a potential solution to the trade-off between scale and replication in long-term studies which may be particularly appropriate for patchily-distributed species. Advantages of the recently-developed multivariate analysis method Principal Response Curves (PRC) for interpreting long-term effects of pesticides on arthropods at the community level are discussed.

Historical Perspective

During the 1970s, pesticide use on arable farmland in the UK increased rapidly [*1*, *2*]. The number of active ingredients used, frequency of applications per season and incidence of the use of pesticide mixtures all increased at a time when the abundance of several taxa of non-target farmland arthropods was found to be in decline in southern England [*3*]. Most studies of the effects of pesticides on arthropods were single-season short-term investigations of individual pesticide applications. These provided valuable information on the risks posed by certain pesticides but could not detect long-term cumulative effects of repeated applications or effects of complex tank mixes. In 1979 a report to the UK Government by the Royal Commission on Environmental Pollution expressed concern "about the scale of pesticide use" and "the possibility of unforeseen and unforeseeable effects". The report gave a number of recommendations as to how pesticide effects research should progress to address these issues and proposed that reducing pesticide usage to the minimum consistent with agricultural objectives should be Government policy [*1*]. The report lent support to the concept of a large-scale study to examine the potential long-term effects of overall regimes of pesticide use on farmland wildlife which had been originally proposed in the mid-1970s [*4*]. Following a period of consultation to develop the best methodological approach, the Boxworth Project was initiated in 1981 under funding from the UK Ministry of Agriculture, Fisheries and Food (MAFF) [*4*].

Boxworth: a 1980s Farm-Scale Approach

The design of the Boxworth Project has been reported previously [*5*] and compared with those of other farming systems studies in western Europe [*6*] so here I refer only to aspects of the design which have a bearing on interpretation of pesticide effects. During the 'treatment phase' (1983-88), effects of three regimes of pesticide use were investigated upon birds, small mammals, invertebrates and plants. Each regime was applied to a discrete block of contiguous fields on a commercial winter wheat farm in eastern England. Two of the pesticide regimes ('Supervised', SUP and 'Integrated', INT) had managed inputs, applied in response to pest, weed or disease thresholds, whilst the third regime ('Full Insurance', FI) comprised prophylactic insurance use of pesticides, intended to mimic high-input commercial farming practice; FI had substantially higher pesticide inputs than SUP or INT [*7*]. The grouping of fields together under SUP (three fields, total 45.7 ha), INT (three fields, total 22.6 ha) and FI (four fields, total 53.2 ha) was necessary for realism of scale but precluded orthodox replication of the treatments [*7*]. A two-year 'baseline' period of pre-treatment monitoring (1981-82) was included in the Project to allow assessment of existing populations and permit the detection of any substantial changes in wildlife abundance at the start of the contrasting regimes.

The Project's design was geared towards detecting gross overall effects of pesticide use on wildlife in the long term rather than statistical testing of specific hypotheses concerning the pesticide regimes [*7*]. It was assumed that, given the large scale of the study, any substantial effects of pesticide use on wildlife would be detectable through intensive monitoring.

Boxworth's Limitations and Lessons

When the Project concluded in 1988 it was evident that the most substantial effects of the pesticide regimes had been upon arthropods. Negative and also some positive effects of the FI regime on abundance of several arthropod species were detected [*8, 9*]. Counts of some species of Carabidae and Collembola had declined sharply in FI fields relative to SUP and INT when the FI regime was initiated and these remained close to zero for the remainder of the treatment phase [*9*]. This endorsed the suitability of the Project's contiguous-field design for detecting at least major effects of pesticides. But the Project was limited in its geographical and agricultural coverage, raising questions as to whether the observed pesticide effects would also occur in other arable crops at other locations. The realism of the relatively inflexible FI regime at the end of the Project was also questioned as it had not kept pace with changes in farming practice during the 1980s. Lessons that emerged from the Boxworth Project included: (1) Different experimental scales are necessary for the study of environmental and economic impacts of pesticide use [*10, 11*]; (2) 'Baseline' monitoring is essential in an unreplicated long-term study [*9*]; and (3) The arthropod species most vulnerable to FI pesticide use were those characterised as having poor dispersal ability [*8*].

Implementing the Lessons : SCARAB in the 1990s

From the Boxworth Project came a need to determine whether the negative effects of intensive pesticide use on arthropods in wheat in eastern England would be likely in other crops and locations. In line with UK Government policy objectives there was also a need to investigate the economic consequences of reducing pesticide use. These questions required different scales of study and were addressed by two different MAFF-funded long-term projects during the 1990s: SCARAB (Seeking Confirmation About Results At Boxworth) was an environmental study with subsidiary economic monitoring whilst TALISMAN (Towards A Low Input System Minimising Agrochemicals And Nitrogen) was primarily an economic study with limited environmental monitoring [*6, 10-12*]. The focus of SCARAB was on arthropods; this permitted the use of smaller experimental areas than those used in the Boxworth Project. From an agronomic perspective, realistic experimental units for studying arthropods would be individual fields. However, heterogeneity of arthropod populations among fields at Boxworth indicated that within-field comparison of treatments would be preferable in follow-up studies [*11*]. SCARAB accordingly used a split-field approach for comparing effects of two pesticide regimes on arthropods. The regimes were: current farm practice (CFP), representing conventional practice but flexible in response to changes in pesticide use as indicated by pesticide usage surveys; and a reduced input approach (RIA), in which insecticides were avoided and use of other pesticide types minimised where possible.

SCARAB comprised eight fields, sited at three locations in England (Table 1) (two of the fields, ON and OS, were adjacent, with identical cropping and pesticide inputs; some previous references to SCARAB consider these as one field). Cropping in each field followed a rotation typical of the locality (Table 1). During a 'baseline' pre-treatment monitoring year (1989-90), both halves of each field received identical pesticide inputs (which approximated the RIA regime) to permit comparison of arthropod abundance in each half of the field before the contrasting CFP and RIA regimes were initiated at the start of the 1990-91 crop season. Pesticide inputs varied between years and fields according to cropping but to ensure long-term continuity each regime was applied to the same half of each field every year during 1991-1996. Arthropods were monitored on up to four occasions per month throughout the Project using pitfall trapping and suction sampling at matched locations in the CFP and RIA halves of each field; sampling details have been given previously [*13, 14*].

Table 1. Design of the SCARAB Project: during 1991-1996 each of the RIA and CFP pesticide regimes was applied consistently to one half of each field

Farm	Drayton		Gleadthorpe			High Mowthorpe		
	52.2°N 1.8°W calcareous clay		53.2°N 1.1°W stony sand			54.1°N 0.6°W calcareous loam		
Field (area)	F1 (11ha)	F5 (8ha)	BA (12ha)	NK (8ha)	SO (12ha)	BU (19ha)	ON (17ha)	OS (17ha)
Year:								
89-90	grass	grass	w-bar	beet	w-bar	w-bar	w-bar	w-bar
90-91	w-wht	grass	beet	s-bar	pots	rape	beans	beans
91-92	w-wht	w-wht	s-wht	w-bar	s-wht	w-wht	w-wht	w-wht
92-93	grass	w-wht	w-bar	beans	w-bar	s-bar	w-bar	w-bar
93-94	grass	grass	pots	w-wht	beet	beans	rape	rape
94-95	grass	grass	s-wht	w-barl	s-wht	w-wht	w-wht	w-wht
95-96	grass	grass	w-bar	beet	w-bar	w-bar	s-bar	s-bar

Note: w- = winter-sown; s- = spring-sown; wht = wheat; bar = barley; beet = sugar beet; pots = potatoes; beans = spring beans; rape = winter oilseed rape.

Results from SCARAB

During the period 1991-1996, overall RIA inputs of herbicides, fungicides and insecticides were, respectively, 48%, 53% and 100% lower than those of the CFP regime [*12*]. Abundance of several arthropod taxa, and the similarity of CFP and RIA arthropod communities, declined after individual CFP applications of broad-spectrum insecticides in most of the Project fields but recovery usually occurred within the same season [*15*]. Only Collembola exhibited persistent long-term differences between CFP and RIA regimes. This occurred in only two of the eight fields (BU and F5), where differences in abundance and species richness persisted from 1991 to the Project's end in 1996 (Figure 1; [*15*]).

In addition to effects of pesticides the project provided information on spatial and temporal distributions of arthropods (Table 2). For example, *Bembidion obtusum* Serville, *B. tetracolum* Say and *Pterostichus cupreus* (L.) (Carabidae) were absent from some fields, reflecting these species' preferences for different soil types, whilst *Pseudosinella octopunctata* Börner (Collembola) occurred only in one field (BU) throughout the Project (Table 2). These patterns of distribution are important for the interpretation and prediction of pesticide effects (see below).

Table 2. Summer occurrence of *Bembidion obtusum* (A), *B. tetracolum* (B), *Pterostichus cupreus* (C) (Carabidae) and *Pseudosinella octopunctata* (D) (Collembola) in the SCARAB project (May-July inclusive)

Field	F1	F5	BA	NK	SO	BU	ON	OS
Year:								
89-90	A C	A C	B	B	B	A	A	A
90-91	A C	A C	B	B	B	A D	A	A
91-92	A C	A C	B	B	B	A D	A	A
92-93	A C	A	B C	B	B	A D	A	A
93-94	A C	A C	B	B	B	A D	A	A
94-95	A C	A C	A B	B	B	A D	A	
95-96	A C	A C		A		A D	A	A

Note: Data are overall summer presence-absence records from two 7-day pitfall trap catches per month for Carabidae (n=16 traps) or one suction sampling per month for Collembola (n=8 samples); species were absent unless indicated.

Interpreting the SCARAB Findings

The long-term differences in the abundance of Collembola under the CFP and RIA halves of fields BU and F5 (Figure 1) at first appear consistent with long-term effects of the pesticide regimes. In F5, a link with pesticide use is evident when pre-treatment and treatment-phase data are compared (Figure 1a, 1b). Differences in collembolan abundance between the regimes in F5 were largest in years when the broad-spectrum OP insecticide chlorpyrifos was applied in winter or early spring under the CFP regime [*15*]. Patterns in the catches of individual collembolan species suggested that the repeated application of OP insecticides in consecutive seasons impeded recovery of some species [*14*]. In BU field however, long-term differences in collembolan distribution are less easily interpreted because the species which exhibited differences in abundance between the CFP and RIA halves of the field were not present in pre-treatment samples (Figure 1c, 1d). Furthermore, unlike the situation in F5, relatively few broad-spectrum OP insecticides were applied to BU under the CFP regime [*14*]. Obvious links between collembolan distribution and use of individual pesticides in BU were therefore lacking. At the end of the treatment phase of SCARAB in 1996, the possible influence of pesticide use on the spatial distributions of *Lepidocyrtus* spp. (Figure 1c) and *P. octopunctata* (Figure 1d) in field BU thus could nether be confirmed nor discounted.

Beyond SCARAB: the Recovery Study

In response to the results outlined above, the SCARAB project was continued in a modified form, the Recovery Study, to address two questions: (1) were the CFP and RIA pesticide regimes responsible for the long-term patterns of collembolan abundance observed in BU; and (2) how long would recovery take for species negatively affected by CFP pesticide use in F5, if the pesticide inputs were reduced ?

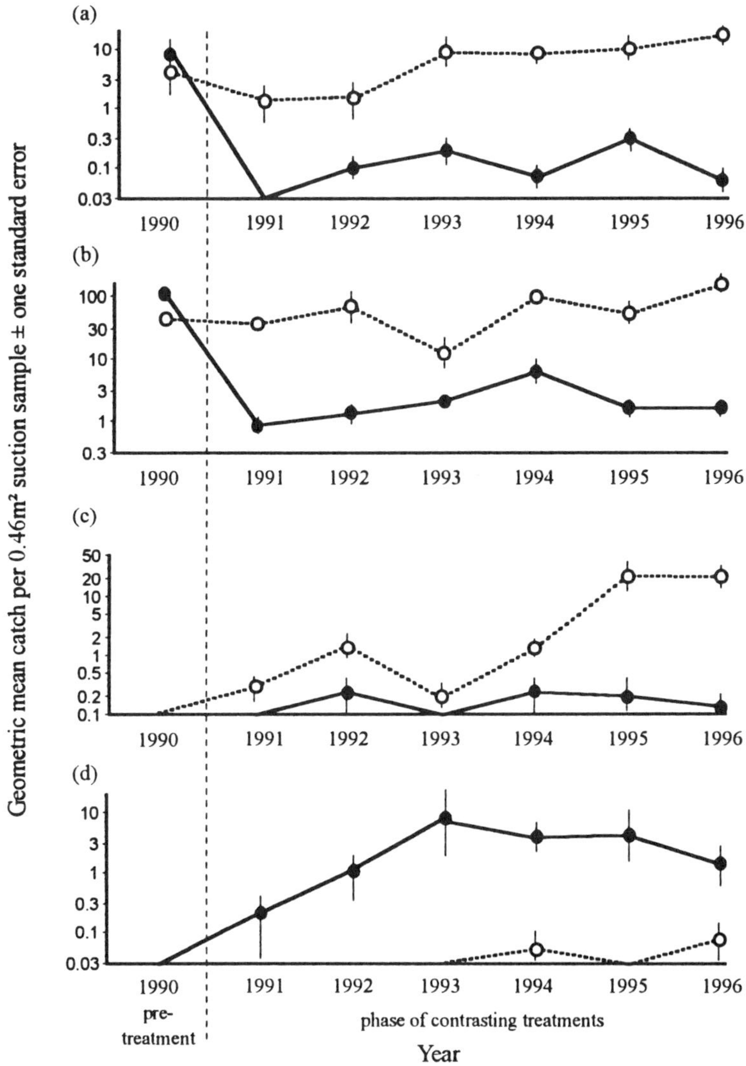

Figure 1. Collembola counts under CFP (•) and RIA (○) pesticide regimes in two SCARAB Project fields in summer (May-July inclusive), 1990-1996: (a) Entomobrya nicoleti in F5; (b) Lepidocyrtus spp. in F5; (c) Lepidocyrtus spp. in BU; (d) Pseudosinella octopunctata in BU (data from 8 to 24 samples per year).

An experimental approach to answer these questions commenced in autumn 1996. This involved spatial reversal of the CFP and RIA pesticide regimes in BU and F5 such that the half of each field formerly under CFP was switched to RIA pesticide inputs and *vice versa*. The rationale was that if the distributions of Collembola in BU (Figure 1c, 1d) resulted from pesticide use, changing the spatial arrangement of the pesticide regimes would cause changes in the collembolan distributions, though not necessarily immediately if effects were indirect. In F5, collembolan recovery would be expected when CFP pesticide inputs were replaced with the lower RIA inputs; results from SCARAB had raised the hypothesis that recovery rates would differ among species [*14*]. The Recovery Study ended in autumn 1999, allowing three years in which to detect any changes in arthropod distributions following reversal of the pesticide regimes. Here I report the latest results from the ongoing data analysis.

Manipulation of the pesticide regimes in F5 led to changes in the distribution of Collembola. When the RIA regime (no insecticides) was replaced with the higher inputs of CFP (which included insecticides), counts of several species declined (e.g. Figure 2a, 2b; [*16*]). Replacement of the former CFP regime with reduced inputs (RIA) led to little change in the abundance of *Entomobrya nicoleti* (Lubbock), indicating no discernible recovery by summer 1998 (cf. Figures 1a, 2a). However, recovery was evident for *Lepidocyrtus* spp. (cf. Figures 1b, 2b) and other species [*16*].

In field BU, counts of *Lepidocyrtus* spp. and *P. octopunctata* were generally low after 1996 but the previous distributions of these species (Figure 1c, 1d) did not exhibit any obvious change in response to spatial reversal of the pesticide regimes in autumn 1996 (Figure 2c, 2d). Highest counts of these species in BU were consistently in the same half of the field during 1991-1998, irrespective of which pesticide regime was applied to it. On the evidence of this preliminary data, the restricted spatial distributions of *Lepidocyrtus* spp. and *P. octopunctata* in BU thus appear to be independent of pesticide use.

Community Analysis: Principal Response Curves

The foregoing discussion has centred on the responses of individual species or genera to pesticide use but it would be helpful to know whether such lower taxa are important for the response of the arthropod community as a whole. Measures of species richness, similarity, diversity, trophic composition and total arthropod abundance differed between the CFP and RIA regimes of F5 [*15*] but such indices hide the relative contribution of individual species.

Multivariate analysis methods such as Redundancy Analysis (RDA) provide a means of summarising complex data sets and the resulting ordination diagrams show the relative contributions of species to the overall variance in the data [*17*]. However, changes in the size of treatment effects can be difficult to follow in an ordination diagram [*15*]. Principal Response Curves (PRC) analysis, which is based on RDA, has been developed recently to optimally display temporal changes in invertebrate communities in aquatic microcosm and mesocosm tests [*18-20*]. As the experimental design of the SCARAB Project was appropriate for PRC analysis, this technique was used to elucidate the response of the terrestrial arthropod communities to the CFP and RIA pesticide regimes. A full description of the PRC methodology is not possible here but is available in the literature [*18-20*].

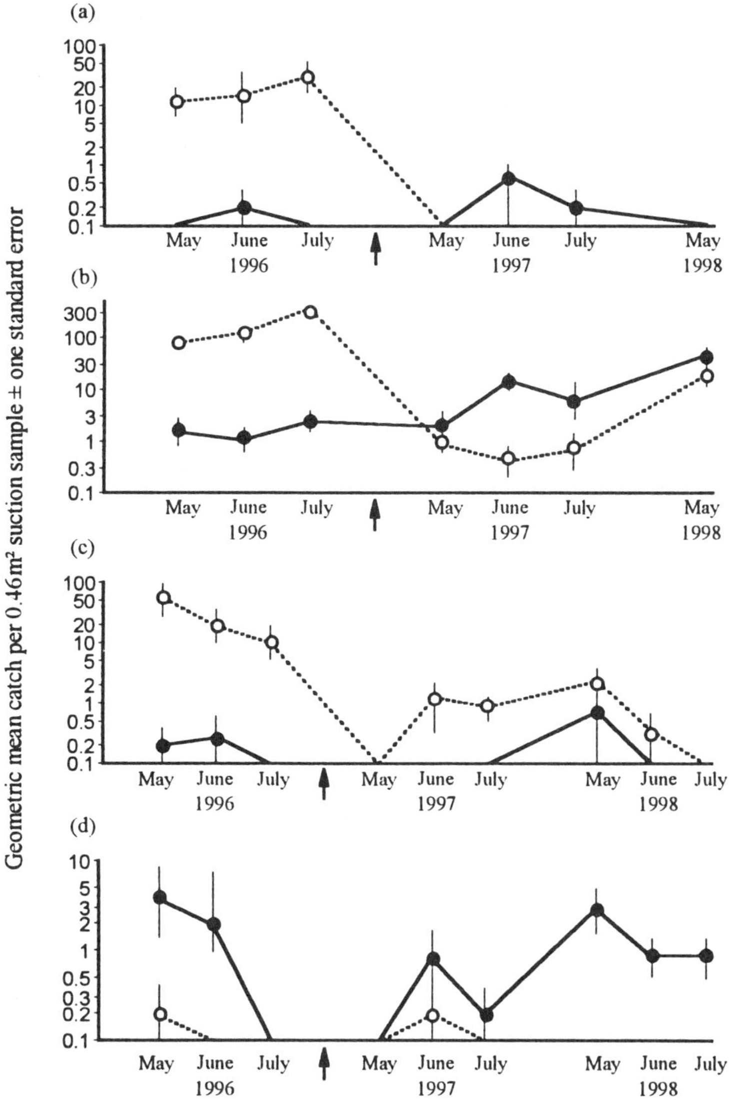

Figure 2. Collembola counts in two SCARAB Project fields under two patterns of pesticide inputs: CFP in 1991-1996 changed to RIA in 1997-1998 (•); RIA in 1991-1996 changed to CFP in 1997-1998 (∘); arrows indicate timing of the changes. Species and fields (a)-(d) are the same as in Figure 1 (data are from 8 samples per month).

The model fitted by PRC analysis is: $y_{d(j)tk} = 0_{tk} + b_k \cdot c_{dt} + e_{d(j)tk}$, where $y_{d(j)tk}$ is the log abundance of species k in sample j of treatment d at time t, 0_{tk} is the mean log abundance of species k in the control (d=0), c_{dt} is the score of the dth treatment at time t, b_k is the weight of the kth species and $e_{d(j)tk}$ is an error term with mean zero and variance 6_k^2. For the SCARAB Project data, the RIA pesticide regime was nominated as the control treatment. By definition, $c_{0t} = 0$ for every t, i.e. the abundance counts are modelled as a count in the control plus a deviation, which is calculated for each treatment level at each sampling date. The time course of the deviations from the control forms the PRC for each level of treatment. The weight of a species b_k indicates how closely the actual abundance of a species relative to the control treatment resembles the fitted relative abundance predicted by the PRC diagram. In more formal ordination terms, a PRC diagram displays the first principal component (canonical coefficient) of treatment effects (c_{dt}) obtained algebraically from a partial RDA (*18-20*). By plotting each treatment d against sampling date t, temporal changes in the overall arthropod community response to experimental treatments are displayed. With PRC analysis the contribution of individual species to the overall community response may be interpreted numerically using the species weights obtained from the analysis [*18-20*].

After ln(2x+1)-transformation of suction-sampled arthropod counts x [*21*], PRC analyses were performed using the software program CANOCO 4 [*18*]. Means of transformed counts for May, June and July in each year were used. PRC diagrams are included here for three data sets obtained by suction sampling in F5: (1) all arthropods; (2) Collembola only; and (3) arthropods excluding Collembola (Figure 3). The PRC diagrams show that Collembola made an important contribution to the overall community response; negative values of c_{dt} for the CFP pesticide regime indicate an overall negative effect of the regime on arthropod abundance relative to the RIA regime, which was nominated as the reference treatment. For brevity, no attempt is made here to numerically interpret the contribution of individual taxa to the treatment effects displayed in Figure 3 (hence, species weights b_k are not given); PRC diagrams (Figure 3) are included here to illustrate the value of the PRC method in concisely summarising temporal changes in arthropod responses to the pesticide regimes, for which purpose the values of c_{dt} may be interpreted simply as indicating the deviation in relative abundance between CFP and RIA pesticide regimes. With Collembola data omitted (Figure 3b), the effects of the CFP regime were not persistent, indicating recovery had occurred following individual OP insecticide applications in each year.

Discussion

Three key findings from the Boxworth and SCARAB projects are: (1) Collembola are more vulnerable than other arthropods to effects of pesticide use in the long term; (2) spatial and temporal distributions of species may not be properly represented in small-scale studies; and (3) the most harmful type of pesticide use for arthropods is repeated application of organophosphorus insecticides in consecutive seasons

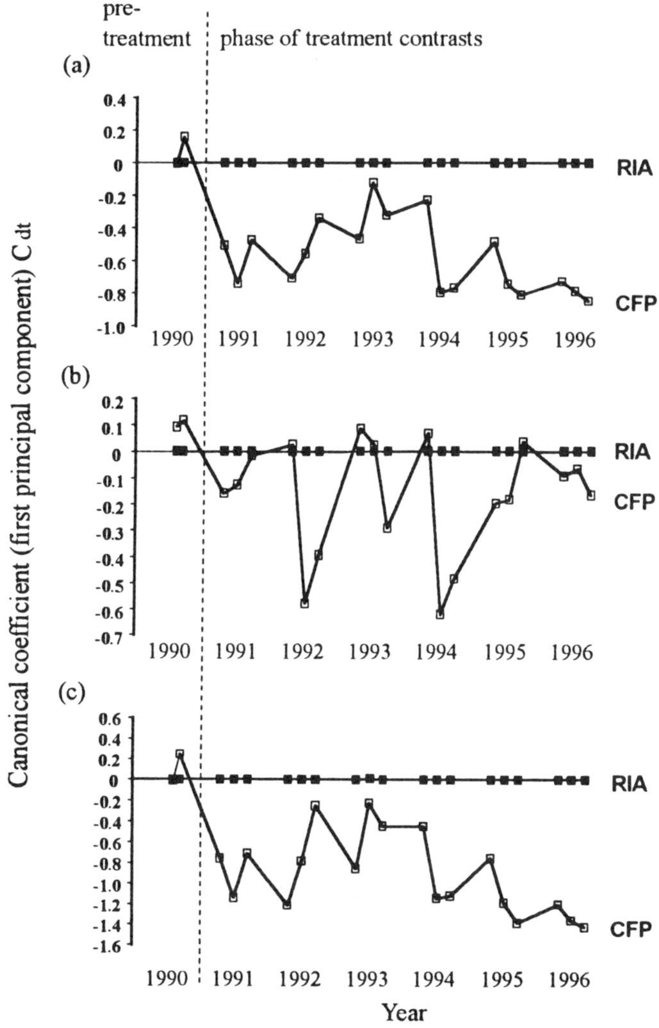

Figure 3. Principal Response Curves (PRC) diagrams summarising the response of arthropod communities in field F5 to Reduced Input Approach (RIA, reference treatment) and Current Farm Practice (CFP) pesticide regimes: (a) all arthropods; (b) arthropods excluding Collembola; (c) Collembola only. Values of C_{dt} indicate the magnitude of the impact of CFP on abundance relative to RIA.

Vulnerability of Collembola

The extensive literature on terrestrial arthropod responses to non-target effects of pesticides is biased heavily towards predatory species. Until recently hardly any work in Britain had investigated effects on Collembola of the most widely used insecticides [*22*] and fungicides [*23*]. Interpretation of pesticide effects on Collembola is difficult because, even though they may be the most abundant arthropods in temperate farmland, the ecology of many species is poorly known [*24*]. So, while the seasonal migration of carabid beetles into arable fields is a known to be major determinant of exposure and recovery, and hence vulnerability [*8, 9*], such information is lacking for Collembola. Only recently has the possibility that they may migrate seasonally into arable fields from non-crop habitats been addressed [*25*].

Recovery plays an important part in determining the significance of overall pesticide effects [*26*]; among predatory arthropods, rates of recovery depend upon species' dispersal ability and may be related inversely to the spatial scale of the experimental treatments [*27-29*], as well as other factors such as the distribution of source habitats for recolonisation and refugia from exposure [*30*]. Hypotheses to explain the slow recovery after long-term CFP pesticide inputs, e.g. of *E. nicoleti,* (Figure 2a) have yet to be tested but include: (1) poor dispersal ability at a field scale, resulting in slow recolonisation; (2) lack of a suitable source of recolonists in the vicinity; (3) low physiological potential for recovery once a population is depleted (e.g. low fecundity, either inherently or as an indirect result of pesticide use); or (4) indirect effects of pesticide use on collembolan interactions with other biota. Variation among species in recovery rates [*31*] could also reflect inter-specific variation in initial exposure to pesticide residues [*14*]. In this respect, the timing of applications appears to be important, as species with different temporal patterns of seasonal abundance may receive a different degree of exposure to a particular pesticide [*14*]. It has been established for predatory arthropods that dispersal ability and reproductive potential are important factors influencing recovery rates [*8, 9*] but such information is lacking for Collembola.

Importance of Species Distributions in Space and Time

The spatial and temporal distributions of species directly influence the interpretation of pesticide effects observed in the field. It seems to be a paradigm that species particularly vulnerable to effects of pesticide use are patchily distributed in farmland. The carabid beetle *Bembidion obtusum* is vulnerable to broad-spectrum insecticides [*8, 32*] but has a fragmented spatial distribution in Britain [*33*] and was absent from some fields in the SCARAB project (Table 2). *Pterostichus cupreus* is used as a bioindicator species in regulatory testing with pesticides [*34*] but is also patchily distributed in Britain [*33*] and was absent from most of the SCARAB project fields (Table 2). The collembolan *E. nicoleti* was vulnerable to the CFP regime of pesticide use in SCARAB but occurred almost exclusively in one field (F5). If vulnerable species are absent from a study site [*22*], effects of pesticides on the wider environment may be underestimated. Conversely, effects may be overestimated if vulnerable species present at a study site are not actually widespread in farmland and so would not normally be exposed to particular scenarios of pesticide usage. It is difficult to design field studies which encompass all the possible spatial and temporal distributions of potentially important species; even contiguous fields which shared the same cropping history, soil type and agrochemical inputs differed in their collembolan species composition [*22*].

In the SCARAB Project the absence of two species of interest during the pre-treatment period of monitoring in field BU (Figure 1c, 1d) suggests that a longer period of pre-treatment monitoring would have been desirable. However, the temporal unpredictability of some arthropods makes it difficult to know *a priori* when certain species will occur in a field. For instance, in field NK, *Entomobrya multifasciata* (Tullberg) quickly became abundant in samples during 1995 and 1996 after several years of absence [*14*]. To include a pre-treatment sampling period long enough in duration that such temporally unpredictable species are likely to be present would be impractical in most field studies where site availability and technical resources are at a premium. From an anthropocentric standpoint, availability of funding and the need for researchers to publish research papers do not weigh in favour of long-term studies [*35*], despite the need for such studies in ecology, as illustrated by the results from Boxworth, SCARAB and other projects [*3, 36, 37*].

Conclusions

Results from the SCARAB Project and Recovery Study show that spatial manipulation of pesticide regimes within a single field can provide valuable information on the responses of arthropods to pesticide use. This experimental approach was the only available means of confirming or refuting potential effects of the pesticide regimes on *P. octopunctata,* given that the species was not present during pre-treatment monitoring in the only SCARAB Project field in which it occurred. The manipulative approach as outlined here thus has potential value as a means of confirming effects of experimental treatments upon species which are highly restricted in spatial distribution.

The importance of Collembola in determining the response of the overall arthropod community to pesticide use was shown clearly using PRC analysis. This method has been used mainly in aquatic ecotoxicological studies to date but is also appropriate for summarising the long-term responses of terrestrial arthropods to effects of pesticide use. The next step in the analysis of the SCARAB Project and Recovery Study data will be to apply the PRC method to investigate how spatial reversal of the pesticide regimes affected the overall arthropod community. PRC analysis is well suited to the study of recovery because changes in the size of treatment effects over time are easily discerned from a PRC diagram (*18-20*).

Repeated use of OP insecticides in most years was the only scenario of pesticide use which had a long-term negative effect on arthropods, as a result of effects on Collembola, the most abundant taxon. This probably represents a worst-case scenario of pesticide use in Britain [15] but has direct relevance to cropping systems worldwide in which high usage of OP insecticides occurs.

Acknowledgements

The Boxworth, SCARAB, and TALISMAN projects were funded by the UK Ministry of Agriculture, Fisheries and Food. Staff at the Agricultural Development and Advisory Service managed the experimental sites, pesticide regimes and arthropod sampling. Paul Van den Brink of the Alterra Research Institute in Wageningen, The Netherlands, performed the PRC analyses on SCARAB data. These people, together with members of the SCARAB and TALISMAN Steering Group and colleagues at Southampton University, are gratefully acknowledged for their contributions to the work reported here.

Literature Cited

1. Royal Commission on Environmental Pollution. *Seventh Report - Agriculture and Pollution*; HMSO: London, 1979; 280 pp.
2. Potts, G.R. In *The Ecology of Temperate Cereal Fields*; Firbank, L.G.; Carter, N.; Darbyshire, J.F.; Potts, G.R., Eds; Blackwell Scientific Publications: Oxford, 1991; pp 3-21.
3. Aebischer, N.J. In *The Ecology of Temperate Cereal Fields*; Firbank, L.G.; Carter, N.; Darbyshire, J.F.; Potts, G.R., Eds; Blackwell Scientific Publications: Oxford, 1991; pp 305-331.
4. Greig-Smith, P.W. In *Pesticides, Cereal Farming and the Environment*; Greig-Smith, P.W.; Frampton, G.K.; Hardy, A.R., Eds; HMSO: London, 1992; pp 1-5.
5. *Pesticides, Cereal Farming and the Environment*; Greig-Smith, P.W.; Frampton, G.K.; Hardy, A.R., Eds; HMSO: London, 1992; 288 pp.
6. Holland, J.M.; Frampton, G.K.; Çilgi, T.; Wratten, S.D. *Annals of Applied Biology* **1994**, 125, 399-438.
7. Greig-Smith, P.W.; Hardy, A.R. In *Pesticides, Cereal Farming and the Environment*; Greig-Smith, P.W.; Frampton, G.K.; Hardy, A.R., Eds; HMSO: London, 1992; pp 6-18.
8. Burn, A.J. In *Pesticides, Cereal Farming and the Environment*; Greig-Smith, P.W.; Frampton, G.K.; Hardy, A.R., Eds; HMSO: London, 1992; pp 110-131.
9. Vickerman, G.P. In *Pesticides, Cereal Farming and the Environment*; Greig-Smith, P.W.; Frampton, G.K.; Hardy, A.R., Eds; HMSO: London, 1992; pp 82-109.
10. Cooper, D.A. In *Proceedings 1990 Brighton Crop Protection Conference - Pests and Diseases*; British Crop Protection Council, 1990; pp 153-162.
11. Greig-Smith, P.W.; Griffin, M.J. In *Pesticides, Cereal Farming and the Environment*; Greig-Smith, P.W.; Frampton, G.K.; Hardy, A.R., Eds; HMSO: London, 1992; pp 200-215.
12. Frampton, G.K. In *Ecotoxicology*; Haskell, P.T.; McEwen, P., Eds; Kluwer Academic Publishers: Dordrecht, The Netherlands, 1998; pp 292-300.
13. Hancock, M.; Frampton, G.K.; Çilgi, T.; Jones, S.E.; Johnson, D.B. In *Ecology and Integrated Farming Systems*; Glen, D.M.; Greaves, M.P.; Anderson, H.M., Eds; Wiley: Chichester, 1995; pp 289-306.

14. Frampton, G.K. *Pedobiologia* **1997**, 41, 179-184.
15. Frampton, G.K. In *Reducing Agrochemical Use on the Arable Farm*; Young, J.E.B.; Alford, D.V.; Ogilvy, S.E., Eds; MAFF: 2000 (in press).
16. Frampton, G.K. *Pedobiologia* **2000** (in press).
17. *Data Analysis in Community and Landscape Ecology*; Jongman, R.J.H.; Ter Braak, C.J.F.; Van Tongeren, O.F.R., Eds; Cambridge University Press: Cambridge, 1995; 299 pp.
18. *CANOCO 4*; Ter Braak, C.J.F.; Smilauer, P; Microcomputer Power: Ithaca, New York, 1998; 351 pp.
19. Van den Brink, P.J.; Ter Braak, C.J.F. *Aquatic Ecology* **1998**, 32, 163-178.
20. Van den Brink, P.J.; Ter Braak, C.J.F. *Environmental Toxicology and Chemistry* **1999**, 18, 138-148.
21. Van den Brink, P.J.;Van Donk, E.; Gylstra, R.; Crum, S.J.H.; Brock, T.C.M. *Chemosphere* **1995**, 31, 3181-3200.
22. Frampton, G.K. *Pesticide Science* **1999**, 55, 875-886.
23. Frampton, G.K.; Wratten, S.D. *Ecotoxicology and Environmental Safety* **2000** (in press).
24. Rusek, J. *Biodiversity and Conservation* **1998**, 7, 1207-1219.
25. Alvarez, T.; Frampton, G.K.; Goulson, D. *Pedobiologia* **2000** (in press).
26. Kelly, J.R.; Harwell, M.A. *Environmental Management* **1990**, 14, 527-545.
27. Jepson, P.C.; Thacker, J.R.M. *Functional Ecology* **1990**, 4, 349-355.
28. Thomas, C.F.G.; Hol, E.H.A.; Everts, J.W. *Functional Ecology* **1990**, 4, 357-368.
29. Duffield, S.J.; Aebischer, N.J. *Journal of Applied Ecology* **1994**, 31, 263-281.
30. Halley, J.M.; Thomas, C.F.G.; Jepson, P.C. *Journal of Applied Ecology* **1996**, 33, 471-492.
31. Frampton, G.K. In *Proceedings Fourth International Conference on Pests in Agriculture, Montpellier, January 1997*; Association Nationale de Protection des Plantes (ANPP): Paris, 1997; 129-136.
32. Frampton, G.K.; Çilgi, T. *In Carabid Beetles - Ecology and Evolution*; Desender, K.; Dufrêne, M.; Loreau, M.; Luff, M.L.; Maelfait, J.-P., Eds; Kluwer Academic Publishers: Dordrecht, The Netherlands, 1994; pp 433-438.
33. *Provisional Atlas of the Ground Beetles (Coleoptera, Carabidae) of Britain*; Luff, M.L.; Biological Records Centre, Institute of Terrestrial Ecology: Monks Wood, England, 1998; 194 pp.
34. *Guidance Document on Regulatory Testing Procedures for Pesticides with Non-Target Arthropods*; Barrett, K.L.; Grandy, N.; Harrison, E.G.; Hassan, S.; Oomen, P., Eds; Society of Environmental Toxicology and Chemistry - Europe: Brussels, Belgium, 1994; 51 pp.
35. Findlay, S.E.G.; Jones, C.G. In *Long-term Studies in Ecology*; Likens, G.E., Ed.; Springer-Verlag: New York, 1988; pp 201-202.
36. Eaton, J.S.; McDonnell, M.J. In *Long-term Studies in Ecology*; Likens, G.E., Ed.; Springer-Verlag: New York, 1988; pp 189-191.
37. Woiwod, I.P. In *The Ecology of Temperate Cereal Fields*; Firbank, L.G.; Carter, N.; Darbyshire, J.F.; Potts, G.R., Eds; Blackwell Scientific Publications: Oxford, 1991; pp 275-304.

Chapter 6

The German Environmental Specimen Bank: Application in Trend Monitoring of Chlorinated Hydrocarbons

P. Marth[1,2], D. Martens[1,2], K.-W. Schramm[1], J. Schmitzer[1], K. Oxynos[1], and A. Kettrup[1,2]

[1]GSF-National Research Center for Environment and Health, Institute of Ecological Chemistry, Ingolstaedter Landstrasse 1, D-85764 Neuherberg, Germany
[2]Technische Universitaet Muenchen Lehrstuhl für Oekologische Chemie und Umweltanalytik, D-85350 Freising-Weihenstephan, Germany

In addition to real-time monitoring of environmental concentrations of chemicals, environmental specimen banks (ESBs) present an appropriate instrument for retrospective monitoring of chemicals in the future. ESBs contain representative and systematically collected biological samples. As a result of low temperature storage (< -150 °C) the sample are not subject to chemical changes during long-term storage. The German ESB is described here, and selected results related to environmental trend monitoring of chlorinated hydrocarbons are presented.

The growing number of environmental pollutants necessitate far-reaching control and monitoring strategies that aim toward the early detection and tracking of pollutants. This is necessary for precautionary environmental protection. Effective risk assessment and management strategies can only be implemented if they are based on reliable data. Against this background, environmental specimen banking (ESB) has gained international recognition as an important component of long-term environmental monitoring and ecotoxicological research over the last decades *(1-4)*. The banking system in Germany developed from a pilot ESB program initiated in 1979 by the Federal Ministry for Research and Technology. During this period, the technical feasibility with respect to the sampling of different species, handling and shipping of samples, deep freezing, homogenization, ultra trace analysis, packing materials, logistics, storage temperature, and documentation was confirmed *(5)*. In 1985, the decision was made to establish a permanent ESB under the responsibility of the Federal Ministry for the Environment, Nature Conservation and Nuclear Safety (BMU), coordinated by the Federal Environmental Agency (UBA). The German ESB with its two branches, the Human Tissue Specimen Bank and the Environmental

Specimen Bank (non-human samples), is a national archive of selected specimens for real time monitoring and retrospective analysis.

Definition

The German ESB is a systematic archive of regularly collected representative environmental specimens from all over Germany *(6-7)*. Because of extremely low storage temperatures (< –150 °C), samples are not subject to chemical changes during long-term storage. The samples preserved in this manner provide the basis for real-time, retrospective and prospective analysis, evaluation of pollutants and identification of the sources of contaminants and their entry points into biological matrices.

Tasks of ESB

The concept of ESB to archive environmental samples for retrospective analysis has been recognized as an integral part of systematic environmental monitoring and covers the following purposes *(8,9)*:

- ESB specimens provide an early warning system of well characterized samples taken according to stringent sampling protocols capable of establishing historical trends of previously unidentified contaminants (retrospective monitoring). They can be analyzed immediately whenever indications on a new chemical hazards become evident. If suitable specimens are available in EBSs, analyses related to the specific time periods and locations can be carried out to elucidate the current and previous environmental burden. Thus rapid action can be taken to reduce environmental impacts, and costs can be cut down considerably.
- Specimens stored in ESBs for a long period of time can be used to evaluate long-term temporal and spatial trends of key contaminant concentrations (real-time contaminant monitoring). They provide a long-term database to determine trends in contaminant burdens and serve as a basis for the improved prediction of various environmental variables. Based on these data, the effect of legislative measures to protect the environment can be evaluated and further risk mitigation strategies can be recommended.
- Specimen banking has an inherent role in assuring the quality of analytical results. The storage of samples under cyrogenic conditions according to standard protocols may serve as reference samples for the documentation of the improvement of analytical efficiency and for the verification of previously obtained monitoring results.
- An ESB is an useful tool for providing reliable and well-documented data on pollutant burden of earlier times. Differences in sampling procedures and analytical methods have made interpreting results difficult. Often only little

information about the condition of the animal and life history data is available, although these data are important factors for interpreting contaminant data.

Sampling Areas and Specimen Types

A range of different sampling areas, representing typical ecosystems in Germany have been chosen by a committee of experts under the auspices of the BMU. Since the reunification of the two German states, representative areas in the former East Germany have been integrated into the ESB routine sampling program.

Continuous sampling is now performed in the following areas:

- The national parks of mud flats in Schleswig-Holstein and Lower Saxony (North Sea) and the Baltic Sea as marine ecosystems.
- River Elbe and river Rhine as limnic ecosystems.
- Saarland and the Halle/Leipzig/Bitterfeld area as urban-industrialized regions.

The number of sampling areas will be further expanded. The final state is shown in Figure 1. For screening purposes, individual samples are also collected in the National Parks of Berchtesgaden and Bavarian Forest in order to determine regional differences in the biology and contaminant exposure of species.

Representative specimens have been selected which reflect the pollution situation of different ecosystems (Figure 1). These species represent different trophic levels of the food chain. Their chemical characterization indicates the amount of contaminants that is potentially bioavailable.

Standard Operating Procedures

Standard operating procedures (SOPs) are the basis for the comparability and reliability of the banked samples. A manual on methods of collecting, processing, storing and analyzing biological samples stored in the German ESB has been produced *(11)*. They contain detailed instructions for:

- selection of sampling sites and specimens
- sampling procedures
- providing cover for repeatability of sampling
- area and sample characterization
- sample treatment and long-term storage
- documentation of sampling and storage conditions
- chemical analysis
- data processing and evaluation
- quality assurance *(12)*.

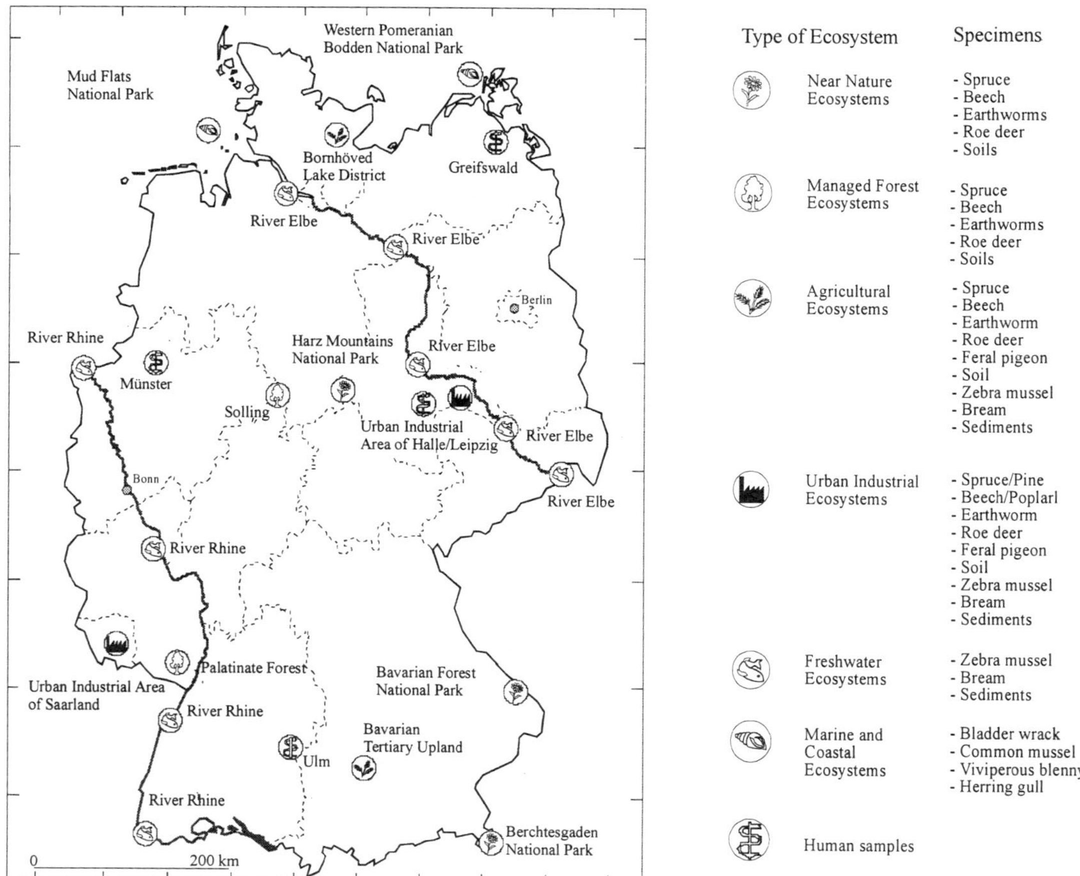

Figure 1. Sampling areas and specimens of the German ESB program (10).

Specimen Sampling, Processing and Storage

Samples are generally sampled once a year during the same season in quantities of about 2.5 kg (wet weight) material per specimen and sampling site. During sampling, a variety of biological data (e.g., length, weight, age) are determined. Biometric sample characterization is a prerequisite for the correct interpretation of results from chemical analysis. The majority of environmental specimens are immediately deep-frozen on site in the gas phase over liquid nitrogen to avoid chemical alterations. An exception to this are earthworms which are kept alive in Petri dishes for five days at 10-12 °C prior to deep-freezing. This procedure assures complete defecation. Thus the contaminant residues represent the biologically available fraction only. The samples are kept in a permanent cyro-chain until further processing. The single samples are homogenized using a cyrogenic grinding procedure designed to provide a homogeneous frozen fresh powder. In this procedure, samples remains frozen during the operation, thus minimizing losses of volatile compounds and avoiding sample degradation due to thawing and refreezing. The powders are divided into sub-samples of about 10 g fresh mass. Most sub-samples are stored in cyro-containers at liquid nitrogen temperatures whereas some are used for an initial chemical characterization.

Analytical Sample Characterization

The choice of pollutants or classes of pollutants for the analytical sample characterization was made according to ecotoxicological importance:

Chlorinated hydrocarbons (CHCs): Pentachlorobenzene (PCBz), Hexachlorobenzene (HCB), α-,β-,γ-Hexachlorocyclohexane (HCHs), Heptachlor (HC), Heptachlorepoxide (HE), Aldrin, Dieldrin, o,p-DDT, p,p'-DDT, p,p'-DDE, p,p'-DDD, Octachlorostyrene (OCS)

- Polychlorinated biphenyls (PCBs) (B28, B52, B101, B138, B153, B180)
- Chlorinated phenols (CPs)
- Polychlorinated dibenzo-p-dioxins and dibenzofurans (PCDD/F)
- Polycyclic aromatic hydrocarbons (PAHs)
- Inorganic compounds (elements, metal organic substances).

The results presented here are restricted to CHCs, PCBs, CPs, and PCDD/F.

The ESB methodology for CHCs and PCBs determination has been described in detail elsewhere *(13)*. Briefly, six sub-samples are mixed with sodium sulfate and sea sand to form a free flowing product, which is extracted with *n*-hexane/acetone (2:1 v/v) in an extraction column. The majority of lipids are removed by size exclusion chromatography (SEC). The CHC fraction is isolated by normal-phase liquid chromatography (LC) on a semi-preparative-scale silica gel column. The fraction is analyzed by high resolution gas chromatography (HRGC) equipped with electron capture detection using two columns of different polarity.

The PCDD/F analysis is carried out by isotope dilution methodology. The *n*-hexane/acetone extract of one sub-sample is purified by several liquid chromatography steps. Clean-up procedures are performed with alumina, sulfuric acidified silica gel, and florisil columns. The identification and quantification is done with capillary HRGC/MS using the EI mode (for details see *(14)*).

For the CPs determination, the samples are hydrolysed with KOH, steam-distilled and extracted with *n*-hexane. CPs are separated from other impurities by solvent-solvent extraction with 0.2 M K_2CO_3 and acetylated by acetic anhydride. CP concentrations are determined by GC-MS and isotope dilution (for details see *(15)*).

Safety Considerations

Organochlorine compounds may be harmful to humans by inhalation, ingestion and skin adsorption. Some of these substances are possible carcinogens and may have toxic effects upon human reproduction. Any contact and inhalation of these substances is avoided. Therefore, gloves, coats and safety glasses are used when processing and handling the samples in order to avoid any contact with and inhalation of these substances.

Results and Discussion

One of the tasks of the German ESB is to monitor the organochlorine concentration in various environmental matrices representing the marine, limnic, and terrestrial ecosystem, and to detect trends in environmental pollution.

Toxic substances such as heavy metals, pesticides, and persistent chlorinated organic chemicals have been found in many species, particularly in top predators of the marine food chain like fish-eating birds. Eggs of herring gull *(Larus argentatus)* have been sampled since 1988 and can be used to assess the chemical contamination of coastal areas along the islands of the North Sea (Wadden Sea) and Baltic Sea (Figure 2). The islands are located in the estuaries of river Weser (Island Mellum), Elbe (Island Trischen), and Oder (Island Hiddensee, sanctuary Heuwiese). In addition, eggs are sampled on the Island Poel (sanctuary Walfisch). This sampling area is not affected by a large river estuary. Female herring gulls feed predominantly on fish species living in the vicinity of the bird sanctuary.

Although a decrease in PCDD/F concentration has been observed until 1992, the further trend is not clear as yet (Figure 3). Further analyses are necessary to confirm the decline of PCDD/F in the environment and to verify the successful implementation of federal risk mitigation programs, e.g., the banning of sea burning of hazardous waste.

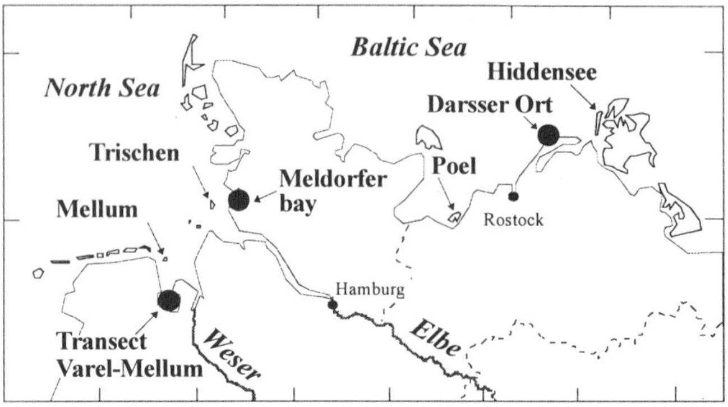

Figure 2. Sampling areas of herring gull eggs and eel-pouts for the German ESB program.

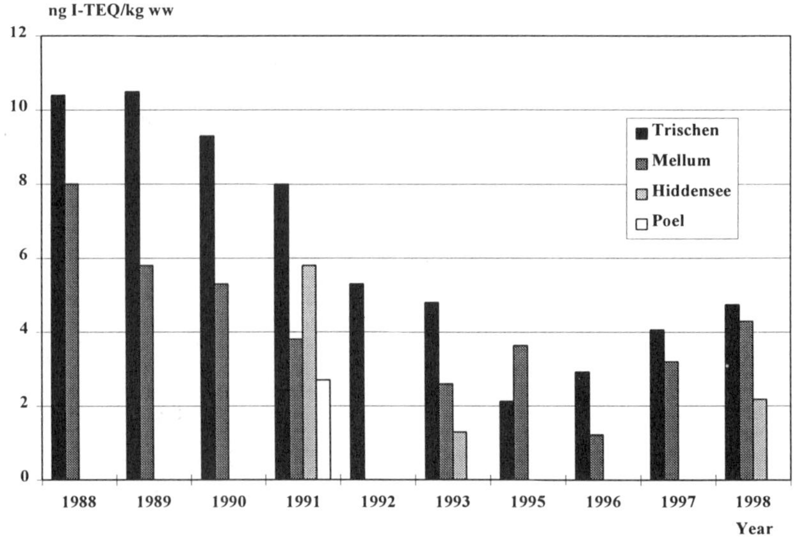

Figure 3. Time series of herring gull eggs from four sampling locations in the North and Baltic Sea (sub-sample of pooled sample, n=1) (ng I-TEQ/kg ww : ng International Toxicity Equivalents/kg wet weight).

The eel-pout (*Zoarces viviparus L.*) has a wide distribution in marine areas of Europe. It is found in the mud flats of the German Wadden Sea and the shallow waters of the Baltic Sea. Eel-pout is sedentary during the entire lifecycle and is, therefore, an appropriate indicator to monitor chemical substances and their environmental impact in particular areas *(16)*.

The highest PCB concentrations were found in eel-pouts liver samples from transect Varel-Mellum which starts at the river Weser (North Sea) where PCBs are the dominant pollutants (Figure 4). Relatively low values of PCBs were determined in fish samples from the Baltic Sea compared to the North Sea, which appears to be due to the limited use of PCBs in the former German Democratic Republic (GDR) (20,000 t between 1955-1985 (*17*)).

High levels of DDT metabolites in samples at the station Darsser Ort may be caused by the continued application of DDT in the former GDR after the banning of DDT in Western Europe *(18)*. Elevated levels of DDD at the Meldorf Bay sampling area could be due to the influence of the river Elbe. This observation is consistent with high concentrations of DDT metabolites found in bream (*Abramis brama*) muscle tissue from the river Elbe *(19)*.

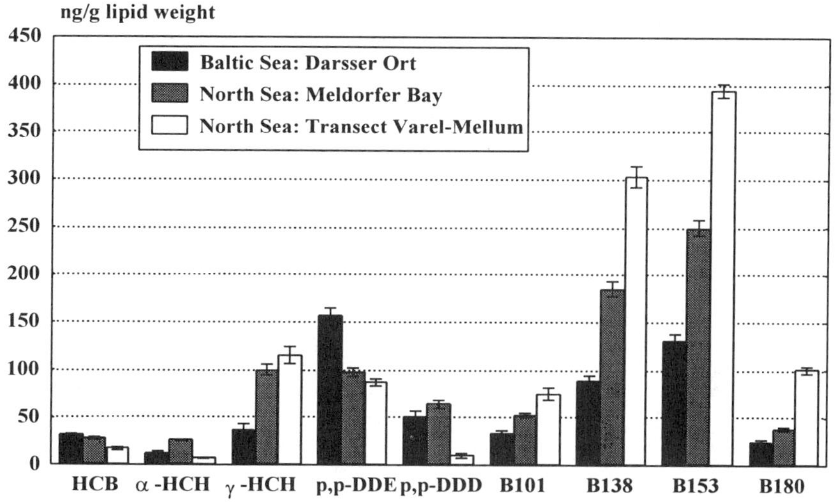

Figure 4. Spatial comparison of mean CHC concentrations in eel-pout liver from North and Baltic Sea in 1997. The bar denotes SD. (Sub-samples of pooled sample, n=6,6,3)

Bream are an ideal organism to monitor freshwater and sediment contamination owing to their small migration radius and their habit to feed in direct contact with the sediment. Bream were caught from different sampling sites along five German rivers (Elbe, Rhein, Saar, Mulde, Saale). The sampling stations were selected with regard to various types and magnitudes of industrial pollution (Figure 5).

In order to characterize local variations of the CHC burden of bream from different limnic ecosystems, comparisons of the pollution situation are drawn in Figure 6.

It is obvious that the sampling sites of the river Elbe exhibit the highest concentrations of HCB and DDE. High values of these contaminants were also found in samples from the river Mulde, a tributary of the river Elbe. This is the result of considerable pollution of both rivers from industrial wastewater discharges of the chlorine industry in the former GDR *(18)*.

In West-Germany, both HCB and DDE concentrations were considerably lower. Elevated levels of PCB were detected in bream from the lower Rhine at Bimmen and from the lower Saar at Rehlingen. The increasing pattern along the river Rhine is partly a result of the inflow of the river Mosel (upstream of Koblenz), containing high amounts of PCB *(20)*. The river Saar is strongly influenced by the emission of the mining industry (e.g., hydraulics oil containing PCB) situated in the Saar region *(21)*.

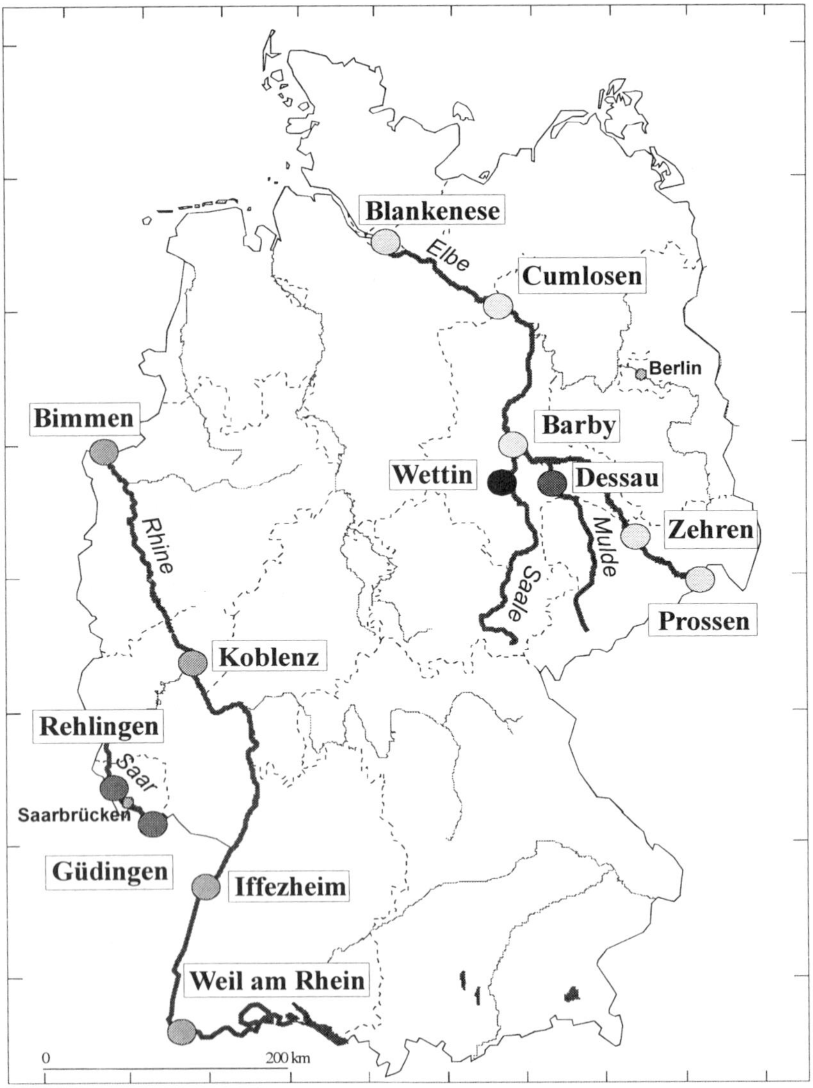

Figure 5. Sampling sites of bream samples, 1997.

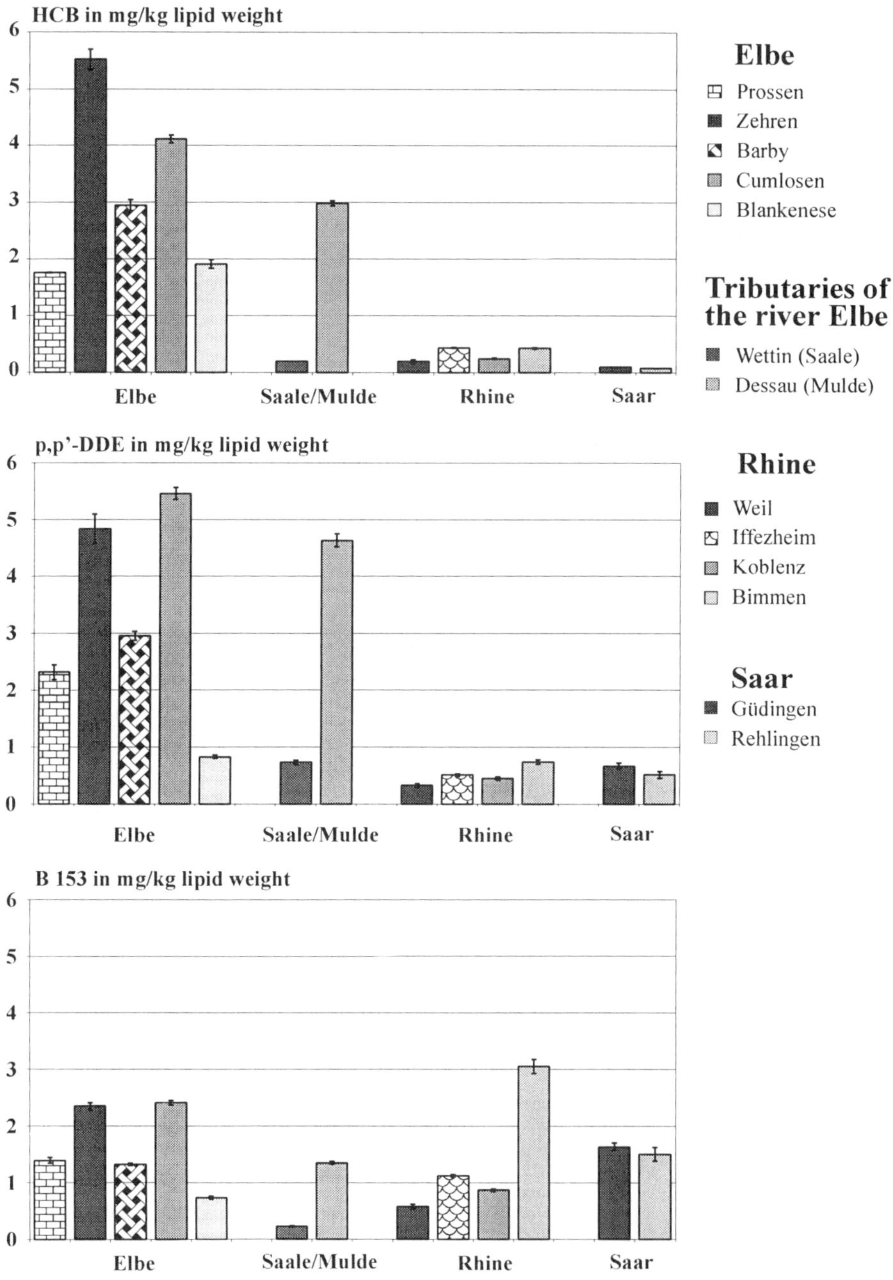

Figure 6. Spatial comparison of mean CHC concentrations in bream muscle from different German rivers in 1997. The bar denotes SD. (Sub-samples of pooled sample, n=6,6,6,6,6,5,6,5,6,6,6,5,6)

Earthworms are virtually ubiquitous and ecologically important soil organisms. As decomposers they are involved in nutrient cycling and are exposed to soil pollutants *(22)*. Deeply burrowing species (*Lumbricus terrestris, Aporrectodea longa*) reflect the biologically available soil contaminants in terrestrial environments.

Sub-samples of pooled samples representing terrestrial ecosystems of typical industrial areas in West-(Saarland) and East (Leipzig) Germany have been analysed for CPs, and allow an evaluation of the environmental pollution over a period of four years (Table I).

High CP concentrations were found in earthworms from both sampling sites in contrast to the corresponding CHC levels *(23)*. Other methylated phenols, guaiacols and catechols were also detected. The CP concentrations exhibit no or only a slight decline over a period of 4 years, except pentachlorophenol (PCP), which shows a strong decrease in samples from Saartal over the last 3 years.

CPs were determined in nearly all samples analyzed which demonstrates the ubiquitous distribution of CPs, especially PCP, in the environment of Germany, without great differences between West- and East Germany.

Conclusion

Based on reliable and well-documented analytical procedures, the results offer the opportunity for long-term monitoring of environmental pollution and detection of spatial differences and time-dependent trends. Apart from long-term storage , the ESB serves as an effective database on organic and inorganic substances in specimens from widely separated geographical areas. The integration of real-time contaminant monitoring with specimen banking provides important baseline data that can be used to plan and manage banking activities. This includes identifying appropriate specimens that are useful in assessing temporal trends and increasing the utility of the banked samples in assessing chemical contaminant accumulation and relationships to biological effects. Specimen banking will enable future investigators to extend their research into the past, thereby increasing their time-line of study. Therefore, it is important to maintain and expand specimen banks or tissue archives associated with long-term environmental monitoring.

References

1. Iyengar, G.V. and Subramanian, K.S. In *Environmental Biomonitoring, Exposure Assessment and Specimen Banking, ACS Symposium Series 654;* Subramanian, K.S; Iyengar, G.V.; Ed.; American Chemical Society : Washington, DC, 1997, pp. 220-245.
2. Zeisler, R. In *Environmental Biomonitoring, Exposure Assessment and Specimen Banking, ACS Symposium Series 654;* Subramanian, K.S; Iyengar, G.V.; Ed.; American Chemical Society : Washington, DC, 1997, pp. 246-253.

Table I. Mean chlorophenol concentrations in ng/g wet weight in earthworms from Saartal and Leipzig (1994-1997)

Location	*Year*	*2.3.4-Trichlorophenol*			*2.3.5-Trichlorophenol*			*2.3.6-Trichlorophenol*			*2.4.5-Trichlorophenol*			*2.4.6-Trichlorophenol*		
		$\overline{x}$	*SD*	*n*	$\overline{x}$	*SD*	*n*	$\overline{x}$	*SD*	*n*	$\overline{x}$	*SD*	*n*	$\overline{x}$	*SD*	*n*
Saartal	1994	0.9		1	0.8		1	2,0		1	11,8		1	217,1		1
	1995	0.9	0.01	2	0.9	0.03	2	2,5	0,02	2	12,2	0,02	2	262,4	4,06	2
	1996	0.8	0.04	5	0.8	0.03	5	1,6	0,07	5	9,1	0,39	5	154,8	7,46	5
	1997	1.0	0.06	4	0.7	0.08	4	2,1	0,08	4	10,2	0,25	4	162,3	3,35	4
Leipzig	1994	1.2	0.31	2	1.1	0.34	2	2,0	0,73	2	7,9	1,72	2	209,3	1,01	2
	1995	1.6	0.24	2	1.3	0.20	2	2,7	0,43	2	9,9	1,55	2	280,5	0,89	2
	1996	1.1	0.03	5	0.9	0.02	5	1,4	0,11	5	7,6	0,26	5	149,4	0,33	5
	1997	1.2	0.03	4	0.7	0.02	4	1,4	0,05	4	6,2	0,23	4	123,1	1,07	4

Location	*Year*	*2.3.4.6-Tetrachlorophenol*			*2.3.5.6-Tetrachlorophenol*			*Tetrachloroguaiacol*			*Pentachlorophenol*		
		$\overline{x}$	*SD*	*n*	$\overline{x}$	*SD*	*n*	$\overline{x}$	*SD*	*n*	$\overline{x}$	*SD*	*n*
Saartal	1994	6.7		1	1.3		1	2,0		1	33,0	2,16	2
	1995	19.7	0.51	2	1.6	0.04	2	2,0	0,17	2	112,6	21,53	3
	1996	25.6	2.75	5	2.4	0.27	5	1,2	0,11	5	66,6	2,76	5
	1997	6.3	0.39	4	1.2	0.07	4	2,0	0,08	4	26,0	0,54	4
Leipzig	1994	3.6	0.52	2	1.4	0.42	2	1,8	0,50	2	23,6	4,54	3
	1995	4.8	0.75	2	2.0	0.31	2	2,7	0,43	2	17,9	3,93	3
	1996	8.6	0.75	5	2.7	0.18	5	1,2	0,09	5	28,4	0,90	5
	1997	2.2	0.13	4	0.9	0.04	4	1,2	0,07	4	17,3	0,13	4

3. Kluge, E.-H. and Subramanian, K.S. In *Environmental Biomonitoring, Exposure Assessment and Specimen Banking, ACS Symposium Series 654;* Subramanian, K.S; Iyengar, G.V.; Ed.; American Chemical Society : Washington, DC, 1997, pp. 254-270.
4. Zhang, P.Q.; Chai, C.F.; Lu, X.L.; Qian, Q.F.; Feng, W.Y. In *Environmental Biomonitoring, Exposure Assessment and Specimen Banking, ACS Symposium Series 654;* Subramanian, K.S; Iyengar, G.V.; Ed.; American Chemical Society : Washington, DC, 1997, pp. 271-276.
5. Boehringer, U. R. *Umweltprobenbank: Bericht und Bewertung der Pilotphase*; Bundesministerium für Forschung und Technologie, Umweltbundesamt, Springer-Verlag: Berlin, 1988.
6. Emons, H.; Schladot, J. D.; Schwuger, M. J. *Chemosphere* **1997,** *9/10*, pp. 1875-1888.
7. Schramm, K.-W.; Kettrup, A.; Schmitzer, J.; Marth, P.; Oxynos, K. *TEN* **1996,** *3*, pp. 43-49.
8. *Biological Environmental Specimen Banking;* Stoeppler, M.; Zeisler, R., Eds.; Sci. Tot. Environ., BESB special issue, Vol. 139/140, 1993.
9. *Biological Environmental Specimen Banking;* Emons, H., Eds.; Chemosphere, BESB 2 special issue, Vol. 34 (9/10), 1997.
10. Klein, R.; Paulus, M.; Wagner, G.; Müller, P. *UWSF - Z Umweltchem Ökotox* **1994**, *6*, pp. 223-221.
11. *Federal Environmental Specimen Bank: Standard Operating Procedures for Sampling, Transport, Storing, and Chemical Characterization of Environmental Specimens and Human Organ Specimens (in German);* Umweltbundesamt, Eds.; Erich Schmitt Verlag: Berlin, 1996; ISBN 3-503-03963-5.
12. Wagner, G. In *Ecoinforma '94;* Alef, K.; Blum, W.; Schwarz, S.; Riss, A.; Fiedler, H.; Hutzinger, O., Ed.; Bayreuth, 1994, Vol. 5, pp. 457-462.
13. Oxynos K.; Schmitzer J.; Dürbeck H.W.; Kettrup, A. In *Specimen Banking;* Rossbach, M.; Schladot, J.D.; Ostapczuk, P., Ed.; Springer-Verlag: Berlin, 1992, pp. 127-137.
14. Schramm, K.-W.; Henkelmann, B.; Kettrup, A. *Wat Res* **1995,** *29*, pp. 2160-2166.
15. Martens, D.; Prachar, V.; Amberg, S.; Oxynos, K.; Schramm, K.-W.; Kettrup, A. *Int J Environ Anal Chem* **1997,** *68*, pp. 415-427.
16. Jacobsen, A.; Neumann, E.; Thoresson, G. *Ambio* **1986**, *15*, pp. 236-238.
17. Von Schaaffhausen, J.; Gramenz J. P. *Aufkommen und Verbleib von PCB-Abfällen in der ehemaligen DDR sowie Herkunft und Entsorgung von PCB-Abfällen aus wehrtechnischen Geräten, aus Geräten bei der Bundesbahn (Reichsbahn) und Bundespost und von PCDM-Abfällen aus dem Bergbau*; BSM-Beratungsgesellschaft für Umweltfragen, final report of the UBA project 103 50 204, Düsseldorf, 1993.
18. Heinisch, E. *Umweltbelastung in Ostdeutschland-Fallbeispiele: Chlorierte Kohlenwasserstoffe*; Wissenschaftliche Buchgesellschaft: Darmstadt, 1992.

19. Marth, P.; Oxynos, K.; Schmitzer, J.; Schramm, K.-W.; Kettrup, A. *Chemosphere* **1997**, *34*, pp. 2183-2192.
20. Internationale Kommission zum Schutze des Rheins (IKSR) Statusbericht Rhein, Koblenz, 1993.
21. Breitung, V. *Vom Wasser* **1992**, *79*, pp. 39-47.
22. Weigmann G. *Water Air Soil Pollut* **1991**, *57/58*, pp. 655-663.
23. *Umweltprobenbank*, Annual Report of the Bank of Environmental Specimens 1997 (in German), Jülich 1998.

Chapter 7

Locust and Tsetse Fly Control in Africa: Does Wildlife Pay the Bill for Animal Health and Food Security?

Ralf Peveling and Peter Nagel

Institute of Environmental Sciences–Biogeography, University of Basle, St. Johanns-Vorstadt 10, 4056 Basle, Switzerland

Broad spectrum insecticides are used to control tsetse flies in the more humid parts and locusts in the arid and semi-arid parts of Africa. Control operations are concentrated in natural ecosystems and affect a wide range of wildlife communities. The large scale use of pesticides adds to other important pressures on wildlife such as habitat loss and fragmentation and often conflicts with conservation policies. In this paper we present results from several environmental monitoring studies conducted in Africa over the last two decades and evaluate the risk of these control strategies to wildlife, in particular birds, reptiles and small mammals.

Africa presents its human population with the worst set of environmental constraints and problems of any of the major land masses of the world (*1*). Drought and desertification, deforestation and soil erosion, tropical diseases of man and livestock are just a few of the most frequently used keywords to illustrate the African case. Africa faces problems which are unique to the continent, either in quality or in extent. In the drier areas, which occupy about two thirds of the continent, the limitations to food security are manifold, with unpredictable and highly variable rainfalls being the most important. However, green pastures and flowering fields of millet and sorghum are by no means a guarantee of secured food since a multitude of pests may demand their share, with locusts and grasshoppers being among the most feared visitors (*2*). Locust control has become an incentive of national and international concern, not only to reduce crop losses but also to sustain traditional cropping systems and to motivate farmers not to abandon agricultural production (*3*). In the more humid parts of Africa, agriculture is more secure, but meat is a resource in short supply. Vector-borne diseases such as animal trypanosomiasis (nagana) limit livestock production and threaten the health of the rural population (sleeping sickness). Despite some 50 years of large scale tsetse fly control, the tsetse-transmitted diseases nagana and sleeping sickness are still considered a major obstacle to the agricultural development of whole regions. For example, the FAO estimates that crop production in countries like Ethiopia could be doubled if infested areas became tsetse-free (*4*).

After the Second World War, chemical pesticides such as DDT and other organochlorines (OCs) were considered the ultimate tool to combat both locusts and disease vectors. The expectations were high: Africa's cattle stock would, it was predicted, increase by 120 million head if trypanosomiasis were brought under control (*5*). While no longer used in African agriculture, DDT still plays a role in the control of disease vectors such as mosquitoes, and its continued use is strongly advocated by public health professionals (*6*).

Yet Africa is not only home to a variety of unique pests and diseases but also hosts one of the most spectacular and diverse faunas of the world, including many endemic species of restricted distribution (*7*). Apart from the Orient, the Afrotropical realm harbours the highest number of threatened vertebrates among all zoogeographic regions of the world (*8*). Thus there is great concern that excessive and indiscriminate use of insecticides against migratory pests and disease vectors could further threaten African wildlife.

Locust and Tsetse Fly Control – Different Objectives, Similar Consequences

Except for the highest mountains and the most arid deserts, there is no place in Africa without either locusts or tsetse. Very often people have to deal with both problems at a time, particularly in the transition zone between semi-arid and sub-humid savannahs (Figure 1).

There are several reasons to address the problems caused by locusts and tsetse flies together – despite their different nature. First, locusts and tsetse flies have a complementary distribution. Second, control operations are usually carried out in natural and often pristine landscapes, thus directly targeting important wildlife habitats and refuges. Third, pesticides are applied on very large scales. For example, more than 250,000 km^2 of land were sprayed during the 1986-89 desert locust plague in Africa and the Middle East (*3*), and the current campaign against migratory locust in Madagascar affected more than 30,000 km^2 since 1997. The joint tsetse control programme in Zambia, Zimbabwe, Malawi and Mozambique covered a fly belt of about 320,000 km^2 since 1985. Finally, until recently, similar pesticides were used in locust and tsetse control.

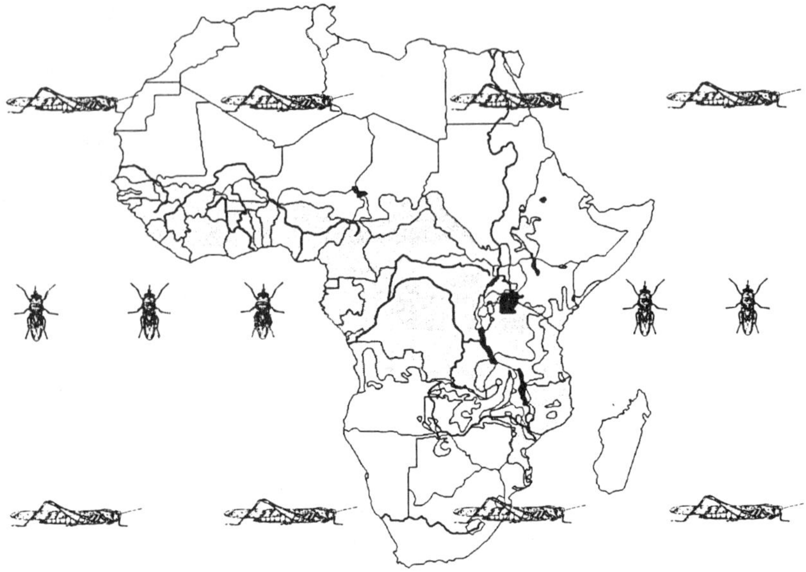

Figure 1. The distribution of tsetse flies (grey area) and locusts (white area) in Africa.

Important Locust and Tsetse species

Locusts

Five locust species occur in the semi-arid and arid zones of Africa. They all belong to the orthopteran family Acrididae. The desert locust, *Schistocerca gregaria* (FORSKÅL), is considered the most destructive species. Its invasion area extends from Senegal to Bangladesh and from Tanzania to Kazakhstan, covering an area of approximately 29 million km^2 (*9*). The African migratory locust, *Locusta migratoria migratorioides* (REICHE & FAIRMAIRE), is distributed in all semi-arid parts of Africa. The most recent plague lasted from 1928-42. However, the Malagasy subspecies *Locusta migratoria capito* (SAUSSURE) represents a permanent threat to grain crops (maize and rice) in Madagascar, and a recent outbreak in 1997 has not yet been brought under control. The red locust, *Nomadacris septemfasciata* (AUDINET-

SERVILLE), is mainly distributed south of the equator but can migrate into northern Somalia and into the Nile valley in Sudan. *Locustana pardalina* (WALKER), which is confined to southern Africa, does not show typical population cycles with an irregular sequence of recession (solitary phase) and invasion (gregarious phase) periods like other locusts but rather maintains gregarious populations in its Karoo outbreak area in South Africa, necessitating permanent control to prevent the spread of swarms into the invasion zone. The Sahelian tree locust, *Anacridium melanorhodon* WALKER, is a major pest of gum arabicum plantations (*Acacia senegal* (L.) WILLD.) in the Sahel, with several subspecies distributed up to southern Europe and the Middle East.

Tsetse Flies

Tsetse flies belong to the dipteran family of Glossinidae which is endemic to Africa (Figure 1). Glossinid flies occupy an area of approximately 10 million km^2 south of the Sahara (*10*). The northern distribution limit coincides with the 450-500 mm isohyet of annual rainfall in the Sahel area, while temperature rather than rainfall limits the distribution in southern Africa (*11*). All of the 23 species are potential vectors of *Trypanosoma* spp. (Protozoa: Trypanosomatidae), the agents of human sleeping sickness and of nagana in cattle. Trypanosomiasis of domestic animals is more widespread than the infection of man (*5*) which is confined to endemic foci (*12*). *Glossina morsitans* WESTWOOD is distributed in the dry savannahs from West to East and southern Africa and is the main vector of nagana. Other important species include *G. longipalpis* WIEDEMANN in riverine forests of West Africa and *G. pallidipes* AUSTEN in the moist savannahs of East and Southeast Africa. Tsetse flies occur mostly at low densities and rest for most of the day in the foliage of trees and shrubs. Host species include a wide range of mammals and reptiles, and both male and female flies take one blood meal every 2-4 days. Infested wildlife are not affected by disease and represent important reservoirs of *Trypanosoma* spp.

Control Agents and Strategies

An overview of control agents used in desert locust and tsetse control is provided in Table I. The quoted dose rates for desert locust are those recommended by FAO (*13*). Effective dosages against other locusts may slightly differ, depending on the target species. The dose rates for tsetse control represent a range of doses used in several African countries over the last 30 years.

Locust control

The early detection of gregarious locust populations in recession areas is paramount to the success of the currently practised preventive control strategy. It involves the use of remote sensing techniques as well as regular surveys on the ground by national locust units. The objective is to prevent the establishment of gregarious populations and to contain outbreaks in the recession areas. Thus control operations are carried out whenever gregarisation is detected, using ultra low volume (ULV) spraying techniques. The aim is to hold locust populations in the recession phase (solitary locusts at low density) indefinitely (*14*). From the 1950s until 1985, the organochlorine dieldrin was extensively used to obtain cheap and effective control (*15*). Dieldrin remains active for several weeks and was sprayed in widely spaced barriers (up to 5 km apart) about 100-150 m wide. Migrating hopper bands which cross a barrier would collect a lethal dose before reaching the adult stage. The low frequency of desert locust upsurges in the period 1960-85 has been interpreted as an indication of the efficacy of this control strategy (*14*). In the late 1980s, OCs were largely replaced by organophosphates (OPs), carbamates and pyrethroids which were applied more frequently and on a larger scale (blanket treatment) to be equally effective. Thus the environmental benefit of using non-persistent insecticides was offset by acute threats to non-target fauna.

Consequently, the barrier spraying technique was resumed in the 1990s using insecticides with moderate persistence such as benzoylureas (insect growth regulators) and phenylpyrazoles. A breakthrough in environmentally friendly locust control was achieved with the development of a mycopesticide, the entomopathogenic fungus *Metarhizium anisopliae* var. *acridum* (Deuteromycotina: Hyphomycetes), which is highly specific to locusts and grasshoppers and can be used in environmentally sensitive areas (*16*). However, the share of mycopesticides in the locust pesticide market, which is dominated by organophosphates and pyrethroids, is as yet very low because commercialisation only began in 1999.

Tsetse fly control

Residual application of organochlorine insecticides was the preferred control technique until the 1980s (*11*). Owing to the high dose rates and the long persistence of OCs, complete eradication of tsetse could theoretically be achieved with single applications. Following the withdrawal of DDT (for aerial treatment) and dieldrin, non-residual sequential aerosol application of low doses of non-persistent (e.g. deltamethrin) or moderately persistent insecticides (e.g. endosulfan) over a period of several months during the dry season was used in large-scale tsetse eradication campaigns. Spraying operations are conducted with low-flying fixed-wing aircraft during temperature inversions at dusk or during the night. A spray cloud with fine droplet sizes (< 40 μm) is produced with ULV rotary atomisers. The cloud remains airborne and slowly drifts into the vegetation and to the resting sites of tsetse. Since pupae of various ages in the soil are not exposed to the spray, the treatment has to be repeated several times to eliminate the surviving generations. Ground spraying of DDT with knapsack compression sprayers is still practised in Zimbabwe (*17, 18*) and can be as effective as aerial control. In this technique, DDT is selectively applied to tsetse resting sites and refuges which usually cover about 20% of the total surface. Alternative control strategies include the sterile male technique (SMT) and the use of

Table I. Insecticides and Dose Rates Used in Desert Locust (Hoppers) and Tsetse Fly Control.

	Effective dose rate (g a.i. ha^{-1})				
	Desert locust		*Tsetse flies*		
			Residual		*Non-residual*
Insecticide	*Full cover*	*Barrier* [a]	*Full cover*	*Ground* [b]	*Full cover*
Organochlorine					
DDT	-	-	330-720	≤1,000	-
(Dieldrin)	-	10-20	150-1,500	≤4,000	-
(Lindane)	-	-	40-100	-	-
Endosulfan	-	-	250-1,200	-	6 -56
Organophosphate					
Chlorpyrifos	225	-	-	-	-
Fenitrothion	450	-	-	-	-
Malathion	925	-	-	-	-
Carbamate					
Bendiocarb	100	-			
Pyrethroid					
Deltamethrin	12.5	-	11- 30	12.5	0.1- 0.4
λ-Cyhalothrin [c]	20	-	-	-	-
Permethrin	-	-	50- 300	-	1.9- 4.3
Benzoylurea					
Diflubenzuron [d]	60	100	-	-	-
Teflubenzuron [d]	30	n.d.	-	-	-
Triflumuron [d]	25	75	-	-	-
Phenylpyrazole					
Fipronil	5	12.5	-	-	-
Fungal pathogen					
M. anisopliae	100	-	-	-	-

NOTE: Products in parenthesis are phased out. DDT is still used in ground control against tsetse and other disease vectors

a = dose within treated barriers, spacing varies with vegetation density and hopper movements
b = treatment of resting sites (≈ 20% of the total area), residual dose within treated sites
c = used in cattle dips and to treat baits for tsetse control
d = tested as sterilising agents on artificial targets for tsetse control
n.d. = not determined; - = insignificant or no use; Sources: (*11, 13, 18, 19, 66, 67*)

stationary attractive devices such as insecticide-impregnated and odour-baited traps and screens of cloth. SMT is restricted in use to eliminating isolated tsetse populations (e.g. on the island of Zanzibar). Traps and odour-baited targets are now widely used in West and in southern Africa. They combine visual (shape and colour) and olfactory cues (e.g. a mixture of 3-n-propyl phenol, 1-octen-3-ol, and 4-methyl phenol) to attract flies. They are fabricated from special material and can be impregnated with pyrethroids or sterilising agents such as benzoylureas (*19*). Apart from SMT, these devices are the most selective and environmentally acceptable control method but require regular maintenance.

Impact of Locust and Tsetse Control on Wildlife

The database on the risk of pesticides to African wildlife is limited. Most studies have focused on birds (*20, 21, 22, 23, 24*) and fish (*11, 20, 24, 25*), whereas observations on mammals, amphibians and reptiles are sparse and often anecdotal (*24, 26, 27*). In general, side-effects of tsetse control have been better studied than side-effects of locust control. This imbalance of scientific effort seems to reflect the different population strategies of locusts and tsetse flies. Tsetse are considered as K-selected which makes them a more predictable target not only for control operators and environmental scientists but also for funding agencies. Agencies are more reluctant to support "R-selected" locusts whose occurrence is unpredictable in time and space.

Examples of locust and tsetse control environmental impact studies are presented below. Four individual case studies have been selected to illustrate effects of former and current locust and tsetse control operations on non-target vertebrates. Two of these studies address the problem of wildlife exposure to organochlorine pesticides. One was carried out in Niger one year after the 1986-89 desert locust control campaign. This campaign was probably the last one in which OCs such as lindane and dieldrin – though largely replaced by OPs, carbamates and pyrethroids – were still used on a significant scale against locusts. Likewise, the study on residues in wildlife following residual application of dieldrin against tsetse flies in Cameroon demonstrates effects of control techniques practised from the 1950s to the early 80s. The other two studies focus on the field toxicity of fenitrothion-esfenvalerate and triflumuron to reptiles in Madagascar and of endosulfan to fish in Zambia. OP-pyrethroid mixtures and benzoylureas have been used against the Malagasy migratory locust since the early 1990s (*28*). Large-scale aerosol applications of endosulfan against tsetse were conducted from the late 1970s to the mid 90s. In most African regions, these tsetse control tactics have recently been replaced by stationary attractive devices, livestock baits and SMT. However, residual applications with deltamethrin and aerosol applications with endosulfan or deltamethrin are still realistic options.

The four case studies are part of different environmental monitoring programmes which focussed on non-target invertebrates. In this paper we only present results for vertebrates. Results for invertebrates and more detailed descriptions of the field studies have been published elsewhere (*11, 29, 30, 31, 32, 33*).

Organochlorine Residues in Desert Wildlife (Niger 1990)

Methods

The study was conducted in a desert locust recession area near In-Abangharit in the Tamesna desert. The site is characterised by sand dunes interspersed with granite outcrops. Widely scattered patches of *Schouwia thebaica* WEBB vegetation were the predominant habitat of the invertebrate fauna and also attracted birds and reptiles. The main objective was to study the impact of desert locust control agents on non-target arthropods. However, since there was evidence that OCs such as dieldrin, which was still on stock in In-Abangharit, had been used in previous years despite the official ban, we also sampled wildlife to assess the OC burden of the fauna. Ten specimens of nine different bird and reptile species were sampled by hand or with nets. Two samples of domestic animals were also taken. The animals were sacrificed and immediately dissected in the field station. Internal organs were removed and stored individually over liquid nitrogen in a dewer. OCs were extracted in Germany (University of Saarbrücken) and analysed by gas chromatography/electron capture detection, using previously described methods (*34*). Residues presented here are for liver (wet weight), except in the fringe-toed lizard, *Acanthodactylus boskianus* (DAUDIN) (internal organs without intestines). Total DDT residues (Σ DDT) are the sum of all o,p' and p,p' isomers of DDT, DDD and DDE. No correction was made to account for post-mortem conversion of DDT to DDD (*17*) since the samples were stored over liquid nitrogen to minimise biodegradation.

Results and Discussion

The concentration of OCs in desert wildlife generally low (Table II). DDT was detected in all specimens and dieldrin and lindane (γ-HCH) in the majority. β-HCH was only found in the European teal, *Anas crecca* L., a palearctic migrant which had obviously failed to cross the Sahara and was stranded in the study area (0.075 mg kg^{-1}). The highest OC body load was found in two insectivorous species, the red-breasted chat, *Oenanthe heuglini* (FINSCH & HARTLAUB) (Σ DDT = 3.95 mg kg^{-1}, γ-HCH = 0.1 mg kg^{-1}), and the desert swallow, *Hirundo obsoleta* (CABANIS) (Σ DDT = 3.21 mg kg^{-1}, γ-HCH = 0.09 mg kg^{-1}). The ratio DDT/DDE ranged from 0.04 to 0.8. Ratios of ≥ 0.5 are considered as an indication of new inputs of DDT into the environment (*35*). In the present study, higher ratios were only recorded in two top predators, the little owl, *Athene noctua* (SCOPOLI) (0.75), and the desert monitor, *Varanus griseus* (DAUDIN) (0.8), providing evidence of recent exposure to DDT. Possible sources were old DDT stocks used against desert locust during the 1986-89 campaign or atmospheric transport from areas treated against disease vectors in the Guinea zone further south. However, the exact spray history of the Tamesna area could not be traced back.

The overall residues in desert wildlife and domestic animals were below reported hazard levels (*24, 27, 36, 37*). For example, reproductive failure in birds due to eggshell-thinning (*18*) requires DDE concentrations about one order of magnitude higher than those observed in the present study (≤ 3.3 mg kg^{-1}). The relatively low

Table II. Organochlorine Residues in Desert Wildlife and Domestic Animals, Tamesna Desert, Niger 1990

Insecticide	*n*	Concentration in liver ($mg\ kg^{-1}$ freshweight) *γ-HCH*	*Dieldrin*	*Σ DDT*	*ratio DDT/DDE*
Mammals					
Goat	1	0.002	0.001	0.008	0.3
Birds (granivorous & herbiv.)					
Streptopelia turtur	1	0.001	0.002	0.330	< 0.1
Anas crecca	1	n.d.	0.001	0.185	< 0.1
Domestic chicken	1	0.002	0.005	0.033	< 0.1
Birds (insectiv. & carnivorous)					
Oenanthe heuglini	1	0.1	0.02	3.950	< 0.1
Sylviidae indet.	1	0.035	n.d.	0.070	< 0.1
Athene noctua	1	0.015	0.02	0.037	0.7
Hirundo obsoleta	1	0.09	n.d.	3.210	< 0.1
Alaudidae indet.	1	0.02	0.01	0.230	0.1
Reptiles (insectiv. & carniv.)					
Acanthodactylus boskianus[a]	2	0.042	0.025	0.240	0.4
Varanus griseus	1	0.005	0.015	0.018	0.8

a = mean of two specimens; n.d. = not detected

environmental concentration of OCs only one year after their possible use in the area reflects the rapid breakdown under the Tamesna desert conditions, where temperatures can exceed 50°C in summer. Most species sampled for residue analysis were resident to the area. However, birds have wide home ranges in desert environments where their food resources may be dispersed over large areas. Thus, residues in birds can hardly be associated with particular locust treatments in a particular region. In contrast, reptiles have more limited powers of migration and dispersal (*27*) and are considered representative indicators of the state and pesticide burden of the study site.

Effects of Fenitrothion-Esfenvalerate and Triflumuron on Lizards (Madagascar 1995)

Methods

The study was conducted towards the end of the rainy season in a wooded savannah in the migratory locust recession area of Southwest Madagascar. It was part of a large-scale field trial on the side-effects of the benzoylurea triflumuron (TFM) and the organophosphate-pyrethroid mixture fenitrothion-esfenvalerate (FE) on non-target arthropods (*29*). Two plots of 800 ha were sprayed, one with TFM (34 g a.i.

ha^{-1}) and one with FE (206.8 g fenitrothion + 4.2 g esfenvalerate ha^{-1}) Each treatment plot was paired with an untreated control of similar size at a distance of 1-2 km. The relative abundance of the endemic lizard *Chalarodon madagascariensis* PETERS (Iguanidae) (*38*) was monitored for four weeks before and twelve weeks after treatment (March to June 1995). Mean monthly temperatures (min-max) decreased from 21-36°C in March to 19-33°C, 17-31°C, and 13-29°C in April, May, and June, respectively. The total rainfall was 87.2 mm. Counts were made along fixed transects in the centre of each plot. The five transects, each 0.5 km in length, were oriented East-West at a distance of ≈ 100 m from each other. Numbers of *C. madagascariensis* were recorded twice weekly by walking the 2.5 km route at constant speed and counting all lizards seen in a belt of approximately 20 m. Lizards were most conspicuous when basking or perching on termite mounts but were also visible on the ground due to the relatively open vegetation (35-50% cover). Adults and juveniles were distinguished by size. Treated and paired control plots were always monitored on the same day by the same person, starting at 1100 h in the treatment plot and at 1400 h in the control. The order was reversed on the next sampling occasion to avoid temporal sampling bias, even though a preliminary survey showed that the variation between late morning and early afternoon counts was very low compared to the variation among plots. The mean of the two weekly counts (mean sightings) was taken as a measure of relative abundance.

Prior to statistical analysis, mean sightings were log (n+1) transformed, and the *difference (D)* treatment minus paired control was calculated for each sampling date (week). These derived values were analysed as pseudo-replicates in time (*39*) with one-way ANOVA by comparing the mean pre-treatment difference (D_{before}) and the mean post-treatment differences (D_{after}) of three different time phases, each comprising four successive weeks: 1-4 (phase I), 5-8 (phase II), and 9-12 (phase III). Differences among means of D were analysed with Tukey HSD test. The *effect (E)* (= change in relative population density corrected for control fluctuations) was calculated for each of the three phases according to $E (\%) = 100 \cdot (1 - (C_b \cdot T_a/C_a \cdot T_b))$, where C_b and T_b are the mean sightings *before* treatment in the control (C_b) and treatment plot (T_b), and C_a and T_a are the mean sightings *after* treatment in the control (C_a) and treatment plot (T_a). A significant effect was assumed if D_{before} and $D_{I,II,III}$ proved significantly different at $P \leq 0.05$. For the graphical presentation, we also calculated the percent *change (C)* in relative abundance, or mean weekly sightings, compared to mean pre-spay levels. Data for TFM and FE were analysed independently.

Results and discussion

The overall relative population density of *C. madagascariensis* was always higher in the south (TFM and paired control) than in the north (FE and paired control) of the experimental area, resulting in the following order: FE < FE-control < TFM < TFM-control (Figure 2). This sequence, which did not change despite strong fluctuations over time, reflects microhabitat and food preferences of *C. madagascariensis*. Though mainly insectivorous, *C. madagascariensis* occasionally feeds on flowers of low growing legumes such as *Mundulea genistoides* (DUMAZ-LE GRAND) and *Indigofera mouroundavensis* (BAILL) which occurred more frequently in the TFM plot and, in particular, in the TFM-control than in the other plots.

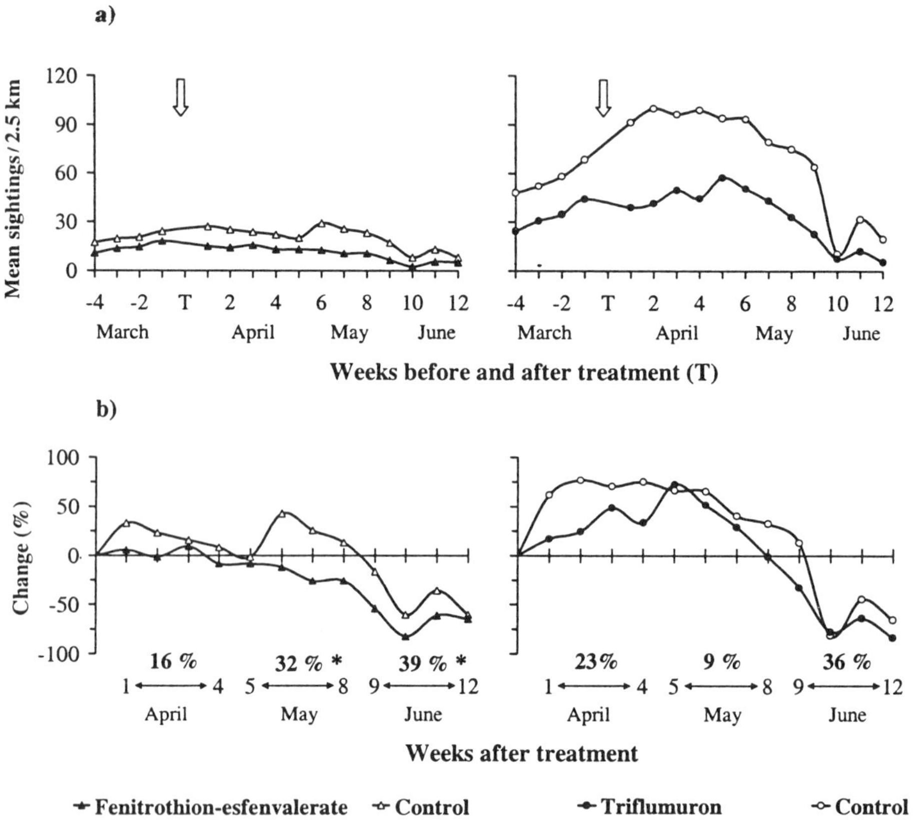

Figure 2. Effect of fenitrothion-esfenvalerate and triflumuron on *Chalarodon madagascariensis*. a) Mean sightings per 2.5 km before and after treatment (arrow); b) Change in relative abundance compared to pre-spray levels. The figures are effects (%) during three different periods: 1-4, 5-8 and 9-12 weeks post-spray; * = $P < 0.05$.

During the four weeks before treatment (March), mean sightings of *C. madagascariensis* increased in all plots, owing to the hatching of juveniles (Figure 2a). This trend continued during the first phase after treatment, and an overall maximum of 100 (2nd week), 58 (5th week), and 29 (6th week) lizards was recorded in the TFM-control, TFM and FE-control, respectively, whereas no further increase occurred in the FE treatment compared to the pre-spray maximum of 18 lizards. Sighting rates in the other plots remained high until 5-6 weeks post-treatment (mid May), but dropped to levels lower than before treatment in June. The decline was due to the tendency of adult and sub-adult *C. madagascariensis* to stay in their burrows during most of the day at the onset of the winter season (*38, 40*). A high proportion of juveniles, on the other hand, was still active during daytime from May to mid June. The change in age structure is reflected by the proportion of juveniles among sighted (= active) lizards which increased from 53.1% pre-spray (March) to 69.1, 72.5, and 98.3% post-spray in April (phase I), May (phase II), and June (phase III), respectively. However, the real age structure of the population and the proportion of non-active juvenile or sub-adult lizards was not know.

Differences among treatments are more evident when comparing the change in abundance relative to pre-spray levels (Figure 2b); up to 77% (2nd week), 73% (5th week), and 43% (6th week) more lizards were encountered after spraying in the TFM-control, TFM and FE-control respectively. In contrast, the highest percentage recorded in the FE treatment was only 9% (3rd week). The seasonal population decline (negative change) in the FE treatment started earlier than in the other groups (4 weeks *versus* 9-10 weeks post-treatment). This temporal shift indicated a moderate effect of FE on *C. madagascariensis*. Effects were 16% (n.s.) in April, 32% ($P < 0.05$) in May, and 39% ($P < 0.05$) in June. Similar effect levels were observed in the TFM treatment (23%, 9%, 36%). However, these were not statistically significant.

The impact of pesticides on vertebrates can be through direct toxicity or, indirect, through food depletion. For example, non-breeding insectivorous birds can respond to food shortages by moving into untreated areas. Thus decreases in bird abundance have been associated with the depletion of insect food rather than direct mortality (*21, 22*). In contrast, lizards such as *C. madagascariensis* are unlikely to move into untreated areas since they are not only highly territorial but also have a low dispersal capacity. The parallel study on non-target invertebrates did not reveal relevant food shortage at all (*29*). The reduction of epigeal arthropods such as ants, termites, spiders and grasshoppers, which are the preferred prey of *C. madagascariensis*, was only short-lived and certainly not limiting. Therefore, direct toxicity, in particular to juvenile lizards, appeared to be the more likely cause for the moderate population reduction in the FE treatment, even though no victims were found in the field. The relative toxicity of some organophosphorus pesticides to iguanid lizards is similar to that in mammals (*41*). Chlorpyrifos has a high oral and dermal toxicity to the West African fringe-toed lizard *Acanthodactylus dumerili* MILNE EDWARDS (Peveling and Demba, unpublished data). However, the evidence of side-effects of OPs on African lizards in the field is scarce (*27*). Chlorpyrifos caused heavy mortality in *A. boskianus* and *A. dumerili* in Mauritania during desert locust treatments (authors' observation). Such kills are not exceptional but were confirmed on several occasions by Mauritanian control teams.

Dermal absorption of pesticide deposits on the ground rather than secondary poisoning via contaminated prey appeared to be the main route of exposure. *C. madagascariensis* occupies the same niche in Madagascar as fringe-toed lizards in Mauritania and shows similar morphological and behavioural traits. Therefore, we assume a similar exposure regime for this species.

In conclusion, the current study showed that *C. madagascariensis* can be adversely affected by fenitrothion-esfenvalerate applied at a comparatively low field dosage. Given a maximum reduction of 39%, single applications are unlikely to have lasting effects on the population of this relatively abundant iguanid lizard. However, a higher risk can be anticipated in species with more restricted distribution, in particular if the same area is treated several times, and when using pure OP formulations which are applied at higher dosages than OP-pyrethroid mixtures (Table I). Therefore, the opinion that locust control with OPs does not result in serious mortality of non-target vertebrates (*22*) cannot be upheld when taking the risk to reptiles into account.

Dieldrin Residues in Savannah Wildlife (Cameroon 1979-84)

Methods

A tsetse eradication programme was conducted during 1976-84 in the highlands of Adamaoua, Cameroon, covering about 10,000 km^2 (*32*). Control operations were concentrated on riverine forests, the dry season habitat of *G. morsitans*, in the western part of the plateau. Dieldrin was applied aerially by helicopters as a residual spray at 750-900 g a.i. ha^{-1}. The sprayed gallery forests represented 15-20% of the total area to be cleared from tsetse (forests plus savannah). Wildlife was sampled for residue analysis in a zone sprayed in 1979. Samples were collected once before and 0.5, 8, 12, 24, 36, 48 and 60 months after treatment, respectively. Birds and bats were caught with mist nets, and small mammals such as rodents and shrews with live traps, using peanut butter as bait. Some of the animals were sacrificed and further processed for residue analysis as described above. However, liquid nitrogen storage was only available from the 3rd sampling cycle. Earlier samples were preserved in 10% formalin and deep frozen until analysis. Residues presented in this study are for liver (wet weight).

Results and discussion

Intensive searches for victims during the first two weeks post-spray provided no evidence of acute poisoning in vertebrates. This contrasts with observations in Nigeria, where dead or dying amphibians, reptiles, birds and mammals were frequently found after a similar spraying operation (*42*). However, there was indirect evidence that insectivorous mammals such as the giant leaf-nosed bat, *Hipposiderus commersoni gigas* (WAGNER), as well as musk shrews, *Crocidura* spp., were severely affected by the dieldrin treatment. Insectivorous bats were not found 8-12 months post-spray, and shrews were not observed for 3 years. Initial median dieldrin residues in liver ranged from 3.90-5.28 mg kg^{-1} in insectivorous bats, and the only shrew found on the first sampling date post-spray showed a concentration of 11.3 mg kg^{-1} (Table III). This level was in the range of residues in warm-blooded animals known to have been killed by dieldrin (*42, 43*). The highest initial median concentration was detected in the lesser epaulet fruit bat, *Micropteropus pusillus* (PETERS) (136 mg kg^{-1}). High and probably toxic levels (9.81 mg kg^{-1}) were also found in the Angolan fruit bat *Lissonycteris angolensis* (BOCAGE) and might explain why no bats were found 8 months after treatment. However, seasonal migrations might also have contributed to the population decline. With regard to their herbivorous feeding ecology and their low position in the food chain, fruit bats were unlikely to accumulate high quantities of dieldrin in the longer term. However, living in the canopy of trees, they were particularly exposed to initial spray deposits on leaves and fruits. Fruit bats trapped one year post-spray and later showed concentrations of < 0.1 mg kg^{-1}. It was assumed that these specimens represented immigrants from untreated areas. Contrary to bats and shrews, residues in the ground-dwelling, seed and fruit-eating soft-furred rat *Praomys jacksoni* (DE WINTON) remained low throughout the 5 year study.

Dieldrin concentrations in birds varied with species and time. Resident insectivorous birds such as the West African thrush, *Turdus pelios* BONAPARTE, the white-crowned and the snowy-crowned robin-chat, *Cossypha albicapilla* (VIEILLOT) and *C. niveicapilla* (LAFRESNAYE), and the blue-breasted kingfisher, *Halcyon malimbica* (SHAW), showed peak concentrations 8-12 months after treatment (Table III). Possible lethal or sub-lethal biological effects were inferred from median residues of 12.2 mg kg^{-1} in *C. niveicapilla* and 6.5 mg kg^{-1} in *H. malimbica*. These levels were in the same order of magnitude as residues in birds found dead or dying after dieldrin applications in Nigeria (*42*). In contrast, the pygmy kingfisher, *Ispidina picta* (BODDAERT), did not accumulate high levels of dieldrin, owing to different feeding and habitat preferences compared to forest-bound kingfishers. *I. picta* mainly feeds in open savannah which was left untreated in the tsetse eradication campaign. Reptiles were not regularly sampled for residue analysis but the results obtained for the skink *Mabuya affinis* (GRAY) suggested accumulation rates similar to those observed in some birds. The minimum residues of dieldrin in liver associated with field mortality in mammals and birds range from 2.4-44 mg kg^{-1} (*37*). Thus the relatively high levels found in the present study clearly indicated that substantial proportions of the fauna – in particular insectivorous species – had been exposed to lethal concentrations. Top predators were not monitored but were expected to be at risk from accumulating lethal concentrations as well.

In conclusion, this type of dieldrin application has to be considered harmful to a wide range of smaller vertebrates and certainly aggravates the risk of (local) extinction of isolated and/or endemic populations.

Fish Kills due to Endosulfan in Miombo Woodland (Zambia 1987)

Methods

Environmental monitoring was conducted during a tsetse control campaign NW of Choma in southern Zambia. The operation aimed to prevent immigration of tsetse into

Table III. Median and range of dieldrin residues in liver in riverine forest wildlife in Adamaoua highland, Cameroon, 1979-84

Period	*Pre-spray*		*Post-spray*													
Month	*½*		*½*		*8*		*12*		*24*		*36*		*48*		*60*	
Species	*n*	*mg kg⁻¹*	*n*	*mg kg⁻¹*	*n*	*mg kg⁻¹*	*n*	*mg kg⁻¹*	*n*	*mg kg⁻¹*	*n*	*mg kg⁻¹*	*n*	*mg kg⁻¹*	*n*	*mg kg⁻¹*
Mammals																
Hipposiderus commersoni gigas (bat, i)	∅		4	3.90 0.64-5.82	0		0		4	0.11 0.01-0.37	2	0.19 0.08-0.29	∅		∅	
Other insectiv. bats (Vespertilionidae)	∅		7	5.28 0.82-7.28	0		0		2	0.35 0.10-0.60	∅		∅		1	0.26
Lissonycteris angolensis (bat, f)	∅		9	9.81 n.d.-94.3	0		2	0.10 n.d.-0.20	1	0.02	3	0.01 0.01-0.02	2	n.d.	∅	
Micropteropus pusillus (bat, f)	∅		11	136 1.48-174	0		3	0.14 0.02-0.2	6	0.08 0.02-0.62	4	0.07 0.01-0.12	2	0.004 0.004 (2x)	[c]2	0.05
Crocidura spp. (shrew, i)	5	n.d.	1	11.3	2	2.33 2.0-2.66	0		0		0		[b]3	0.04 0.03-0.05	∅	
Praomys jacksoni (rodent, g, f)	19	n.d.	13	0.37 n.d.-1.20	7	0.35 0.02-0.99	3	0.29 0.04-0.74	4	0.07 0.03-0.12	1	0.09	2	0.002 n.d.-0.004	∅	
Birds (insectivorous)																
Halcyon malimbica	4	n.d.	4	3.46 2.21-4.30	6	5.0 2.34-7.90	2	6.52 5.44-7.70	5	2.24 1.54-2.57	3	0.6 0.14-0.78	1	1.75	1	0.24
Ispidina picta	6	n.d. n.d.-0.04	10	0.60 0.28-1.29	0		2	0.58 0.48-0.68	5	0.62 0.43-1.47	4	0.43 0.16-1.17	3	0.06 0.005-0.09	1	2.0
Turdus pelios	10	n.d. n.d.-0.04	8	0.44 0.20-0.91	2	2.08 1.39-2.77	2	3.0 n.d.-6.0	6	0.08 0.06-0.13	1	0.05	2	0.003 0.003 (2x)	[c]2	0.002
Cossypha albicapilla	2	n.d.	4	0.27 0.04-0.34	[a]2	12.2 4.3-20.0	0		1	1.3	1	0.33	[d]1	0.17	1	0.36
Reptile (insectivorous)																
Mabuya affinis	∅		7	0.93 0.03-1.63	∅		∅		1	0.03	∅		∅		∅	

f = frugivorous, g = granivorous, i = insectivorous; n.d. = not detected, ∅ = no samples taken; 0 = no samples found; *a* = *Cossypha niveicapilla*; *b* = sample comprises two shrew species: *C. denti* & *Suncus infinitesimum*; *c* = samples pooled prior to analysis; *d* = *Cossypha polioptera* adjacent rangeland. 4,500 km^2 of miombo woodland were sprayed five times between July and October 1987 with endosulfan, using fixed-wing aircraft (sequential aerosol treatment). Nominal dosages were 22 (1st cycle), 18 (2nd cycle) and 14 g a.i. ha^{-1} (3rd to 5th cycle), but amounts actually applied varied due to imprecise spraying of the aircraft. The fish fauna was studied in several dams located in the centre and in the western part of the spray block (*44*). The dams had been constructed between 1967 and 1982 along temporary river courses to secure an all-year water supply for man and livestock. The size of the dams varied between 1.7-9.0 ha in June and 0-4.5 ha in October, and in depth, respectively, between 2.0-4.1 m and 0-3.5 m. Water temperatures were 15-29°C on the surface, 17.5-25.5°C at 1 m depth and ≈ 22°C at a depth of ≥ 2 m. Shallow dams usually dry out during the dry season and often do not contain fish. However, 2-7 fish species were present in all permanent water bodies studied. *Barbus* spp. and *Oreochromis* spp. were the most abundant species. Since spray aircraft can hardly avoid treating small surface waters, these were expected to be exposed several times to endosulfan. Mortality in fish was recorded after each spraying cycle by collecting and counting all dead and debilitated fish washed ashore along a stretch of 100 m on the windward shore of the dam. The search for victims was continued daily until no more fish were found. The overall mortality was assessed by extrapolating to the total shoreline covered with dead fish. The surviving fish population was estimated after the last cycle by catching all fish from 0.2-0.3 ha of water, using gillnets of various mesh widths (2-4 cm).

Results and discussion

After the first treatment, high mortality of *Tilapia sparrmanii* A. SMITH was observed at Mandala dam (Figure 3), whereas the effects on *Barbus* spp. (including *Barbus paludinosus* PETERS) were not as serious. About 50% of all victims were < 4 cm in length, another 50% between 4-8 cm. Mortality continued until 15 days post-spray in both species. No dead fish were found after the 2nd treatment. However, the 3rd cycle caused heavy mortality again, with more than 200 dead fish per 100 m. Only 23% of the victims belonged to size classes ≤ 8 cm which indicates that abundance of smaller fish had been considerably reduced after the 1st cycle. The remaining treatments did not cause relevant mortality in *T. sparrmanii* since the stock had already been largely eliminated. The total number of dead *T. sparrmanii* was estimated at 26,000 and the surviving population at 150, resulting in an overall mortality of > 99%. About 5,700 *Barbus* spp. survived the five endosulfan treatments, compared to a total kill of 1,700. Thus the overall mortality was only 23%, but smaller fish (< 4 cm) had almost completely disappeared.

Effects on fish in the other dams were equally dramatic. About 34,000 *Oreochromis macrochir* (BOULANGER) and *O. mossambicus* (PETERS) perished in Chikwangala dam (98%) along with 20,500 *Barbus* spp. (97%). The highest toll in absolute numbers was recorded in Macha dam where about 700,000 *Barbus* spp. were killed (> 99.9%). Macha dam was the largest (9 ha in the beginning) but also the

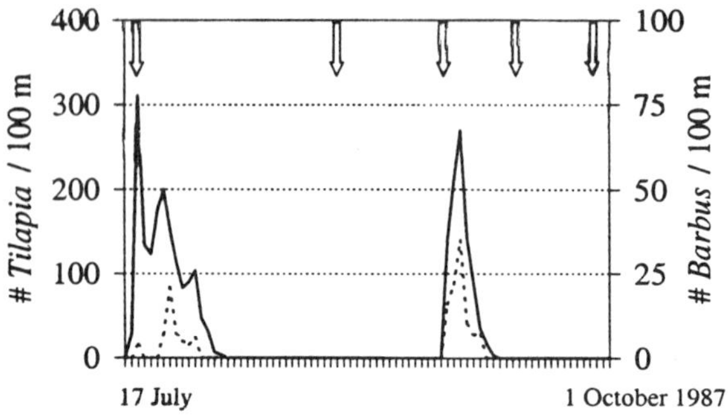

Figure 3. The number of dead or debilitated Tilapia *spp. (solid line) and* Barbus *spp. (broken line) along 100 m shoreline of Mandala dam during a tsetse eradication campaign in Zambia. The arrows denote the days of treatments.*

shallowest (2 m) water body in this study. Therefore, relative pesticide concentrations were higher than in deeper ponds and fish could not escape into deeper water. The median residue of endosulfan in the total body of dead fish from different dams was 1.05 mg kg^{-1} (n = 24, *Tilapia* spp. and *Barbus* sp.).

In order to assess the level of fish mortality in a treatment conducted according to standard operating procedures, a test treatment was conducted at Namuse dam, which lay within a bloc to be treated later in the season. All spray parameters were controlled and the droplet deposition was monitored with Magnesium-Oxide (MgO) coated glass slides. The field dosage was confirmed to be 20 g a.i. ha^{-1}. *T. sparrmanii* and *Barbus* spp. were the only fish species present. In total, less than 50 dead fish were collected per 100 m shoreline during several days post-spray (median endosulfan residue of 5 specimens of *Tilapia* sp. and *Barbus* sp.: 0.51 mg kg^{-1}). In contrast, > 3,600 *Barbus* spp. and ≈ 130 *T. sparrmanii* per 100 m shoreline were recorded in Namuse dam within three days after the next spraying operation at the same dose. In Mandala dam, for example, massive kills were recorded after the 1st and 3rd spraying cycles but not after the 2nd one. These results indicated that the real dose rates were highly variable between cycles and within spray blocks, resulting in a patchwork of pesticide cover sometimes higher and sometimes lower than the intended dose. No major fish kills would have been expected if the spray blocks had been treated at nominal dose rates. Such irregular spray patterns usually result from inadequate application technology, maintenance and spray performance (e.g. calibration and/or navigation errors) and pose a risk of primary and secondary poisoning to wild and domestic animals as well as man. The study also illustrates the importance of environmental monitoring to check compliance (or detect non-compliance) with technical and environmental standards during tsetse control campaigns.

A study on indirect effects on piscivorous birds at Mandala dam revealed that the pied kingfisher, *Ceryle rudis* (L.), was more successful in catching fish after spraying, and that the preferred prey shifted from *T. sparrmanii* to *Barbus* spp. (*45*). The three individuals did not show any sign of secondary poisoning and did not desert the area during the observation period. In contrast, at Chikwangala dam, all resident kingfishers disappeared after the 2nd spray cycle. Long-term effects on the reproduction were not investigated.

Risk Assessment

The four environmental impact studies demonstrate exposures of wildlife and fish to insecticides used in locust or tsetse control. The evidence of unacceptable adverse effects was variable, depending on the ecotoxicity of the pesticides and, in case of endosulfan, on the quality of the application. The examples also reflect the evolution of pesticide use since the late 1970s. The tsetse studies (1979-87) and the desert locust study in Niger (1990) were conducted when OCs were still in use, whereas the migratory locust study in Madagascar (1995) dealt with products which have replaced OCs. The different history has to be taken into account when discussing the effects on wildlife presented in this paper.

Our results showed that tsetse control operations with dieldrin were particularly harmful. All specimens sampled for residue analysis were alive when collected. Median concentrations of dieldrin, which ranged between 0.1-136 mg kg^{-1} in the first year, were in the range of those reported from field-collected victims (*37*) and certainly high enough to indicate lethal effects in a range of different wildlife species. The real population decline of sensitive species was not monitored. However, shrews seemed to have disappeared for several years and were only sampled again four years after treatment. These observations provided clear evidence of intolerable losses of some vertebrate species, and confirmed that the withdrawal of dieldrin in the mid 1980s was a necessary step to reduce wildlife exposure to this highly toxic and persistent insecticide. Dieldrin concentrations measured in specimens from Niger were much lower ($\leq$ 0.025 mg kg^{-1}) than in Cameroon. This might indicate that the ban on dieldrin during the 1986-89 campaign was largely respected, despite unofficial reports on the continued use of dieldrin and other OCs. On the other hand, the dose rates used in desert locust control are lower than those used in tsetse control (Table I), and the breakdown is more rapid in the desert than in humid savannahs. These factors would also result in reduced exposure of desert wildlife compared to their savannah counterparts. Residues of DDT and its metabolites were higher than residues of other OCs, but as indicated before, below critical levels (*36*).

However, while residue analyses may be more reliable measurement endpoints than estimates of relative abundance, difficulties arise in linking particular concentrations with certain biological effects and implications for wildlife populations and communities (*46*). For example, chlorinated hydrocarbons including pesticides such as

DDT and endosulfan can act as endocrine disruptors (*47*) and have been associated with developmental and reproductive abnormalities in fish (*48*) and wildlife (*49, 50*). It is difficult to determine if these factors play a role in Africa. There is at present little, information to base the discussion on. However, despite possible inputs from distant areas, environmental concentrations of hormonally active pesticides and other endocrine disruptors such as PCBs are certainly lower in Africa than in industrialised countries. For example, in the beginning of the 1990s, Africa shared only about 4% of the global crop protection market and 7% of the insecticide market, compared to 31% (insecticides: 19%) in Western Europe and 26% (insecticides: 19%) in North America (*51*). Nonetheless, intense pesticide use in vector control programmes (tsetse fly, black fly, mosquito and snail control) and agriculture can overlap in some areas (*52*), leading to high environmental concentrations of different pesticides. In these areas, there might be an elevated risk of lethal and sub-lethal effects on wildlife, including endocrine disruption. Furthermore, while the consumption of insecticides has dropped in industrialised countries, it is gradually increasing in the developing world (*51*).

The case study in Madagascar revealed a moderate decline of *C. madagascarienesis* exposed to a product composed of fenitrothion and esfenvalerate. The risk of fenitrothion to terrestrial vertebrates at dose rates used against desert locust has been classified as medium and the risk of pyrethroids as low (*13*). OP-pyrethroid mixtures were not included in this assessment, however, they would be expected to be less hazardous since OP dosages are generally lower than with pure OP formulations. For example, the field dose of fenitrothion against migratory locust in Madagascar is 500 g ha^{-1} when used alone and 245 g ha^{-1} when combined with esfenvalerate (*28*). The real dose applied in our study was even lower (207 g). We conclude that, while the moderate decline seems to be acceptable on the first look, more serious effects on lizards can be expected when using pure formulations of fenitrothion and other OPs such as chlorpyrifos. Thus the risk to reptiles certainly deserves more attention in future monitoring programmes, not only in Madagascar, where 88% of the nearly 300 reptile species are endemic (*7, 53*), but also on the African continent.

Among the tsetse control agents used for sequential aerosol treatment, none seems to be dangerous to terrestrial vertebrates (*11*) but endosulfan is moderately toxic to fish (*54*). Our study demonstrated that high mortality may occur if standard operating protocols are not respected. The severity of the ecological side effects of residual application of DDT and pyrethroids is a matter of debate (*11, 18, 55*). It is hoped that improved spraying and track guidance technologies, a mosaic-like spraying pattern with different insecticides (*11*) and the better understanding of the necessary width of buffer zones (*56*) will help to further reduce the risk of contamination of aquatic ecosystems in tsetse and locust control.

An important aspect of the risk assessment of locust and tsetse control operations is their different temporal and spatial pattern. Locust control applications are usually conducted in relatively small areas of 1-12 km^2 (*22*) resulting in a mosaic of treated and untreated land. Moreover, the frequency and intensity of control operations is highly variable between years. Thus, wildlife is only partially exposed to pesticides which improves the chances for recovery if negative effects occur.

In contrast, applications for tsetse control often cover huge continuous areas within a single season (*11*). Moreover, control often aims at the complete eradication of isolated tsetse populations (pockets) which are either completely separated from the main population or are connected by ecological corridors (bottle-necks). In some areas large-scale programmes were launched to eliminate certain tsetse species from complete belts, i.e. large parts of the total range, isolated from other parts by natural barriers such as unsuitable habitat. Therefore, tsetse habitats such as riverine forests, forest islands in savannahs or savannah woodland were almost completely treated with insecticides, leaving few refuges for wildlife. Apart from chemical control, extremely hazardous host game eradication and large-scale bush clearing operations were launched into the 1960s and 70s. Furthermore, since tsetse flies present a permanent threat to livestock and humans, control operations were often conducted in the same areas for many years (*18*). Thus exposure of wildlife to pesticides was very high in the past. It was not until recently that control programmes aimed at the management rather than full eradication of tsetse populations, using more environmentally acceptable control techniques such as artificial (traps and screens) and living targets (treated cattle).

The risk to wildlife does not only depend on the pattern and scale of the treatments or the ecotoxicity of the insecticides but also on the magnitude, quality of and interaction with other human impacts. These have to be brought into perspective when assessing the relative importance of pesticide hazards compared to other threats.

Other Human Impacts on African Wildlife

About 73% of agriculturally used African drylands, comprising approximately 14.3 million km^2 of irrigated lands, rainfed croplands and rangelands, are degraded (*57, 58*). Biological, physical and socio-economic factors are the driving forces in a "cycle of unsustainability" which leads to ever-increasing and accelerating human pressure on the fauna and flora of both favourable and marginal lands (*59*). Slash and burn agriculture, deforestation, bush fires and intensive hunting result in a continuous loss and fragmentation of wildlife habitats and in the segregation of isolated and often non-viable populations.

Locust and tsetse species can be positively or negatively affected by these modifications of landscapes. For example, most of the former dry forests in south-western Madagascar have been transformed into vast grasslands, providing new feeding and breeding habitats for the graminivorous Malagasy migratory locust and aggravating the pest status of this species. In contrast, the African migratory locust has greatly suffered from the drainage of the Niger flood plain in Mali, the core recession area in continental Africa (*9*), for agricultural irrigation schemes. Deforestation and agricultural development may also lead to a reduction in abundance of tsetse flies. While domestic animals can replace wildlife as hosts, most tsetse species prefer undisturbed woody vegetation and need shaded places for reproduction. Therefore, certain tsetse flies can be regarded as indicators of habitat quality.

To illustrate human pressures on biological resources in tsetse-affected countries, we will focus on two impact indicators, (a) livestock production and (b) charcoal and fuelwood production for domestic use (Figure 4). The cattle stock grew from 96.6 million heads in 1963 to 121.9 million in 1973 and to 182.6 million in 1998 (statistical database of FAO). The real increase of 60.7 million cattle between 1973 and 1998 is far below the previously mentioned expected increase of 120 million heads which was based on the assumption that trypanosomiasis was controlled (*5*). This might be interpreted as an indication that tsetse control, on a regional scale, was not very successful. However, the constraints to livestock production are manifold, and factors such as the escalating degradation and loss of pastureland (*59*) and marketing restrictions are equally important causes for the low growth rates. Charcoal and firewood in Africa usually stem from unmanaged forests and woodlands. Therefore, production figures directly reflect the degradation of these important wildlife habitats. The evidence is clear: the production nearly tripled from 188 million m^3 in 1963 to 500 million m^3 in 1998 (statistical database of FAO). The increase is an unambiguous indication of the strong pressure on the remaining forest and woodland resources in Africa.

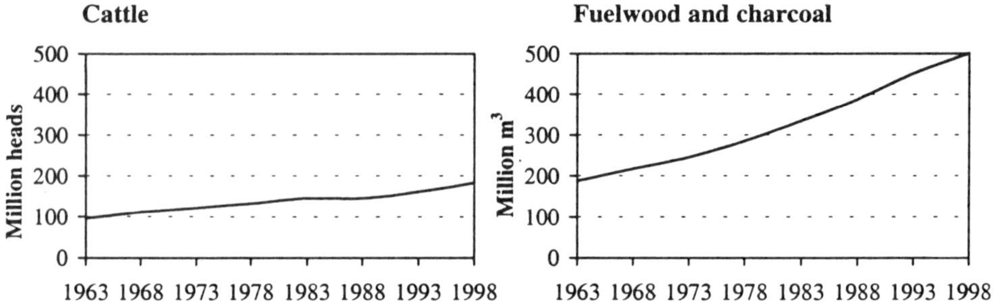

Figure 4. Cattle production (left) and fuelwood and charcoal production (right) in tsetse-affected countries in Africa.

One of the most controversial issues in the discussion about the risks and benefits of tsetse control is the role of nagana as a factor controlling human migration and limiting rural development (*60*). Tsetse flies, so is the argument of conservationists, slow down the transformation of natural landscapes into farmland and halt the destruction and fragmentation of wildlife habitats and refuges. The counter-argument is that this transformation takes place anyway as long as a growing human population is in need of new land. Neither position can be fully rejected. For example, areas cleared of tsetse in the 1970s in Nigeria were found to be altered dramatically through human activity (*43*). The tsetse eradication campaign in Zimbabwe and adjacent countries clearly aimed to transform natural woodland into agricultural land and involved the settling of human populations in the framework of a rural development programme. On the contrary, our own studies in central and northern Ivory Coast (*61, 62, 63*) showed that changes in land use were not directly related to tsetse control (Figure 5). For example, in Mbengue (Sudano-Guinean transition zone) deforestation rates and farmland extensions were highest from 1950-70, i.e. before the first tsetse control campaign was launched in the late 1970s. The same phenomenon was observed in Tortiya (central Guinea zone), where ≈ 70% of the forest cover in 1950 had already been cleared for agricultural use at the beginning of the first tsetse control operation in 1991. In contrast, the land use pattern in Bouaké (southern Guinea zone) in 1991 was nearly the same as in the 1950s, despite several years of tsetse control. We conclude that tsetse flies might delay, but cannot prevent the transformation of wildlife habitats into agricultural land. In Africa, the rural population has little choice but to exploit all accessible resources to make their living however inconvenient and sometimes dangerous this may be.

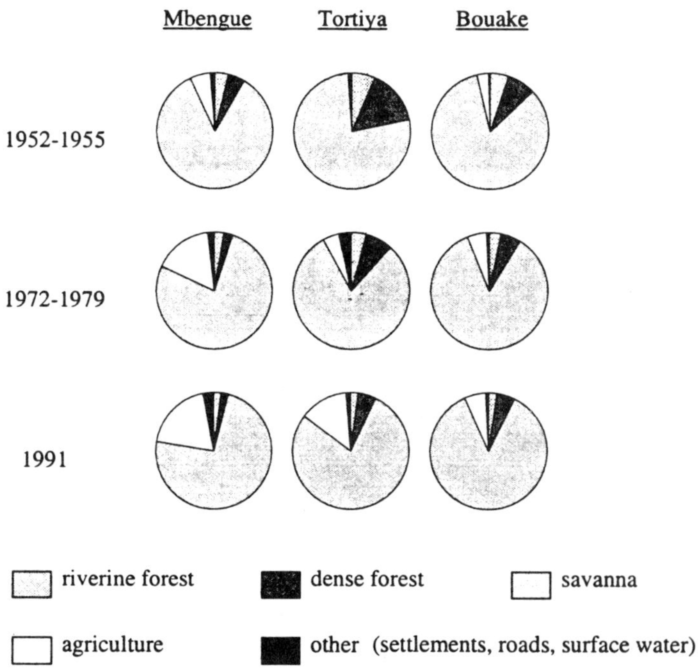

Figure 5. Land use between the 1950s and 1991 in three different regions in Ivory Coast. Tsetse control started in 1978 (Mbengue), 1990 (Bouaké) and 1992 (Tortiya), respectively. Analysis based on aerial photographs (1:20,000 to 1:80,000) covering 1,812 km² (1952-55), 2,310 km² (1972-79) and 550 km² (1991).

Who Pays the Bill?

Homo sapiens has the highest biomass of all animal species on earth (*1*). His biological success is closely linked with the evolution of agriculture and has almost always been to the detriment of wildlife. Pesticides against locusts and tsetse flies have certainly been an important item on the bill, however, the most important single threat to African biodiversity is habitat clearance and alteration (*7*). Bearing in mind the multitude and scale of other direct and indirect human pressures, locust and tsetse control can hardly be considered as a major cause for the decline of wildlife in Africa unless on a local scale. Selective techniques such as baited targets are now widely used and make the control of tsetse more environmentally friendly. Similarly, new insecticides such as entomopathogenic fungi and selective chemicals are now integrated into locust control programmes (*64*).

On the other hand, broad-spectrum insecticides are still used on fairly large scales in emergency situations and do of course threaten wildlife. If massive wildlife kills were acceptable during the first decades of chemical control, they are certainly no longer acceptable today. The reason is that wildlife subjected to a range of other pressures is more vulnerable to added pesticide stress. Furthermore, the seemingly infinite faunal reservoirs to compensate for local extinctions no longer exist. Thus the art is to carefully weigh the risks and benefits of different pest management scenarios, including the no control option, and to perform necessary control to the highest technical standards so as to minimise the risk to the non-target fauna (*65*).

Acknowledgements

Most of the research presented here has been funded by the German Development Cooperation (GTZ) and the Commission of the European Communities. We would like to thank these organisations for their support. We are also grateful to the following persons for their contributions in the laboratory or in the field: J. Krüger, Saarbrücken, (residue analysis) and J. Rafanomezantsoa, Antananarivo, (lizard study). J. Kohl and D. Jäger, Basle, are acknowledged for preparing two of the figures and two reviewers for their helpful comments on an earlier draft of this paper.

References

1. Simmons, I. G. *Changing the face of the earth;* 2nd ed.; Blackwell, Oxford, 1996.
2. *New strategies in locust control;* Krall, S.; Peveling, R.; Ba Diallo, D., Eds.; Birkhäuser: Basel, 1997.
3. Showler, A. T. In *New strategies in locust control;* Krall, S.; Peveling, R.; Ba Diallo, D., Eds.; Birkhäuser: Basel, 1997; pp 461-465.
4. FAO *Press Release* **1997**, *67*, 1-2.
5. Gratz, N. G. In *Chemie der Pflanzenschutz- und Schädlingsbekämpfungsmittel;* Wegler, R., Ed.; Springer: Berlin, 1976; Vol. 3, pp 85-181.

6. Stolberg, S. J. *New York Times* 1999, Vol. CXLVIII (51,629).
7. Stuart, S. N.; Adams, R. J.; Jenkins, M. D. In *Biodiversity in Sub-Saharan Africa and its islands;* Occasional Papers of the IUCN Species Survival Commission 6; IUCN: Gland, Switzerland, 1990.
8. World Conservation Monitoring Centre *Global biodiversity: status of the earth's living resources;* Chapman & Hall: London, 1992.
9. *Locust handbook;* Steedman, A., Ed.; 3rd ed., Natural Resources Institute: Chatham, UK, 1990.
10. Leak, S. G. A. *Tsetse biology and ecology. Their role in the epidemiology and control of trypanosomosis;* CABI Publishing: CAB International, Wallingford, UK, 1999.
11. Nagel, P. *Environmental monitoring handbook for tsetse control operations;* Margraf Verlag: Weikersheim, Germany, 1995.
12. Jordan, A. M. *Trypanosomiasis control and African rural development*; Longman: London, 1986.
13. FAO *Evaluation of field trial data on the efficacy and selectivity of insecticides on locusts and grasshoppers;* 7th Meeting of the Pesticide Referee Group, Rome, 2-6 March 1998; FAO: Rome, 1998.
14. Lecoq, M.; Duranton, J. F.; Rachadi, T. In *New strategies in locust control;* Krall, S.; Peveling, R.; Ba Diallo, D., Eds.; Birkhäuser: Basel, 1997; pp 467-473.
15. *Lutte contre le criquet pèlerin par les techniques existantes;* van Huis, A., Ed.; Wageningen Agricultural University: Wageningen, Netherlands, 1994.
16. Lomer, C. J. In *New strategies in locust control;* Krall, S.; Peveling, R.; Ba Diallo, D., Eds.; Birkhäuser: Basel, 1997; pp 159-169.
17. Douthwaite, R. J. *J. Appl. Ecol.* **1995**, *32*, 727-738.
18. *DDT in the tropics – The impact on wildlife in Zimbabwe of ground-spaying for tsetse fly control;* Douthwaite, R. J.; Tingle, C. C. D., Eds.; Natural Resources Institute: Chatham, UK, 1994.
19. Langley, P. A. *Bull. Entomol. Res.* **1995**, *85*, 495-500.
20. *Environmental effects of chemical locust and grasshopper control – A pilot study;* Everts, J. W., Ed.; FAO: Rome, 1990.
21. Mullié, W. C.; Keith, J. O. *J. Appl. Ecol.* **1993**, *30*, 536-550.
22. Keith, J. O.; Bruggers, R. L. *J. Raptor Res.* **1998**, *32(2)*, 151-158.
23. Keith, J. O.; Bruggers, R. L.; Matteson, P. C.; Abderrahim, El H.; Ghaout, S.; Fiedler, L. A.; El Hassan, A.; Gillis, J. N.; Phillips, R. L. *An ecotoxicological assessment of insecticides used for locust control in Southern Morocco;* DWRC Research Report No. 11-55-005; USDA, APHIS, Denver Wildlife Research Center: Denver, CO, 1995.
24. Wiktelius, S.; Edwards, C. A. *Rev. Environ. Contam. Toxicol.* **1997**, *151*, 1-37.
25. Dejoux, C. *La pollution des eaux continentales Africaines;* Travaux et Documents n° 213; Editions de l'ORSTOM: Paris, 1988.
26. Lambert, M. R. K. *Environ. Pollut.* **1993**, 82, 231-237
27. Lambert, M. R. K. *Rev. Environ. Contam. Toxicol.* **1997**, *150*, 31-73.

28. Randriamantsoa, M. *Manuel sur la lutte antiacridienne;* DPV-GTZ, Promotion de la Protection Intégrée des Cultures et des Denrées Stockées: Antananarivo, Madagascar, 1998.
29. Peveling, R.; Rafanomezantsoa, J.-J.; Razafinirina, R.; Tovonkery, R.; Zafimaniry, G. *Crop Prot.* **2000**, *18(10)*, 659-676.
30. Peveling, R., Weyrich, J. In *Biological control of locusts and grasshoppers;* Lomer, C. J.; Prior, C., Eds.; CAB International, Redwood Press: Melksham, 1992; pp 321-336.
31. Peveling, R.; Weyrich, J.; Müller, P. In *New trends in locust control;* Krall, S.; Wilps, H., Ed.; Schriftenreihe der GTZ 245; TZ-Verlagsgesellschaft: Rossdorf, Germany, 1994; pp 147-176.
32. Müller, P.; Nagel, P.; Flacke, W. *Oecologia* **1981**, *50*, 187-194.
33. SEMG *Environmental monitoring of tsetse control operations in Zambia and Zimbabwe. Report 1987-1990;* RTTCP, 5th EDF, SEMG Project, Saarbrücken, 1993.
34. Ejobi, F.; Kanja, L. W.; Kyule, M. N.; Mueller, P.; Krueger, J.; Nyeko, J. H. P.; Latigo, A. A. R. *Bull. Environ. Contam. Toxicol.* **1996** *56 (4)*, 551-557.
35. Kucklig, J. R.; Hravey, H. R.; Ostrom, P. H.; Ostrom, N. E.; Baker, J. E. *Environ. Toxicol. Chem.* **1996**, *15*, 1388-1400.
36. Blus, L. B. In *Environmental contaminants in wildlife – Interpreting tissue concentrations;* Beyer, W. N.; Heinz, G. H.; Redmon-Norwood, A. W., Eds.; SETAC Special Publication Series; CRC Lewis Publishers: Boca Raton, 1996; pp 49-71.
37. Peakall, D. B. In *Environmental contaminants in wildlife – Interpreting tissue concentrations;* Beyer, W. N.; Heinz, G. H.; Redmon-Norwood, A. W., Eds.; SETAC Special Publication Series; CRC Lewis Publishers: Boca Raton, 1996; pp 73-97.
38. Blanc, C. P. *Oecologia* **1969**, *2*, 292-318.
39. Stewart-Oaten, A.; Murdoch, W. W.; Parker, K. R. *Ecology* **1986**, *67*, 929-940.
40. Raholijaona, M. In *Symposium sur la lutte antiacridienne à Madagascar;* Scherer, R.; Fong Hong, L., Eds.; DPV-GTZ Promotion de la Protection Intégrée des Cultures et des Denrées Stockées; Antananarivo, Madagascar, 1997; pp 353-366.
41. Hall, R. J.; Clark, D. R. Jr. *Environ. Pollut.* **1982**, *(series A.) 28*, 45-52.
42. Koeman, J. H.; den Boer, W. M. J.; Feith, A. F.; de Iongh, H. H.; Spliethoff, P. *Environ. Pollut.* **1978**, *15*, 31-59.
43. Koeman, J. H.; Rijksen, H. D.; Smies, M. *Neth. J. Zool.* **1971**, *21 (4)*, 434-463.
44. Böhme, Ma.; M.S. thesis, Institute of Biogeography, University of Saarbrücken, Germany, 1988.
45. Böhme, Mo.; M.S. thesis, Institute of Biogeography, University of Saarbrücken, Germany, 1990.
46. Keith, J. O. In *Environmental contaminants in wildlife – Interpreting tissue concentrations;* Beyer, W. N.; Heinz, G. H.; Redmon-Norwood, A. W., Eds.; SETAC Special Publication Series; CRC Lewis Publishers: Boca Raton, 1996; pp 1-47.

47. Gillesby, B. E.; Zacharewski, T. R. *Environ. Tox. Chem.* **1998**, *17 (1)*, 3-14.
48. Arcand-Hoy, L. D.; Benson, W. H. *Environ. Tox. Chem.* **1998**, *17 (1)*, 49-57.
49. Guillette, L. J. Jr.; Pickford, D. B.; Crain, D. A.; Rooney, A. .; Percival, H. F. *Gen. Compar. Endocrin.* **1996**, *101*, 32-42.
50. Gray, L. E. Jr.; Ostby, J.; Wolf, C.; Lambright, C.; Kelce, W. *Environ. Toxicol. Chem.* **1998**, *7 (1)*, 109-118.
51. Römbke, J.; Moltmann, J. F. *Applied Ecotoxicology;* CRC Lewis Publishers: Boca Raton, 1996.
52. *Environmental concerns in A.I.D. programs for locust and grasshopper control*; U.S. Agency for International Development (USAID): Washington D.C., 1991.
53. Glaw, F., Vences, M. *A field guide to the amphibians and reptiles of Madagascar;* Zoologisches Forschungsinstitut und Museum Alexander König: Bonn, Germany, 1994.
54. Douthwaite, R. J.; Fox, P. J.; Matthiessen, P.; Russel-Smith, A. *Environmental impact of aerial spraying operations against tsetse in Botswana*. Proc. 6th Int. Agr. Aviation Congr.; Centre for the Application of Pesticides: Cranfield, UK, 1981; pp 221-227.
55. Lambert, M. R. K., Grant, I. F., Smith, C.·L., Tingle, C. C .D., Douthwaite, R. J., *Effects of deltamethrin ground-spraying on non-target wildlife. Environmental impact assessment of ground-spraying operations against tsetse flies in Zimbabwe;* Natural Resources Institute (NRI): Chatham, UK, 1991.
56. Lahr, J.; Gadji, B.; Demba, D. *Crop Prot.* **2000**, in press.
57. Biswas, M. R. *Ambio* **1994**, *23*, 192-197.
58. McNeely, J. A.; Gadgil, M.; Levèque, C.; Redford, K. In *Global biodiversity assessment;* Heywood, V. H., Ed.; United Nations Environment Programme; Cambridge University Press: Cambridge, 1995; pp 711-821.
59. Steiner, K.G. *Causes of soil degradation and development approaches to sustainable soil management;* GTZ, Margraf Verlag: Weikersheim, Germany, 1996.
60. Nagel, P. In *Livestock production and diseases;* Zessin, K.-H., Ed.; Proc. 8th Conf. Inst. Trop. Vet. Med.; Berlin, Germany, 1996; pp 446-452.
61. Erdelen, W.; Nagel, P.; Peveling, R. *Applied Geography and Development* **1994**, *44*, 17-31.
62. Nagel, P.; Erdelen, W.; Peveling, R. *Verhandlungen der Gesellschaft für Ökologie* **1995**, *24*, 67-72.
63. Abel, F. M.S. thesis; Institute of Biogeography, University of Saarbrücken, Germany, 1994.
64. Lomer, C. J., Bateman, R. P., Dent, D., de Groote, H., Douro-Kpindou, O.-K., Kooyman, C., Langewald, J., Ouambama, Z., Peveling, R., Thomas, M. *Agric. Forest Entomol.* **1999**, *1*, 71-88.
65. *Pesticides: Managing risks and optimizing benefits;* Ragsdale, N. N., Seiber,

J. N., Eds.; ACS Symposium Series 734; Amercian Chemical Society: Washington, DC, 1999.

66. Gruys, P. In *Lutte contre le criquet pèlerin par les techniques existantes;* van Huis, A., Ed.; Wageningen, Netherlands, 1994; pp 19-30.
67. Launois-Luong, H.; Launois, M.; Rachadi, T. *La lutte chimique contre les criquets du Sahel;* Collection Acridologie Opérationnelle 3; Département de Formation en Protection Végétaux: Niamey, Niger, 1988.

Chapter 8

Estimation of Red-Winged Blackbird Mortality from Toxic Bait Application

James F. Glahn[1] and Michael L. Avery[2,3]

[1]**Animal and Plant Health Inspection Service, National Wildlife Research Center, U.S. Department of Agriculture, Mississippi Field Station, P.O. Drawer 6099, Mississippi State, MS 39762–6099**
[2]**Animal and Plant Health Inspection Service, National Wildlife Research Center, U.S. Department of Agriculture, Florida Field Station, 2820 East University Avenue, Gainesville, FL 32641**

Application of slow-acting toxic bait is one method of reducing local populations of depredating blackbirds. Estimating numbers of birds killed in such baiting operations is difficult because affected birds die off-site and are seldom recovered. We conducted bioassays and flight pen studies of red-winged blackbirds (*Agelaius phoeniceus*) with a slow-acting, toxic brown rice bait to determine whether bird mortality could be predicted accurately using the Poisson and binomial discrete probability distributions. Bioassays confirmed that brown rice treated with 2% (w/w) 2-chloro-p-acetotoluidide was effective as a 1-particle lethal bait for redwings. Within a 0.2-ha flight pen, we offered this bait diluted 1:99 with untreated particles to three 60-bird flocks and three 20-bird flocks in a simulated baiting operation. Across the 6 test groups, we recorded bird mortality not different from that predicted using the Poisson distribution. We obtained the same estimates using the binomial distribution. Although a number of factors could influence the relevance of our findings to field use, the application of discrete probability distributions appears superior to using estimates of bait consumption as a means for evaluating blackbird mortality due to slow-acting toxic baits.

Introduction

For a number of years, reduction of winter blackbird populations in Louisiana and Texas with toxic bait has been a key component of integrated management strategies for reducing damage to cultivated rice (*1*). Although there are no objective data documenting increased rice crop yields following toxic baiting of blackbirds, surveys of rice producers support the overall effectiveness of this technique (*1*). There is currently

[3]Corresponding author: e-mail: Michael.L.Avery@usda.gov.

U.S. government work. Published 2001 American Chemical Society a Federal section 3 registration for the use of the toxin DRC-1339 (3-chloro-p-toluidine hydrochloride) to bait blackbirds at pre-roost staging areas (EPA registration number 56228-30). Most baiting to date has been conducted in Louisiana and Texas, but plans are under discussion to extend blackbird baiting programs to other areas (*2*).

Development of blackbird baiting techniques has focused on slow-acting toxicants such as DRC-1339 and closely related compounds such as 2-chloro-p-acetotoluidide (CAT). These compounds are attractive as blackbird toxicants because of their high toxicity to blackbirds and starlings but low toxicity to most mammals and predatory birds (*3*). The mode of action of these compounds is not well-understood, but death appears to result from nephrotoxicity (*4*). Time to death can extend to several days following ingestion.

Despite the potential utility of DRC-1339 and CAT for reducing local blackbird numbers, accurate determination of the impacts of baiting operations on target species populations is not straightforward because most poisoned birds die far from the bait site. This precludes direct counts of dead birds as an evaluation option. Alternatively, bait consumption has been used most often to estimate the number of blackbirds killed in baiting trials with CAT and DRC-1339 (*5, 6, 7, 8, 9*). These mortality estimates in field operations have relied upon consumption and mortality estimates from laboratory trials (*10*) or extrapolations from the theoretical number of treated particles taken per unit weight of bait consumed by birds. Independent verification of the validity of these approaches is lacking (*1*).

One potentially useful option not previously considered is application of discrete probability function distributions to predict mortality. Use of the binomial or the Poisson distribution is relevant because (1) in blackbird baiting operations, a single treated bait is formulated to be lethal, and (2) treated baits are presented in a mixture diluted with untreated baits. For a given bait dilution, it is straightforward to apply either function to calculate the proportion of birds that ate zero treated baits. All other blackbirds then consumed ≥1 treated bait and died. The only additional information needed to apply these functions is the mean number of total baits consumed by individual birds on the bait site.

We are unaware of any attempt to estimate blackbird mortality using this approach or to compare the accuracy of estimates obtained with theses functions with those derived from bait consumption. Thus, we conducted feeding trials to: (1) establish the lethality of a single treated bait, (2) establish an appropriate dilution of untreated to treated bait, (3) document feeding behavior of blackbird flocks on simulated bait sites, (4) predict mortality using binomial and Poisson distributions and bait consumption estimates, and (5) compare the mortality estimates to the actual number of birds killed.

The study took place during October-November 1987 when both CAT and DRC-1339 were being considered for registration as blackbird toxicants. Eventually, DRC-1339, not CAT, was registered. Nevertheless, because the two compounds are very similar in their effects on blackbirds (*4*), our findings are applicable to current blackbird baiting programs with DRC-1339 or to any other slow-acting bait formulated to deliver a single-particle lethal dose.

Methods

Test Birds

All test birds were male red-winged blackbirds trapped in the vicinity of Gainesville, Florida and held in captivity for at least 2 weeks at the Florida Field Station of the U.S. Department of Agriculture's National Wildlife Research Center. We maintained birds in group holding cages (1.2 x 1.2 x 1.8-m) in a covered outdoor aviary and allowed them free access to commercial game bird diet and water.

Bioassay

Three days before testing, we removed 10 birds from their communal holding cage and placed them individually in 46 x 46 x 91-cm cages in a covered outdoor aviary. Except during bait presentation, test birds had unlimited access to medium grain brown rice and water. The bait used in all testing was a medium grain brown rice bait formulated at the National Wildlife Research Center to contain 2% (w/w) CAT using 1% Alcolec S as a sticker.

On the treatment day, we fasted test birds for approximately 5 h, and then gave 5 randomly selected birds 1 treated bait mixed with 4 untreated rice baits in their feed cups. The other birds received 5 untreated brown rice grains. During the 4-h bait presentation, we covered the bottom of each cage with a shallow aluminum tray to reduce the possibility of bait loss. After baits had been eaten, birds again received untreated rice for 48 h. Dead birds were necropsied; survivors were released.

Flight Pen Trials, Untreated Bait

Within a 46 x 46-m flight pen at the Florida Field Station, we tilled and smoothed eight 9 x 12-m bait plots. As needed, we covered each baited plot with black polyethylene plastic sheeting to prevent birds or rain from reaching the bait. A 3 x 10 x 2-m drop-in decoy trap located in one corner of the flight pen served to hold and to recapture groups of test birds. In this trial, and in the one with treated bait, we allowed birds access to the baited plot only in the morning and late afternoon to simulate morning and evening activity at staging area bait sites in the field.

This set of trials involved two 20-bird groups and two 60-bird groups. One day before testing, we removed birds from their holding cages, placed the test group in the decoy trap, and provided brown rice and water. We individually identified 15 randomly selected birds in each group with 1.9 x 6.4-cm plastic tail tags inscribed with unique letter-number codes and attached to the base of the tail feathers with hot-melt glue. After the birds fasted overnight in the decoy trap, we released them into the flight pen between 0730 and 0800 to forage on a 9 x 12-m plot hand-broadcast with untreated brown rice at a rate of 28 kg/ha (300 g/plot).

Two observers with binoculars and spotting scopes watched from an observation blind at the north end of the flight pen to quantify foraging behavior of birds using the baited plot. We recorded latency to use of the baited plot, lengths of flock feeding bouts, and the number of rice grains consumed by individually marked birds during 1-min observation periods. We defined a flock feeding bout as starting when more than half of the birds in the test group were foraging on the plot and ending when all birds left the plot and failed to return within 5-min. After morning observations, birds were allowed to return to the decoy trap for food and water. Later the same day, test birds fasted for 2 h and we again released them into the flight pen at approximately 1500 to forage on the same plot. We recorded foraging behavior as before.

Flight Pen Trials, Treated Bait

The trials involved 3 test groups of 20 male redwings and 3 test groups of 60 male redwings each. We tested each group separately in the flight pen following the same procedures used with untreated bait. The 9 x 12-m test plots held 300 g bait/plot, with CAT-treated brown rice diluted 1:99 (3 g treated and 297 g untreated).

When the morning feeding bout ended, observers entered the flight pen, covered the baited plot with plastic sheeting, and opened the decoy trap to allow birds to return to the trap for food and water. Birds then fasted for 2 h and we released them again at 1500. We recorded behavior using the same procedure as the morning trial. Following the afternoon feeding activity, we retrapped birds from the test group and searched the flight pen for dead birds. We maintained the test group on brown rice and water for 7 days to assess mortality. We released survivors and necropsied half of the dead birds in each test group to confirm that death was due to CAT poisoning.

We assessed bait consumption from each 9 x 12-m test plot by establishing 10 (3 test groups) or 25 (3 test groups) randomly located 0.09-m^2 sampling quadrats. We marked the corners of each quadrat with small wooden sticks and set the number of rice baits in each at 12. We recounted the baits on sampling quadrats after exposure to foraging birds. To estimate the mass of treated bait consumed by birds in the test group, we multiplied the overall reduction in mass by the proportion of treated bait applied (1 in 100). We then divided by 19 mg/bait to yield an estimate of number of treated baits consumed by the test group.

Estimation with Binomial and Poisson

We used mean rates of individual bait consumption from birds in each test group to generate mortality predictions from the probability functions. For the binomial distribution, the basic function is $\mathbf{(p + q)^n}$, where p is the proportion of treated baits offered (also termed the bait dilution rate), q (= 1-p) is the proportion of untreated baits offered, and n is the mean number of total baits (treated and untreated) eaten per bird. In expanding the binomial function, the term $\mathbf{q^n}$ is the only one of interest as it represents the proportion of birds that ate no treated bait; all other birds ate at least one and died. The proportion that died is thus $\mathbf{1-q^n}$.

The Poisson distribution is represented by the function, $\mathbf{P(x) = (np)^x / (x!\, e^{np})}$, where e = 2.7183, n = mean total number of baits eaten/bird, p = the bait dilution, and x = mean number of treated baits eaten/bird. When x = 0, that is when no toxic bait is taken, the expression reduces to **P (0) = e** $^{-np}$. Thus, P(0) represents the proportion of the flock that did not consume a toxic bait, and **1-P (0)** is the proportion of the flock expected to die from eating at least 1 treated bait.

We applied Chi-square goodness of fit tests to examine whether observed mortality from baiting trials differed from that predicted using the binomial and the Poisson distribution functions.

Results

Bioassay

Each of the 4 birds that ate treated bait died within 48 h. One bird that was offered treated bait ate nothing. Birds that received only untreated bait survived. Necropsied birds all exhibited signs typical of CAT poisoning (*11*): separated gizzard linings, mottled livers and kidneys, and deposition of urates on internal surfaces, particularly the pericardium.

Flight Pen Trials, Untreated Bait

Flock foraging bouts averaged 4.5 min (SE = 0.6, n = 10) and individual feeding rates averaged 23.9 seeds/min (SE = 1.5, n = 44). The product of these two values, 108 seeds/bout, is an estimate of the individual bait consumption during a daily feeding bout. Based on this, for trials with treated bait, we used a dilution of 1 treated bait per 100 total baits because we intended for a blackbird to obtain only 1 treated bait during a typical daily feeding bout.

Flight Pen Trials, Treated Bait

The latency to start of feeding in the baited plot varied from about 10 sec to over 17 min following release from the decoy trap. Estimated total bait consumption varied among the 6 test flocks from 9.2% (SE = 9.2%) to 74.7% (SE = 7.8) (Table I). Generally, the 60-bird groups fed longer, removed more bait, and had higher estimated individual rates of treated bait consumption than did the 20-bird flocks. The morning foraging bouts tended to be longer than those in the afternoon, but the birds readily fed on the treated bait site the second time.

As suggested by bait consumption estimates, mortality in the 60-bird groups ($\bar{x}$ = 78%) exceeded that in the 20-bird groups (68%). For all groups, approximately 92% of the eventual mortality occurred within 48 h of initial exposure to the treated bait (Fig. 1). No deaths occurred beyond 96 h following exposure. Post-mortem examination of 90 of the 181 dead test birds revealed signs typical of CAT poisoning (*11*) in each case.

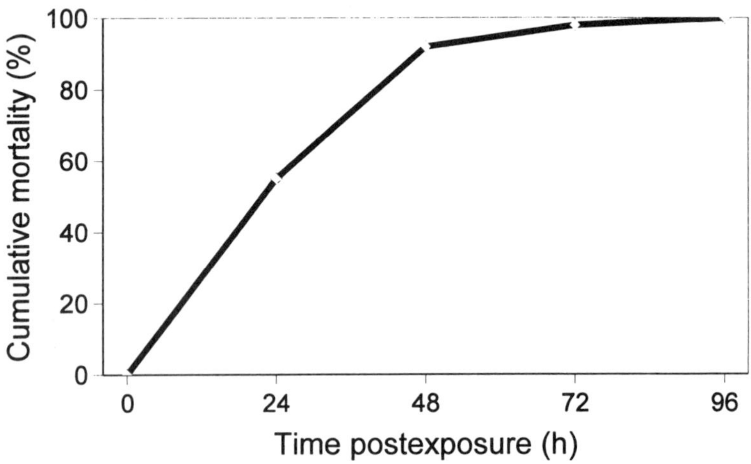

Figure 1. Cumulative mortality among three 20-bird and three 60-bird red-winged blackbird flocks following exposure to brown rice bait treated with 2% (g/g) CAT, a slow-acting toxicant.

Morality Estimation

Chi-square goodness of fit tests showed no difference ($P = 0.39$) between observed mortality and that predicted from the binomial and Poisson distributions (Table I). Only in group 6 did the observed mortality differ substantially from the predicted mortality. In contrast, estimates of treated bait consumption consistently exceeded observed mortality, particularly in the 60-bird groups (Table I). If we assumed 1 treated bait represented 1 dead bird, then mortality based on bait consumption would be greatly overestimated.

Table I. Bait consumption and mortality among groups of red-winged blackbirds during baiting trials in a 0.2-ha flight pen, Gainesville, FL. Each group of birds was exposed to 300 g of brown rice bait, 3 g of which was treated with 2% (g/g) CAT, a slow-acting toxin.

Test group	*Number of birds*	*% total bait consumption* $\bar{x}$	*SE*	*Baits removed*[1] *Total*	*Per bird*	*Dead birds* *Actual*	*Binomial*[2]	*Poisson*[2]
1	20	9.2	9.2	1453	73	10	10.4	10.4
2	60	50.8	10.8	8021	134	44	44.4	44.4
3	20	16.3	3.7	2574	129	13	14.5	14.4
4	60	71.7	12.5	11321	188	45	50.9	51.0
5	60	74.7	7.8	11794	197	51	51.7	51.6
6	20	10.3	24.2	1626	81	18	11.1	11.2

1 - Total bait removal is calculated by multiplying 300 g by the % total bait consumption and dividing by 0.019 g/bait. Then divide by the number of birds to get bait consumption per bird (n).

2 - Predicted from binomial and Poisson distribution functions. There was no difference ($P > 0.5$) between predicted and actual mortality for either distribution ($X^2 = 5.23$, 5 df).

Discussion

In order to evaluate the effectiveness of applying toxic bait for blackbird control, reliable estimates of mortality due to such management programs are needed. Reliable mortality estimates are also needed for the development of realistic environmental assessments and for incorporation into blackbird population models that will help determine long-term management strategies. Attempts to estimate mortality from consumption of single-particle lethal bait are hampered by not knowing how many treated baits are taken per bird. As illustrated in our study, inaccurate and unrealistic overestimates of mortality arise if each treated bait removed is assumed to represent one dead bird. Such an approach will yield an estimate of the maximum possible number killed which could have utility by putting an upper bound on baiting mortality *(1)*. It is unlikely, however, that this type of estimate will accurately reflect mortality because at least some individuals actually ingest >1 treated bait. In our study, the latter proved to be the case, as the estimated ingestion of treated bait among the 6 test groups ranged from 0.9 to 2.5 baits per dead bird. Empirical verification of multiple treated bait ingestion in field studies would be difficult to accomplish.

Given the possibility that blackbirds ingest >1 treated bait, but given the difficulty in documenting this, then determining mortality using bait consumption estimates reduces to guesswork. The best that can be done is to calculate a range of mortality estimates assuming different levels of bait consumption per bird (*1*).

The use of discrete probability distributions offers an alternative that is independent of estimates of total bait consumption. In our flight pen trial, both the Poisson and the binomial distributions yielded predictions very similar to the actual mortality from bait consumption. Assumptions inherent in the use of the Poisson distribution are: (1) the amount of bait distributed is large relative to the amount eaten, (2) the treated and untreated baits are evenly mixed, and (3) birds do not distinguish between treated and untreated baits. Assumption (1) does not apply to the binomial distribution, so the binomial might be preferable to the Poisson for that reason. The only data needed to calculate mortality estimates from the binomial or the Poisson distribution are numbers of baits eaten by individual blackbirds during feeding bouts on the bait site. These observations might be difficult to make, but it is certainly reasonable to do so because the result will be more reliable projections of the impact of toxic baiting on blackbird populations.

The relevance of our findings to field situations depends on the extent to which the flight pen foraging conditions resembled foraging conditions in the field. At this point, we have no field data on individual or flock foraging bout lengths at bait sites, so we cannot compare our experimental results to the birds' behavior under field conditions. We observed nothing to indicate that the test birds' feeding activity in the flight pen was abnormal. The birds readily adjusted to the flight pen environment and were not reluctant to feed from the baited plot. Another question concerns the relatively small flock sizes and sex of redwings tested. Blackbird foraging flocks often number in the thousands and at certain times are dominated by female redwings or other blackbird species (*12*). Because of differences between this study and field conditions in size and composition of blackbird flocks, there remains some question as to the applicability of our results to the field situation. Clearly, more detailed in-depth study of blackbird behavior at bait sites is needed.

Management Implications

If toxic baiting continues to be an important component of blackbird control strategies, then the accurate mortality predictions made in this study using the binomial and Poisson distributions should be of interest. Accurate mortality projections can be made and insights can be gained that might assist in designing baiting programs to meet specific objectives. For example, if mortality is to be maximized in a program using single lethal dose baits, and if multiple toxic bait ingestion is irrelevant, then for an average consumption (n) of 100 baits/bird, a dilution of 1 toxic bait/20 total baits (i.e. $q = 0.95$) would result in virtually 100% mortality ($1\text{-}q^n = 0.994$, or 99.4% mortality). However, based on the overall 75% mortality of birds foraging on bait in this study, we believe the 1:99 dilution is practical for achieving efficacious blackbird control and also for reducing nontarget hazards in the field. Allowing some individuals to survive exposure to bait may actually increase overall efficacy because these birds may likely return to the bait site and serve to decoy in other birds on subsequent days of baiting.

The average bait consumption per bird (n) will have to be obtained empirically during the pre-baiting period through close observation of feeding behavior on the bait site. The feeding bouts of as many birds as possible should be recorded to maximize the confidence in the estimate of feeding bout length. This value is then used with the dilution rate in either the binomial or the Poisson distribution to calculate the proportion of birds killed. A projection of the absolute number of dead birds requires an estimate of the total population using the bait site. This can only be obtained through careful, consistent monitoring of the bait site.

Acknowledgments

P. W. Lefebvre, R. E. Matteson, and C. O. Nelms assisted with various aspects of the study. We thank J. F. Heisterberg, C. E. Knittle, and A. R. Stickley for reviewing earlier drafts of this paper. D. T. Bohdanowicz assisted with preparation of the manuscript.

References

1. Glahn, J. F.; Wilson, E. A. *Proc. East. Wildl. Damage Control Conf.* **1992**, *5*, 117-123.
2. Avery, M. L.; Kenyon, M. J.; Linz, G. M.; Bergman, D. L.; Decker, D. G.; Humphrey, J. S. *J. Wildl. Manage.* **1998**, 62, 388-394.

3. DeCino, T. J.; Cunningham, D. J.; Schafer, E. W., Jr. *J. Wildl. Manage.* **1966**, 30, 249-253.
4. Schafer, E. W., Jr. In *CRC Handbook of Pest Management in Agriculture*; Pimentel, D., Ed.; Boca Raton, FL, **1991**. Vol. II., pp 599-610.
5. Besser, J. F.; Bray, O. E. *Bird Damage Research Report 45*. U. S. Fish Wildl. Serv., Denver Wildl. Res. Cen. 1977.
6. Knittle, C. E.; Guarino, J. L.; Nelson P. C. *Proc. Vertebr. Pest Conf.* **1980**, *9*, 31-37.
7. Glahn, J. F.; Heisterberg, J. F. *Bird Damage Research Report 198*. U. S. Fish Wildl. Serv., Denver Wildl. Res. Cen. 1981.
8. Cummings, J. L.; Schafer, E.W., Jr.; Cunningham, D. J. *Bird Damage Research Report 344*. U. S. Fish Wildl. Serv., Denver Wildl. Res. Cen. 1985.
9. Cummings, J. L.; Schafer, E. W., Jr. *Bird Damage Research Report 357*. U. S. Fish Wildl. Serv., Denver Wildl. Res. Cen. 1985.
10. Schafer, E. W., Jr.; Brunton, R. B.; Schafer E. C.; Chavez, G. *Bird Damage Research Report 93*. U. S. Fish Wildl. Serv., Denver Wildl. Res. Cen. 1978.
11. Lefebvre, P. W.; Matteson, R. E.; Decker, D. G.; Hillman, C. D.; Palmer, J. E. *Bird Damage Research Report 73* . U. S. Fish Wildl. Serv., Denver Wildl. Res. Cen. 1978.
12. Labisky, R. F.; Brugger,K. E. *Technical Report 36*. U. S. Fish Wildl. Serv., Florida Cooperative Fish Wildl. Res. Unit. 1989.

Chapter 9

Exposure Assessment of *Rana catesbeiana* Collected from a Chlorpyrifos-Treated Cornfield

Sean M. Richards, Todd A. Anderson, Steven B. Wall, and Ronald J. Kendall

The Institute of Environmental and Human Health, Texas Tech University, Texas Tech University Health Sciences Center, Lubbock, TX 79416

Global anuran decline is well documented; however, the reason(s) for this are not apparent. Many governmental and private sectors are seeking answers to this dilemma. One limiting factor in determining the cause of anuran decline is available data. Relatively few studies have gathered data to assess the potential for anuran exposure to agriculturally applied pesticides and subsequent risk. For these reasons, bullfrogs (*Rana catesbeiana*) were collected from a cornfield and adjacent pond following application of chlorpyrifos, an organophosphate used in the control of corn rootworms. Frogs were collected from two sites; one received an application of chlorpyrifos at the label rate (1.35 kg A.I./ha or 1.2 lbs. A.I./acre), the other received no insecticide application. Brain and plasma cholinesterase activities of bullfrogs from both sites were measured and compared. Our data indicate that while the inherent biological variation of both frog brain and plasma cholinesterase is large, no bullfrogs were significantly exposed to chlorpyrifos while occupying the cornfield or adjacent pond.

Recently, the scientific community has been inundated with questions about frog (Order Anura) population decline. Numerous studies have documented dwindling populations and discussed the probable extinction of many species of anurans (1-10). Many areas worldwide have chronicled an incidence of frog decline, often in numerous genera (11). Yet, a scarcity of data impairs the ability of regulatory agencies and researchers in assessing risk and offering a solution to this colossal puzzle.

Many theories have been developed in an attempt to explain the seemingly high rate of anuran disappearance. These theories include events such as natural disasters, climatological shifts, competition, and succession of habitat (12). Other contributing factors may include breeding habitat alteration, reduction, or destruction (3). Heyer (9) as well as Hayes and Jennings (10) propose that non-indigenous species of fish and frogs may be removing amphibians from their natural habitat via predation and/or competition. Acid rain may be a culprit by lowering the pH of breeding water so that amphibian egg mortality drastically increases (13). Increased levels of UV-B radiation, due to depletion of the ozone layer, may contribute to amphibian decline (2). Even seemingly natural events such as pathogens (14, 15) and climatologic disturbances (16) have been suggested as possible explanations to the ongoing anuran loss.

Agricultural pesticide application has also been scrutinized as a potential cause for anuran decline (2, 17). Although few studies have directly linked pesticide use to anuran decline (18), frogs have been shown to be sensitive to a myriad of herbicides and insecticides (19-24).

Chlorpyrifos [*O*,*O*-diethyl *O*-(3,5,6-trichloro-2-pyridyl) phosphorothioate] is one organophosphate (OP) for which little anuran field exposure data exists (25). Of all the organophosphates applied in the US, chlorpyrifos is used on the most acreage, accounting for approximately 7 million acres per year (26). In order to maximize land use, chlorpyrifos-treated row crops are often planted close (< 3 m) to water sources that serve as anuran habitat. This practice increases the chance for anuran exposure to chlorpyrifos via direct contact in the field, rainfall drainage, or irrigation runoff. The US EPA (27) classifies chlorpyrifos as highly toxic to freshwater fish; the toxicity of chlorpyrifos is such that applications as low as 0.01 lb active ingredient (A.I.)/acre may cause fish mortalities. In the absence of toxicity data, it is sometimes assumed that frogs are likely to be as sensitive to OPs as fish (28).

Despite the evidence of frog decline and their potential sensitivity to OPs such as chlorpyrifos, the USEPA has yet to establish or require toxicity testing with the order Anura. However, many groups have recognized the need for anuran protection and expansion of the small toxicological database (29-31). Governmental agencies are recognizing needs and calling for researchers to determine the rank of tolerance of anurans, identify sensitive species, define an indicator species, and evaluate all avenues of exposure (31).

In the spring of 1998, while collecting specimens for a concomitant study, numerous bullfrogs (*Rana catesbeiana*) were noticed occupying a cornfield recently treated with chlorpyrifos. Cognizant of anuran decline and the lack of data pertaining to frogs and insecticide exposure, we collected bullfrogs and analyzed them for biomarkers of chlorpyrifos exposure. Our original study design did not call for collection of frogs, as we did not expect to find any in the cornfield. It is not known why the frogs were in the field, although it is suspected that they were feeding on an abundance of Junebugs (genus *Phyllophaga*). The data from this opportunistic study are reported herein.

Methods and Materials

South central Iowa (93.3° longitude/40.9° latitude), the area from which the *Rana* specimens were collected, is characterized by rolling croplands interspersed with ponds, rivers, and associated tributaries. During May of 1998, chlorpyrifos was applied at the label rate (1.35 kg A.I./ha or 1.2 lbs. A.I./acre) to multiple 16.2 ha (40 acre) cornfields. Two cornfields, one treated with chlorpyrifos (Treatment Site), the other treated with blank clay granules (Reference Site), partially encircled ponds. These ponds, approximately 0.2 ha (0.5 acre) in size, contain water all year and receive precipitation run-off. Following chlorpyrifos application (Day 0), rain events were recorded (in inches) on Day 3 (0.04), Day 7 (0.85), Day 8 (0.10), Day 9 (1.32), Day 10 (0.01), Day 11 (1.30), Day 12 (0.05), Day 13 (0.02), and Day 14 (0.01)

Within a 0.5 mile radius of the treatment site, the farming regimes included hay, pasture, soybeans, and oats. The field containing soybeans received one application of glyphosate (RoundUp®), an herbicide, prior to planting, while the remaining fields were pesticide free. Farming regimes surrounding the reference site within a 0.5 mile radius included hay, oats, and pasture, all were maintained without pesticide applications. Approximately 4 miles separated the treatment and reference site. To our knowledge, neither of these fields received an OP application for two years prior to this study.

On two different occasions, 3 and 15 day post application (DPA), bullfrogs were collected from corn rows and pond margins of both sites. Upon collection, frogs were rinsed with tap water, and anesthetized with Tricane (Ethyl m-aminobenzoate). Blood was collected via heart puncture, using a heparinized 27-gauge needle. Bullfrogs were then decapitated while the brain was left intact; head and carcasses were frozen whole at –80°C.

At the time of analysis, brains were removed while frozen. Great care was taken to excise consistent portions of the central nervous system. The entire forebrain (including olfactory bulb), midbrain, and hindbrain regions were removed. The hindbrain was severed from the spinal cord at the posterior portion of the medulla. Following removal, brains were homogenized in 0.05 M Tris buffer (pH 7.4) at 1:4 (w:v). An aliquot of the brain homogenate was further diluted to a final 1:99 in the same buffer and assayed immediately for ChE activity. Plasma samples were thawed, diluted 1:5 in 0.05 M Tris buffer, and immediately analyzed.

Cholinesterase activities were determined colorimetrically using a modification of the Ellman assay (32) as described by Gard and Hooper (33) and using acetylthiocholine iodide (AThCh, Sigma Chemical Co., St. Louis, MO) as the substrate. Brain samples were evaluated for total ChE while plasma samples were evaluated for total ChE, AChE, and BChE. All cholinesterase activities are reported in µmole of AThCh hydrolyzed / min / gram brain or ml plasma.

To determine if the bullfrogs were exposed to an OP, we compared the plasma and brain ChE activity of bullfrogs from the treated site to that from the reference site. In addition, a reactivation assay was conducted by measuring the plasma and brain ChE activity before and after exposure to a dephosphorylating compound, pyridine-2-aldoxime methochloride (2-PAM). Theoretically, if ChE has not been in contact with a ChE inhibitor, the activity should be the same before and after exposure to 2-PAM. If the activity of a sample treated with 2-PAM was 10% greater than that sample without 2-PAM, and a statistically significant increase was shown by a one-tailed Student's t-test ($P < 0.05$) (34), the sample was considered "reactivated". Such a reactivation would suggest the presence of an OP-inhibited enzyme.

In order to confirm that chlorpyrifos was not present on either field (prior to our application), pre-application and post-application soil samples were collected from the top 12 cm (4.72 in) of the treated and reference site. Soil was collected from the treatment site immediately following application of chlorpyrifos (Day 0) and subsequently on Days 2, 4, 8, 16, 32, and 64 post-application. Soil was collected from the Reference Site on Days 0, 4, 16, and 64 post-application of blank granules. Five sampling points were randomly chosen within cultivated areas of each study site. From each point, 8 soil samples were collected, 4 from within the corn furrow and 4 between furrows. Each set of 4 samples was pooled in the laboratory to form a composite sample for analysis. This provided 10 composite samples per site for each time point (5 in furrow samples, 5 between furrow samples) with the exception of the pre-application samples that were collected prior to cultivation of the corn.

All samples were extracted using a mixture of hexane:acetone (3:1), homogenization, and mechanical agitation. Extracts were treated with anhydrous sodium sulfate to remove excess moisture and filtered through a 0.45 μm Acrodisk® filter. Extracts were concentrated under a stream of N_2 prior to analysis.

Each extract was analyzed using a Hewlett-Packard Model 6890 Series gas chromatograph (GC), operating in the splitless mode, equipped with a ^{63}Ni electron capture detector. The column was a 30 m x 0.25 mm i.d. fused-silica capillary, coated with DB-5 stationary phase, film thickness = 0.25 μm (J&W Scientific, Folsom, CA). Oven temperature was programmed from 100°C with a 1 minute initial hold to 200°C at 30°C/minute, and a final hold time of 4 minutes. Inlet and detector temperatures were 200°C and 320°C, respectively.

The chromatographic system was calibrated before each analysis session. All calibrations consisted of 4 standards of analytical-grade chlorpyrifos (0.058 μg/ml , 0.116 μg/ml , 0.58 μg/ml , 1.16 μg/ml) in hexane. Following every 15th sample, one of the calibration standards was re-analyzed to confirm GC performance. By fortifying soil samples with known amounts of chlorpyrifos, and using the described procedure of extraction, cleanup, and concentration, the method limit of detection for chlorpyrifos in the soil samples was 0.025 µg/g in soil. The method analyte recovery was also assessed and all data were corrected accordingly.

Samples were considered positive for chlorpyrifos when the sample chromatogram contained a peak eluting within the retention time window of the chlorpyrifos standard. However, positive detections were not confirmed by mass spectroscopy.

Statistical Methods

Analysis of Variance (SAS Version 6.12, SAS Institute) was used to compare differences in frog blood plasma and brain cholinesterase activities (α=0.05). Total ChE and BChE activity values were log-transformed, as necessary, to meet ANOVA assumptions. Levene's test was used to examine homogeneity of variance. Factorial models were utilized to examine the presence of significant interactions between main effects. Models consisted of dependent variables; Total ChE, AChE, and BChE activity, and independent variables; site, frog weight, and date of collection. Plasma and brain activities and mean body weights were compared using a two-tailed t-test (α=0.05). A diagnostic threshold, calculated by subtracting 2 standard deviations from the mean ChE activity of Reference Site frogs, was established and used to evaluate frog enzyme activity levels from Treatment and Reference Sites (35). Samples that exhibited activities below the diagnostic threshold were considered (based on weight of evidence) to have been exposed to a ChE-inhibiting insecticide.

Results

Brain and plasma total ChE activities were variable, but did not indicate exposure to chlorpyrifos. Although plasma samples were analyzed for BChE activity, we found that AChE is the predominant ChE in bullfrog plasma, accounting for approximately 95% of the total activity. Therefore we chose to report the total ChE of the bullfrog plasma instead of AChE and BChE separately. Brain and plasma activities are reported as percent of mean Reference Site activities. Brain ChE activities ranged from 66 – 141% of mean Reference Site values (Table I). Similar variation occurred in plasma ChE activities (65 – 138 % of Reference Site mean) (Table II). Although some brain and plasma samples appeared to be relatively inhibited (i.e., activities 66 and 65% of Reference Site mean), there were no statistically significant ChE activity deficiencies. This conclusion is based on the finding that no samples exhibited activities two standard deviations below the mean of Reference Site samples.

The mean weight (± SE) of the bullfrogs from the Treatment and Reference Site was 113.36 ± 23.95g, ranging from 6.55 to 315.10g. In order to examine the effect of weight on brain and/or plasma ChE activity, frogs were divided into groups according to their body weight and statistically analyzed. Multiple weight class comparisons were made, none of which indicated that *Rana catesbeiana* brain ChE activity is dependent upon body weight. However, the plasma ChE activity did appear to be dependent upon the weight of the bullfrog. Plasma total ChE activity for bullfrogs ≤ 80g (n=11) was significantly higher ($p<0.0001$) than that of bullfrogs ≥ 80g (n=12). If weight may be roughly equated with age (with obvious exceptions such as gravid females), then it may be assumed that the plasma total ChE activity decreases with age in *Rana catesbeiana* adults. Though these are preliminary results based on a small sample size (n=23), an inverse relationship between plasma ChE activity and age has, to our knowledge, been reported just one other time, in the mallard (*Anas platyrhynchos*) (36).

Table I. Brain ChE activity of *Rana catesbeiana*

Collection	*Site*	*n*	*ChE Activity* *Mean*	*SE*	*Range*
3 DPA	Treatment	11	115.3	11.78	76.3 - 124.2
	Reference	6	100	11.89	66.2 - 137.3
15 DPA	Treatment	13	108.0	5.60	81.0 - 141.3
	Reference	17	100	6.22	63.2 - 136.6

NOTE: ChE activity is expressed as % mean activity of controls.

Table II. Plasma ChE activity of *Rana catesbeiana*

Collection	*Site*	*n*	*ChE Activity* *Mean*	*SE*	*Range*
3 DPA	Treatment	7	93.7	8.32	71.9 - 126.9
	Reference	7	100	20.01	73.2 - 124.6
15 DPA	Treatment	9	109.53	7.30	74.9 - 137.9
	Reference	7	100	7.82	65.2 - 106.6

NOTE: ChE activity is expressed as % mean activity of controls.

Mean chlorpyrifos concentrations (± SE) in the Treatment Site furrow ranged from 20.5 ppm (±1.13) on the day of application to 0.67 ppm (±0.23), 64 days post-application. At the time of frog collections, 3 and 15 DPA, in-furrow concentrations were 15.55 ppm (±1.77) and 4.88 ppm (±0.18), respectively. No chlorpyrifos was detected on the reference site. The treatment site received 3.7 inches of rain from Day 0 to Day 15. In a previous study by this laboratory (1997, unpublished data), chlorpyrifos was applied at the same label rate and manner as in this study. Temporal soil concentrations of chlorpyrifos in 1997 were similar to those reported above; however, in 1997 there was no rainfall until 18 days post application.

Discussion

Based on the data presented, it appears that the ChE of *Rana catesbeiana* was not inhibited during their sporadic visits to a chlorpyrifos-treated cornfield, or while residing in the adjacent pond. Activities of brain and plasma ChE, from both Treatment and Reference Site, were highly variable. Such intraspecies variability in ChE levels of birds and mammals has long been acknowledged (35-39). Additionally, Baker (40) reports similar variability in the brain ChE activities of two species of salamanders. The use of a diagnostic threshold theoretically compensated for such variability, so samples that exhibited low activity were not misidentified as being chemically inhibited. If our only means of identifying inhibited samples was to compare individuals to the diagnostic threshold, sample groups with a high degree of variability may cloak OP inhibited individuals. However, by examining each sample for the potential to reactivate, we reduce the potential for the diagnostic threshold to hide samples that may have been OP inhibited. Since no sample fell below the diagnostic threshold or reactivated, a conclusion that the bullfrogs were not significantly exposed to chlorpyrifos was made.

Limited studies have shown adult anurans to be relatively resistant to some OPs (41). The reason for the perceived resistance is not fully known. Amphibians, in general, have a highly vascularized, moist skin surface that is very permeable to water, gases, and ions (42). The highly water permeable skin of frogs means that the surrounding water is essentially an extension of the extracellular space of the animal (43). Water moves across frog skin in three ways: evaporation, diffusion, and osmosis (44, 45). Evaporation is an efflux of water vapor. Diffusion is the equal influx and efflux of water, one molecule at a time. (43). Diffusion results in a net water influx of zero; this is distinctly different from osmosis. For the scope of this discussion, osmosis is the only relative mode of water influx, as this process allows a bulk movement or flow of water to pass into the frog. Osmosis surpasses all other means of hydration by the frog, as amphibians typically do not drink water (46). In fact, Bentley and Yorio (47) conclude from tests with dehydrated *Rana catesbeiana*, *Bufo marinus*, and *Xenopus laevis* that oral uptake of water was, "incidental and unimportant in rehydration". The semipermeable, hydrophilic skin of the anuran may partially explain the perceived resistance of the frog. In order for a lipophilic chemical to pass across the frogs' skin, it may have to bind with water or some other osmotic liquid. Due to the hydrophobic nature of many insecticides, their ability to osmose may be limited. Shah et al. (48) present data that suggest penetration of a pesticide into a frog depends upon the toxicants octanol:water coefficient (K_{ow}) and water solubility. In their study, numerous insecticides (dissolved in acetone) were applied to the back of adult grass frogs (*Rana pipiens*). These insecticides included (in order of increasing water solubility): DDT, dieldrin, permethrin, parathion, and carbaryl. Based on their report, it appears that amount and rate of penetration was generally related to the water solubility (Figure 1) and inversely proportional to the K_{ow}. Dieldrin was the only exception, penetrating less than permethrin and DDT.

The hydrophilic skin of frogs may provide protection from many of today's nonpolar, hydrophobic insecticides. Shah et al. (48) also noted that once the pesticide penetrated the skin of the green frog, the amount detected in the liver and other organs was surprisingly low. This is suggestive of a higher capacity for metabolism and/or excretion as compared to mammals and birds. It may be that the resistance conferred by the hydrophilic skin works in concert with superior metabolism and excretion. Anurans, however, are a very diverse group and skin permeability is quite variable (49). Shah et al. (48) examined the skin of one species of frog at one age. Much more research is needed before making generalizations about frogs and pesticide resistance.

In our study, a granular pesticide was used. This probably afforded the bullfrogs even more protection, due to the dry form. Sprayable formulations typically contain emulsifiers. Such formulations (along with increased heat and pH) may dramatically increase the efficiency by which pesticides penetrate the skin of the frog.

Concentrations of chlorpyrifos in the soil were comparable to those detected in a previous study by this laboratory (unpublished data). Due to the spontaneous nature and limited budget of this study, no pond water or sediment samples were collected from either site. Such data should have been collected, as it would have valuable for determining precise chlorpyrifos exposure rates. However, the high organic carbon partitioning coefficient (K_{oc} = 8500) and low water solubility (1.4 ppm) of chlorpyrifos would lead one to expect concentrations to be highest in the soil, followed by pond sediments, and small amounts in pond water (50). Therefore, estimates of frog exposure based on chlorpyrifos residues in the furrow of the cornfield would be highly conservative, as the bullfrog rarely inhabits cornfields and concentrations of chlorpyrifos elsewhere are probably much lower. The similarity of the temporal chlorpyrifos concentrations between two plantings (1997 and 1998), despite the differences in rainfall, further indicates the tendency for the chlorpyrifos to remain in the corn furrow.

Conclusions

Our study indicates that the adult *Rana catesbeiana* inhabiting one (1) farm pond adjacent to a treated cornfield were not significantly exposed to chlorpyrifos. Due to the spontaneous, opportunistic nature of this study, much caution should be taken when examining the results. This study examined one treated site and one reference site; the *n* of both was very low. Anura and other orders of amphibia are a very

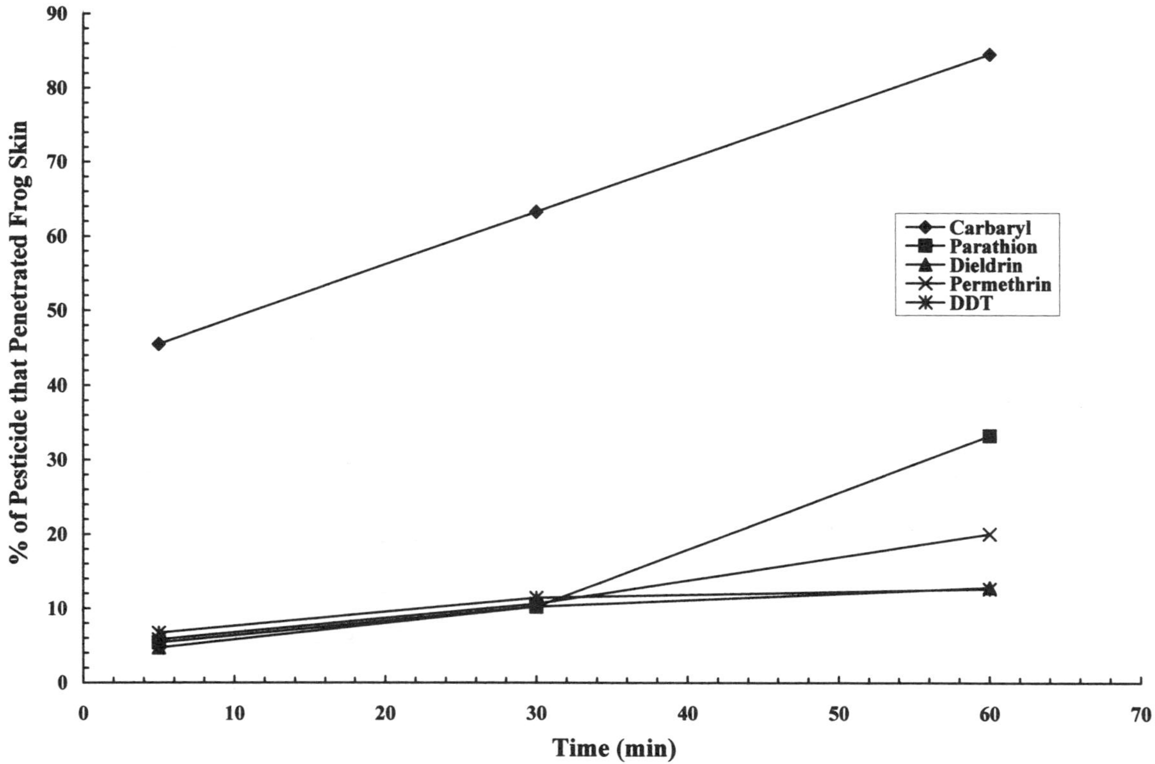

*Figure 1. Relationship between water solubility and penetration of selected insecticides through the skin of Rana pipiens. The legend ranks the compounds in descending order of water solubility. * Created from the data of Shah et al. (47).*

diverse group for which generalization is difficult. Most of the current anuran research is performed on a few species based on ease of obtainment, abundance in the wild, and ease of maintenance in the laboratory, not on how well the organism represents the order Anura (51). These studies must also be reviewed with caution. In order to adequately study, and ultimately preserve these vital organisms, much more research with multiple species from multiple age groups is needed.

References

1. Louma, J. R. *Audubon* **1997**, *99*, 60-69.
2. Blaustein, A. R. *Herpetologica* **1994**, *50*, 85-97.
3. Halliday, T. R. *Environ. Rev.* **1993**, 1, 21-25.
4. Barinaga, M. *Science* (Washington, D.C) **1990**, *247*, 1033-1034.
5. Blaustein, A. R.; Wake, D. B. *Trends Ecol. Evol.* **1990**, 5, 203.
6. Phillips, K. *Biosci.* **1990**, *40*, 422-424.
7. Vitt, L. J. *Biosci.* **1990**, *40*, 418.
8. Wyman, R. L. *Conserv. Biol.* **1990**, *4*, 350-352.
9. Heyer, W. R.; McDiarmid , R. W.; Weigmann, D. L. *Biotropica* **1975**, *7*, 100-111.
10. Hayes, M. P.; Jennings, M. R.J. *Herpetol.* **1986**, *20*, 490-509.
11. Hansen, L. J. 19th Annual SETAC Conference, Charlotte, NC, Abstract #542, 1998
12. Moulton, C. A. Ph.D. thesis, North Carolina State University, Raleigh, NC, 1996.
13. Leuven, R. S. E.; den Hartog, W. C.; Christianns, M. H. C.; Heijligers, W. H. C. *Experientia.* **1986**, *42*, 495-503.
14. Carey, C. *Conserv. Biol.* **1994**, *3*, 355-362.
15. Morell, V. *Science* (Washington, D.C). **1999**, *284*, 728-731.
16. Pounds, J. A.; Crump, M. L. *Conserv. Biol.* **1994**, *8*, 72-85.
17. La Marca, E.; Reinthaler, H. P. *Herpetol. Rev.* **1991**, *22*, 125-128.
18. McAlpine, D. F.; Burgess, N. M.; Busby, D. G. *Bull. Environ. Contam. Toxicol.* **1998**, *60*, 30-36.
19. Guzman, J. A.; Guardia, T. *Bull. Environ. Contam. Toxicol.* **1978**, *20*, 32-38.
20. Berrill, M.; Bertram, S.; Wilson, A.; Louis, S.; Brigham, D.; Stromberg, C. *Environ.Toxicol. Chem.* **1993**, *12*, 525-539.
21. Dumpert, K. *Ecotox. Environ. Safety* **1987**, *13*, 324-338.
22. Marchal-Segault, D.; Ramade, G. *Environ. Res.* **1981**, *24*, 250-258.
23. Paar, K. R.; Ghate, H. V.; Katdare, M. *Bull. Environ. Contam. Toxicol.* **1983**, *31*, 170-176.
24. Sanders, H. O. *J. Water Poll. Cont. Fed.* **1970**, *24*, 1544-1550.
25. Calumpang, S. M. F.; Medina, M. J. B.; Tejada, A. W.; Medina, J. R. *Bull. Environ. Contam. Toxicol.* **1997**, *58*, 909-914.
26. Ollinger, M.; Aspelin, A.; Shields, M. *Agribusi.* **1998**, *14*, 199-212.
27. USEPA ; Water Quality Standards Handbook, OSTP, 823-04-005A: Washington DC, 1994.

28. Berrill, M.; Bertram, S.; McGillivray, L.; Kolohon, M.; Pauli, B. *Environ.Toxicol. Chem.* **1994**, *13*, 657-664.
29. Grillitsch, B.; Linder, G. 19th Annual SETAC Conference, Charlotte, NC, Abstract #458, 1998.
30. Sparling, D. W.; Pickney, A. E. , 19th Annual SETAC Conference, Charlotte, NC, Abstract #454, 1998.
31. Westerman, A. G.; Kerchner, M. D.; Birge, W. J. Platform presentation, 19th Annual SETAC Conference, Charlotte, NC, 1998.
32. Ellman, G.; Courtney, K.; Andres, V.; Featherstone, R. *Biochem. Pharmacol.* **1961**, *7*, 88-95.
33. Gard, N.; Hooper, M. *J. Wildl. Dis.* **1993**, *29*, 1-9.
34. Sokal, R.; Rohlf, F. *Biometry*; W.H. Freeman and Co.: San Francisco, CA, 1981.
35. Hill, E. *J. Wildl. Dis.* **1988**, *24*, 51-61.
36. Fairbrother, A.; Craig, M.A.; Walker, K.; O'Loughlin, D. *J. Wildl. Dis.* **1990**, *26*, 67-77.
37. Edwards, J. A.; Brimijoin, S.J. *Neurochem.* **1982**, *38*, 1393-1403.
38. Bennett, E. L.; Krech, D.; Rosenzweig, M. R.; Karlsson, H.; Dye, N.; Ohlander, A. *J. Neurochem.* **1958**, 3, 153-160.
39. Westlake, G. E.; Martin, A. D.; Stanley, P. I.; Walker, C. H. *Comp. Biochem. Physio.* **1983**, *76C*, 15-24.
40. Baker, K. N. *Arch. Environ. Contam. Toxicol.* **1985**, *14*, 685-691.
41. Anderson, R. A.; Mikalsen, A. *Gen. Pharmacol.* **1977**, *9*, 177-181.
42. Spotila, J. R.; O'Connor, M. P.; Bakken, G. S. In *Environmental Physiology of the Amphibians*; Feder, M. E., Burggren, W. W., Eds.; The University of Chicago Press: Chicago, IL, 1992; Vol. 1, pp 59-80.
43. Boutilier, R. G.; Stiffler, D. F.; Toews, D. P. In *Environmental Physiology of the Amphibians*; Feder, M. E., Burggren, W. W., Eds.; The University of Chicago Press: Chicago, IL, 1992; Vol. 1, pp 81-150.
44. Koefoed-Johnson, V.; Ussing, H. H. *Acta Physio. Scan.* **1953**, *17*, 38-43.
45. Warburg, M. R. *J. Zool.* **1965**, *13*, 317-330.
46. Shoemaker, V. H.; Hillman, S. S.; Hillyard, S. D.; Jackson, D. C.; McClanahan, L. L.; Wither, P. C.; Wygoda, M. L. In *Environmental Physiology of the Amphibians*; Feder, M. E., Burggren, W. W., Eds.; The University of Chicago Press, 1992; Vol. 1, pp 125-150.
47. Bentley, P. J.; Yorio, T. *J. Exper. Biol.* **1980**, *79*, 41-46.
48. Shah, P. V.; Monroe, F. E.; Guthrie, F. E. *Drug Chem. Toxicol.* **1983**, *6*, 155-179.
49. Bentley, P. J.; Baldwin, G. F. *Amer. J. Physio.* **1980**, *239*, 505-508.
50. Racke, K.Rev. *Environ. Contam. Toxicol.* **1993**, *131*, 1-154.
51. Feder, M. E. In *Environmental Physiology of the Amphibians*; Feder, M. E., Burggren, W. W., Eds.; The University of Chicago Press: Chicago, IL, 1992; Vol. 1, pp 1-7.

Chapter 10

Modulation of Teleost CYP1A1 Activity by EBI Fungicides (Propiconazole, Clotrimazole) and Interaction with an Organophosphate Insecticide

S. L. Levine[1], J. S. McClain[2], and J. T. Oris[2]

[1]Department of Veterinary Science, Pennsylvania State University, University Park, PA 16802
[2]Department of Zoology, Miami University, Oxford, OH 45056

This review examines mechanisms by which clotrimazole and a technical grade formulation of propiconazole (TGP) modulate cytochrome P4501A1 (CYP1A1) expression and mechanisms by which TGP may enhance the acute toxicity of the organophosphate insecticide parathion. *In vitro*, clotrimazole was a mixed-type noncompetitive inhibitor and TGP was a noncompetitive inhibitor of rainbow trout hepatic microsomal CYP1A1 catalytic activity. *In vivo*, analytical grade formulations of clotrimazole and propiconazole caused inhibition of CYP1A1 catalytic activity but did not influence hepatic CYP1A1 gene expression. In contrast, rainbow trout hepatic CYP1A1 expression following waterborne exposures to TGP exhibited a mixed-pattern response. The mixed-pattern response was characterized by inhibition of CYP1A1 catalytic activity and concurrent induction of CYP1A1 mRNA levels. Induction of CYP1A1 expression by TGP was determined to result from AhR-active compounds within the formulation. Although exposure to TGP can modulate CYP1A1 expression, it remains to be determined whether modulation of CYP1A1 is related to the enhancement of acute parathion toxicity to fathead minnows.

Introduction

Over the past two decades a number of ergosterol biosynthesis inhibiting (EBI) imidazole and triazole derivatives have been approved for use in agricultural applications. These fungicides were designed to inhibit cytochrome P450-mediated ergosterol biosynthesis and are used as systemic foliar fungicides [1]. A technical grade formulation of the antifungal triazole compound propiconazole (Figure 1) has been shown to be a potent inducer of cytochrome P450 drug metabolizing isoforms in mammalian species [2,3], bobwhite quail [3,4], and recently in rainbow trout (*Oncorhyncus mykiss*) [5]. Currently, technical grade propiconazole (TGP) is widely used on banana plantations and TGP is aerially applied 40 or more times per year in an attempt to control the fungal disease Black Sigatoka [6]. Concerns have been raised in the scientific and popular press over potential toxic interactions between antifungal imidazole/triazole compounds and organophosphate (OP) insecticides [7,8]. Presently, the opportunity exists for concurrent application of TGP along and different OP insecticides on banana plantations [9].

As a result of high precipitation rates, irrigation practices, and an unenclosed design, a large quantity of pesticide applications on banana plantations is lost to surface runoff, leaching, and erosion [6]. Aquatic systems are often incorporated into drainage systems to allow excess water to runoff into local streams and rivers for additional restriction of fungal growth [9,10]. Therefore, aquatic ecosystems near areas of frequent applications could receive high inputs of TGP. In a recent analysis of pesticide concentrations in Costa Rican banana plantations, propiconazole concentrations in nearby streams were as high as 24 μg/L during the dry season [11]. During periods of high precipitation, propiconazole concentrations in aquatic ecosystems could reach concentrations higher than 24 μg/L. In a comprehensive study which surveyed pesticide concentrations in aquatic ecosystems of Central America, propiconazole was found to be the most widely distributed pesticide [12]. Propiconazole was present in 60% of the samples collected in effluents, 56% of the samples collected in creeks, 43% of the samples collected from main rivers, and some riverine sediment samples reached 130 μg/kg [12]. Because of the potential for interactions between TGP and other pesticides, there is the need to develop a biomarker for TGP exposure. One candidate biomarker for TGP exposure is cytochrome P4501A1 (CYP1A1) expression.

This chapter summarizes recent studies which: 1) Characterized the influence of a model EBI compound (clotrimazole; Figure 1) and TGP on CYP1A1 expression; 2) Examined whether CYP1A1 expression could be used as a biomarker of TGP exposure; and 3) Explored the hypothesis that a mechanistic relationship exists between modulation of cytochrome P450 activity by TGP and enhanced toxicity of OP insecticides to fish. This hypothesis was initially tested with the EBI compounds prochloraz and penconazole with red-legged partridge [13-14]. These studies demonstrated that some EBI compounds can enhance the toxicity of OP insecticides.

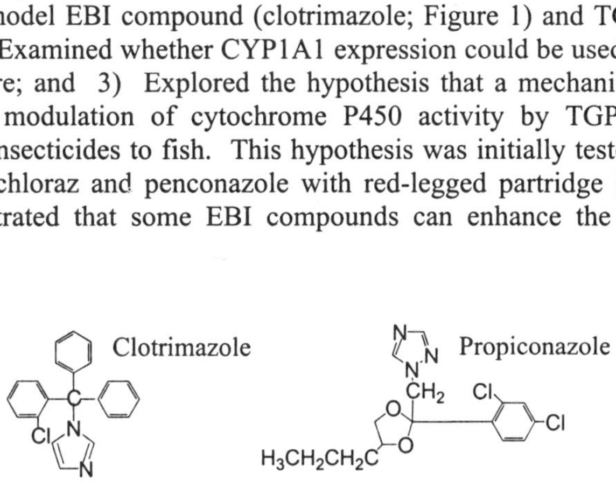

Figure 1. Structure of clotrimazole and propiconazole.

Regulation of CYP1A1 Expression and its use as a Biomarker

The molecular mechanisms involved in induction CYP1A1 expression have been extensively studied in mammalian species. Induction of CYP1A1 expression in mammalian and teleost species is initiated with the binding of the aryl hydrocarbon receptor (AhR) by a ligand. Well-characterized AhR ligands include aromatic hydrocarbons, polychlorinated biphenyls, dibenzodioxins, and dibenzofuran congeners [15]. In mammalian models, the ligand activated AhR translocates to the nucleus [16], binds with the AhR nuclear translocator protein, and then dimerizes to specific DNA sequences termed dioxin responsive elements (DREs) in the regulatory region of the gene [17]. A similar induction process is believed to occur in fish [18]. Currently, there is little information on the functionality of the teleost AhR in terms of binding affinities for nontraditional agonists, particularly at environmentally relevant concentrations.

The primary role of the CYP1A1 enzyme is to enhance the elimination of water insoluble compounds [19]. A single atom of oxygen is added to the foreign molecule resulting in the addition of a hydroxyl group which is the first phase for increasing the water solubility and hence the excretability of the compound. The CYP1A1 enzyme accounts for the majority of benzo[a]pyrene (BaP) hydroxylase activity in mammals [20] and fish [19]. Induction of CYP1A1 expression in fish liver and gill has been successfully used to identify exposure BaP [21] and other planar environmental contaminants [22]. However, the use of CYP1A1 expression and other cytochrome P450 genes as a biomarker for TGP and many other pesticides/fungicides requires further investigation.

Mechanisms of CYP1A1 Inhibition by Clotrimazole and Propiconazole

EBI imidazole and triazole compounds have been shown to either act partially or exclusively as noncompetitive inhibitors of CYP1A1 catalytic activity. Initial work with hepatic microsomes isolated from gizzard shad (*Dorosoma cepedianum*) and difference spectroscopy demonstrated that clotrimazole can directly bind to the heme of cytochromes P450 [23]. In these experiments, addition of 0.1 μM clotrimazole to the reference cuvette resulted in the formation of a type II binding spectrum (Figure 2, curve 2). The type II spectrum was characterized by a gain in absorbance at 427 nm and a loss of absorbance between 400 to 412 nm. Addition of clotrimazole to a final concentration of 1 μM resulted in an increase in the magnitude of the type II binding spectrum (Figure 2, curve 3). The isobetic point for the two binding curves occurred at 417 nm which is approximately the same isobetic point found with mammalian hepatic microsomes. Bubbling the reference cuvette for 30 sec with CO and reducing with sodium dithionite displaced clotrimazole allowing for recovery of the cytochrome P450 difference spectra (Figure 2, curve 4). Recovery of the cytochrome P450 difference spectra demonstrated clotrimazole reversibly binds to the heme of cytochromes P450. In a similar manner, propiconazole was found to directly bind to the heme of cytochromes P450 and can be displaced by CO [5].

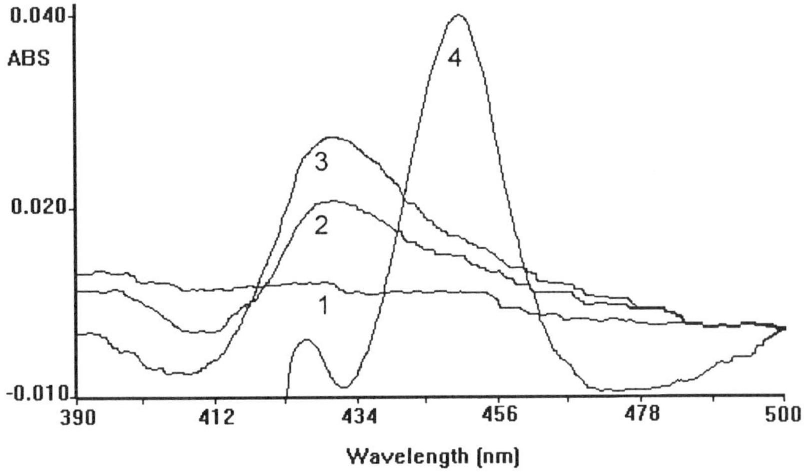

Figure 2. Type II binding spectra for gizzard shad hepatic microsomes after the addition of clotrimazole to the reference cuvette (curves 2 and 3) and recovery of the CO difference spectrum (curve 4). Scan line 1 represents equal absorbance between cuvettes. (Reproduced from reference 23, Copyright 1997 Environmental Toxicology and Chemistry.)

Although, by definition, coordination with the heme of cytochromes P450 is a noncompetitive mechanism, CYP1A1 kinetic assays were performed with rainbow trout hepatic microsomes to determine if inhibition of CYP1A1 catalytic activity, indexed as ethoxyresorufin-*O*-deethylase (EROD) activity, was either purely noncompetitive or noncompetitive with a competitive component. Clotrimazole caused a decrease in apparent V_{max} values and a increase in apparent K_m values [23,24], which is indicative of a noncompetitive mixed-type inhibitor (Figure 3a) [25]. Noncompetitive mixed-type inhibitors have different affinities for the free enzyme and the enzyme-substrate complex ($K_I \neq K_{IS}$) [25]. Secondary plots indicated that K_I and K_{IS} values were significantly different; the value for K_I was calculated to be 3 ± 3 nM and the value for K_{IS} was calculated to be 1900 ± 5 nM [24]. K_{IS} being greater than K_I indicates that clotrimazole has a greater affinity for the free enzyme over the enzyme substrate complex and also indicates that the noncompetitive component is the dominant mechanism of inhibition [25]. Increases in propiconazole concentration also caused apparent V_{max} values to decrease but propiconazole did not cause a significant effect on apparent K_m values (Figure 3b) [5]. The lack of significant differences in apparent K_m values, along with a concentration-dependent decrease in apparent V_{max} values, is indicative of purely noncompetitive inhibition of EROD activity. The IC_{50} values for clotrimazole and propiconazole were 0.51 ± 0.01 μM and 4.7 ± 0.1 μM, respectively [23,5]. Clotrimazole's mixed-pattern of inhibition allows it to be a more effective CYP1A1 inhibitor compared to propiconazole [23,24].

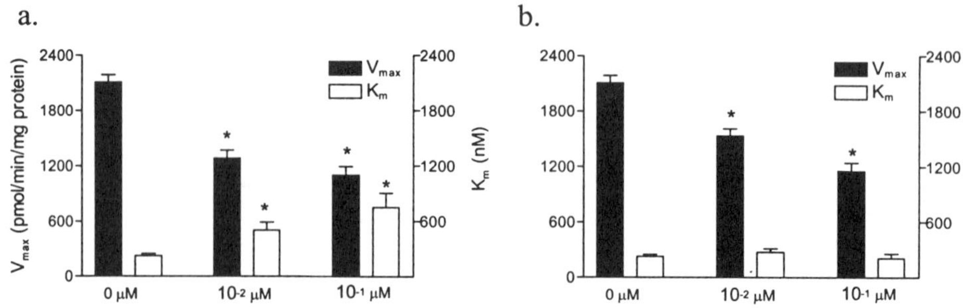

Figure 3. Inhibition of EROD activity + SEM by (a) clotrimazole (b) propiconazole. Values with an asterisk are significantly different from control values (μM) ($p < 0.05$). (Figure 3b is reproduced from reference 5, Copyright 1999 Environmental Toxicology and Chemistry.)

In Vivo Effects of Propiconazole on CYP1A1 Expression

In Vivo, TGP modulated hepatic CYP1A1 expression at the transcriptional and catalytic levels [5]]. Rainbow trout given 24 h waterborne exposures to concentrations ≥ 19 μg TGP/L demonstrated a mixed-pattern response. The mixed-pattern response was characterized by induction of CYP1A1 gene expression but inhibition of EROD activity (Figure 4a). Although EROD activity was elevated following exposure to 90 μg TGP/L, the level of EROD activity was lower than expected based on CYP1A1 mRNA levels. Induction of CYP1A1 mRNA levels suggested that TGP contained a compound(s) that was capable of activating AhR-mediated CYP1A1 expression.

Since 19 μg TGP/L caused induction of CYP1A mRNA levels along with significant inhibition of EROD activity after 24 h of exposure, this concentration was chosen to examine the influence of TGP on hepatic CYP1A1 expression over a 120 h exposure [5]. A 120 h exposure to 19 μg TGP/L produced a biphasic response with CYP1A1 mRNA levels and EROD activity (Figure 4b). The biphasic response at the level of gene expression was characterized by initial induction of CYP1A1 mRNA levels after 24 h of exposure but a return to basal levels between 72 and 120 h of exposure. The biphasic response at the catalytic level was characterized by initial inhibition of EROD activity after 24 h of exposure followed by induction of EROD activity after 72 h of exposure. The decrease of CYP1A1 mRNA levels between 72 and 120 h of exposure may have resulted from first-pass metabolism of an AhR ligand(s) in TGP as it partitioned through the gill membrane. In other words, high deactivation rates of AhR-active compounds by the gill between 72 and 120 h of exposure caused a decrease in the amount of AhR-active compounds that reached the liver through the systemic circulation. Cytochrome P450 monooxygenase systems in gills may have toxicological significance because they are in direct contact with chemicals dissolved in the water, and these chemicals must pass directly through the branchial epithelium before entering the circulation. In a recent investigation, our lab demonstrated that hepatic CYP1A1 mRNA levels also returned to basal levels between 72 and 120 h of continuous BaP-exposure, however, gill CYP1A1 mRNA levels remained maximally induced between 72 and 120 h of exposure [21]. It was hypothesized that significant first-pass metabolism of BaP by the gill, as BaP partitioned into the systemic circulation, modulated hepatic CYP1A1 expression during continuous exposures. Since TGP can induce CYP1A1 expression, measurement of CYP1A1 mRNA levels in a tissue, such as gill, which is in direct contact with the environment could potentially be used to identify exposure to TGP in the field.

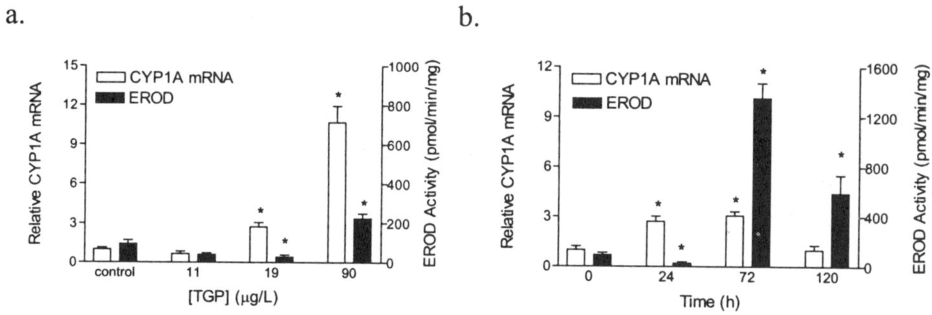

Figure 4. (a) Concentration-response and (b) time-response for rainbow trout hepatic CYP1A1 mRNA + SEM and EROD activity + SEM. Values with an asterisk represent values that are significantly different compared to control values ($p < 0.05$). (Figures are reproduced from reference 5, Copyright 1999 Environmental Toxicology and Chemistry.)

Unidentified Compounds in TGP Cause CYP1A1 Induction

To determine if induction of hepatic CYP1A1 mRNA levels resulted from propiconazole itself being an AhR ligand, or from the presence of an unidentified AhR ligand(s) in TGP, rainbow trout received intraperitoneal (i.p.) injections with either TGP (90.2% purity) or with AGP (99.9% purity). Fish that were treated with 100 mg TGP/kg had an 1100% increase in CYP1A1 mRNA levels and a, lower but, significant 500% increase in EROD activity (Figure 5a) [5]. However, 100 mg AGP/kg did not affect CYP1A1 mRNA levels, but caused a greater than 80% inhibition of EROD activity (Figure 5a) [5]. Comparable results to those reported with AGP, were obtained with rainbow trout that were treated with 50 mg clotrimazole (99.9% purity)/kg [24]. Analytical grade formulations of these compounds only caused inhibition of EROD activity and did not influence CYP1A1 gene expression. The inability of either analytical grade clotrimazole or AGP to activate rainbow trout CYP1A1 gene expression, is consistent with the assumption that TGP contains an unidentified compound(s) which is an AhR ligand.

An additional comparison between AGP and TGP was conducted with a recombinant mammalian HepG2/40-6 cell line [5]. The HepG2/40-6 cell line was created by stably integrating a plasmid that contains the firefly luciferase gene that is under the control of several DREs which respond to AhR agonists with a concentration-, time- and AhR-specific induction of luciferase activity [26]. HepG2/40-6 cells were incubated with either carrier alone, AGP, TGP, or 2,3,7,8-tetrachlorodibenzo-*p*-dioxin (TCDD) as a positive control in a 24-well plate format for 4 h. Incubations with AGP at a maximum concentration of 100 μM caused only a slight increase in DRE-dependent luciferase activity, however, incubations with TGP concentrations ranging from 2 to 100 μM caused a concentration-dependent increase in luciferase activity (Figure 5b). Induction of luciferase activity in the HepG2/40-6 cell line and induction of EROD activity *in vivo* by TGP implies ligand mediated activation of the AhR, but does not directly confirm it. In order to assess the ability of these formulations to transform the AhR into its active DNA-binding form, gel retardation analysis using guinea pig hepatic cytosol was conducted [5]. Although this assay does not directly confirm the ability of an AhR agonist to induce CYP1A1 expression, the positive correlation between induction of luciferase activity in the HepG2/40-6 cell line and the ability of TGP to stimulate AhR transformation and DNA-binding demonstrated that TGP contains an AhR agonist (Figure 5c).

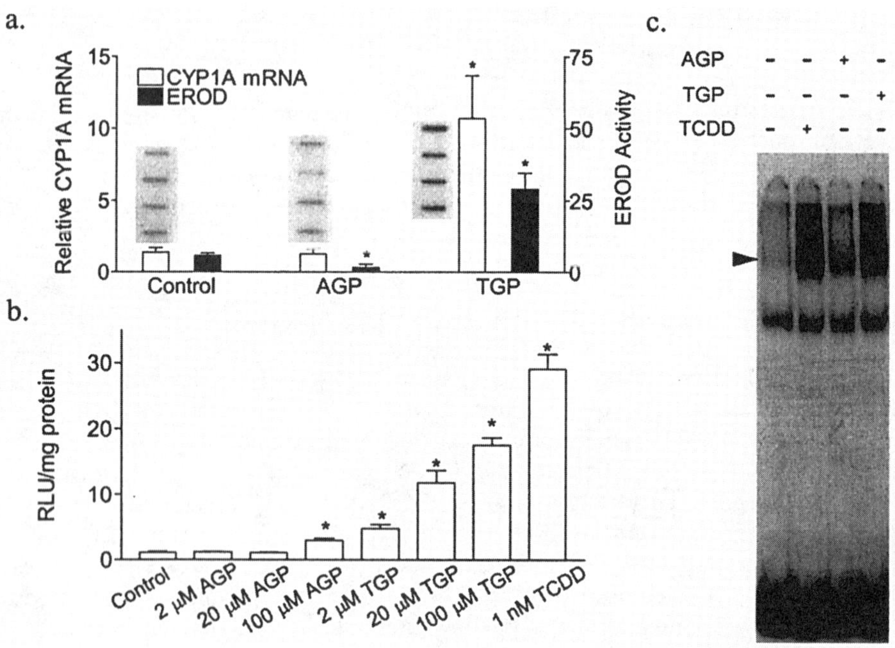

Figure 5. (a) Modulation of hepatic CYP1A1 expression by TGP and AGP; EROD activity is expressed as pmol/min/mg protein and the inset shows slot blots for CYP1A mRNA levels. (b) Influence of AGP, TGP, and TCDD on DRE-driven luciferase activity. Values are expressed as relative light units (RLU) and values with

an asterisk are different from controls ($p < 0.05$) (c) Gel mobility shift induced by 1 nM TCDD, 100 μM AGP, and 100 μM TGP. The arrow indicates the retarded band induced by ligand-AhR-DRE complex. (Figures 5a and 5c are reproduced from reference 5, Copyright 1999 Environmental Toxicology and Chemistry.)

In our initial work with TGP, reverse-phase HPLC with spectrophotometric analysis (λ = 268 nm) was used to detect the presence of AhR-active compounds within TGP that were not found in AGP [5]. It was hypothesized that the compounds that were detected at 268 nm may contain AhR-active compounds. To isolate TGP fractions which contain AhR ligands, 50 μL of 100 mM TGP was injected onto a C18 reverse-phase column (4.6 X 150 mm, 5 micron) and eluted with a linear gradient from 30% acetonitrile:70% water to 70% acetonitrile:30% water over 45 min at 30 ^{o}C with a flow rate of 1.0 mL/min [5]. Fractions were collected at 1 min intervals, vacuum concentrated, resuspended in 5 μL of DMSO, and then added to the HepG2/40-6 cells in a 24 well-plate format and luciferase activity was measured 4 h post-exposure. Many of the TGP fractions contain an AhR active compound(s) (Figure 6). Fractions 6, 16, 19, 36, and 42 showed the highest induction of DRE-driven luciferase activity. Fractions 29-30 contained concentrations greater than 0.1 M propiconazole and propiconazole at a concentrations greater than 200 μM were toxic to the HepG2/40-6 cell line. Therefore, to test for the presence of AhR-active compounds in fractions 29-31 a concentration response experiment was conducted with concentrations ranging from 1 μM to 100 μM TGP.

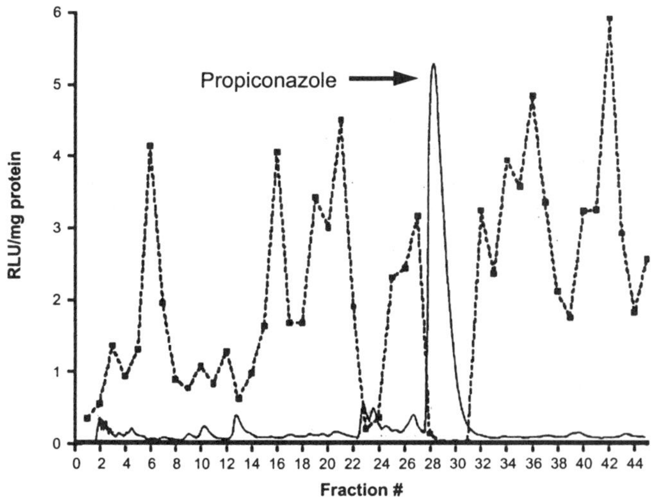

Figure 6. Identification of TGP fractions which cause induction of luciferase activity in the HepG2/40-6 cell line. The dotted line represents relative luciferase activity (RLU) and the solid line is the chromatograph for TGP.

The fractions containing propiconazole (fractions 29-31) do not appear to contain a significant amount of an AhR ligand(s) (Figure 7a). The concentration-response with the combined propiconazole fractions 29-31, were in agreement with results obtained with 100 μM AGP (Figure 5a). To further characterize the inducing potential of AhR active compounds within TGP, several fractions which induced luciferase activity in Figure 6 were combined and retested in the HepG2/40-6 cells in a 24 well-plate format. The selected fractions produced a concentration-dependent increase in luciferase activity (Figure 7b).

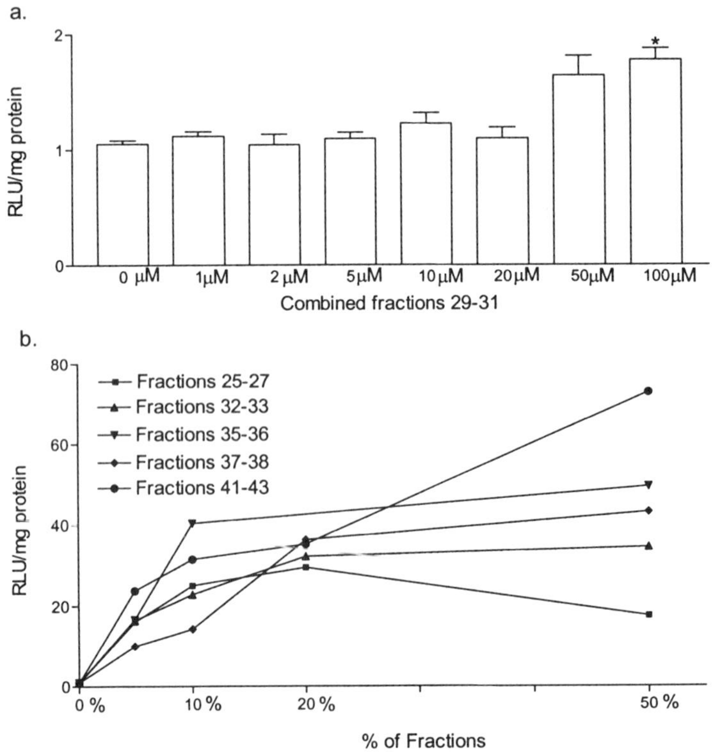

Figure 7. (a) Influence of combined fractions 29-31 on relative luciferase activity (RLU) in the HepG2/40-6 cell line. Values with an asterisk are different from 0% controls ($p < 0.05$). (b) HepG2/40-6 concentration-response for combined TGP fractions that contained AhR-active compounds in Figure 6.

Interactions between EBI Fungicides and Cytochrome P450 Substrates

Preliminary studies were conducted with clotrimazole to determine if pre-treatment with this compound could influence BaP-bioconcentration. BaP is primarily metabolized by CYP1A1, but many other CYP isoforms are capable of metabolizing BaP to more water soluble products [19]. To assess the influence of clotrimazole on BaP-bioconcentration, gizzard shad were injected with 50 mg clotrimazole (99.9% purity)/kg and 48 h later given waterborne exposures to approximately 1 μg BaP/L [23]. Gizzard shad that were pre-treated with clotrimazole had significant inhibition of EROD activity, compared to fish exposed to BaP alone, and bioconcentrated 8- and 11-times more parent-BaP after 1 and 3 d of BaP-exposure, respectively (Table 1). Parent-BaP was separated from BaP-metabolites by reverse phase HPLC with an ion-pairing reagent [23]. It was not determined whether clotrimazole altered the pattern of BaP metabolites or relative proportions of BaP metabolites. However, this work demonstrated that this class of compounds can alter the metabolic fate of a cytochrome P450 substrate during a concurrent exposure.

Table 1. Effect of Clotrimazole on BaP Bioconcentration

BaP exposure	Treatment: BaP [Whole fish parent-BaP] (μg/ kg)	Treatment: BaP + CLOT [Whole fish parent-BaP] (μg/kg)
1 d	3.1 ± 0.2^{a}	26.1 ± 6.7^{b}
3 d	2.8 ± 0.3^{a}	29.9 ± 10.1^{b}

Treatment groups (BaP or BaP + clotrimazole (CLOT)) with the same letters were not significantly different ($p < 0.05$). (Reproduced from reference 23, Copyright 1996 Environmental Toxicology and Chemistry.)

Previously, it was demonstrated with the EBI fungicides prochloraz and penconazole can enhance malathion toxicity to red-legged partridge, and enhancement resulted from increased cholinesterase inhibition [13, 14]. Increased cholinesterase inhibition was believed to have resulted from induction of a novel cytochrome P450 isoform that is capable of desulphuration of malathion to its toxic product malaoxon. Under the same experimental conditions, TGP did not enhance malathion toxicity [14]. It was speculated by this group that TGP did not enhance malathion toxicity because cytochrome P450 inhibition by TGP dominated over cytochrome P450 induction by TGP at the dose of 200 mg/kg. This may be a plausible interpretation of the results, since a mixed-pattern response occurred following injections with rainbow trout that received 100 mg TGP/kg [5]. Although it is presently not known which cytochrome P450(s) are responsible for the desulfuration reaction of OPs, the CYP1A, CYP2B, and CYP3A isoforms of mammalian species are known to catalyze desulphuration reactions. Therefore, there may be drug metabolizing cytochrome P450 isoforms may be involved in OP activation [27, 28]. A proposed mechanism for activation of the OP parathion by a cytochrome P450 isoform which is shown in Figure 8.

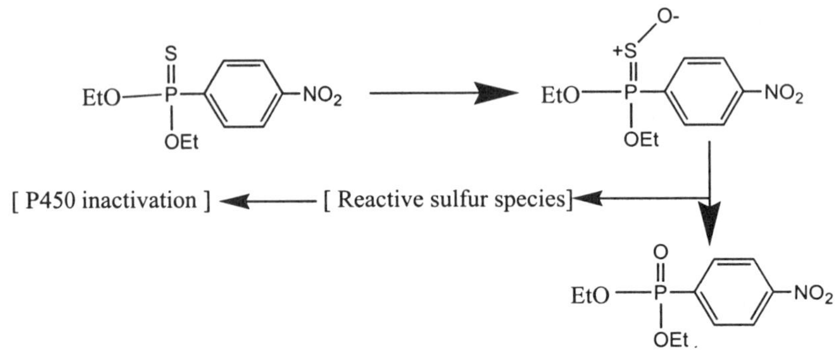

Figure 8. Proposed mechanism of cytochrome P450 activation of parathion.

We recently re-tested the hypothesis that induction of cytochrome P450 activity by an EBI imidazole/triazole would enhance the toxicity of an OP insecticide by increasing activation rates relative to inactivation rates [13, 14]. Our results demonstrated that TGP can enhance the acute toxicity of parathion to fathead minnows (*Pimphales promelas*). This enhancement was uncovered by comparing survival distributions of groups of fish that were pre-exposed to TGP followed by exposure to parathion with survival distributions of groups of fish that were exposed to parathion alone (Figure 9). Survival distributions were compared with the Kaplan-Meier (product limit) test [29]. The concentrations of TGP used in the survival analysis experiments were higher than reported field concentrations for propiconazole. The target exposure concentration of 125 μg TGP/L was chosen as a worst case scenario for propiconazole concentrations during times of high precipitation. Survival times were reduced for fish pre-exposed to 138 ± 9 μg TGP/L followed by exposure to 1.9 ± 0.1 mg parathion/L compared to fish that were exposed to 1.9 ± 0.1 mg parathion/L alone (Figure 7). Although exposure to 1.9 mg parathion/L alone was acutely toxic to fathead minnows, body burdens near this concentration have been measured in freshwater species of fish in El Salvador.

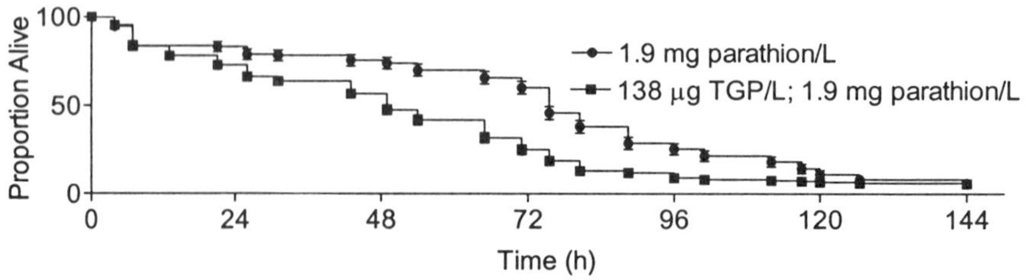

Figure 9. Survival distributions for fathead minnows pre-exposed to TGP and then exposed to 1.9 ± 0.1 mg parathion/L or to 1.9 ± 0.1 mg parathion/L alone. (Reproduced from reference 30, Copyright 1999 Pesticide Biochemistry and Physiology.)

Apparently these species of fish are less sensitive to parathion toxicity compared to fathead minnows [12]. The early life stage maximum allowable toxicant concentration (MATC) for fathead minnows exposed to TGP is above concentrations used in the acute toxicity study. The MATC value for fathead minnows spans the range of 430 to 970 μg/L [30].

Time-to-death experiments were consistent with *in vitro* data that demonstrated that hepatic microsomes isolated from TGP-treated fathead minnows (79 μg TGP/L for 48 h) had a higher rate of paraoxon formation relative to untreated microsomes [30]. The rate of paraoxon formation for control microsomes was 7.0 ± 2.6 pmol/min/mg protein and the rate of paraoxon formation for TGP-treated microsomes was 16.5 ± 0.2 pmol/min/mg protein which was significantly higher relative to the control microsomes [31]. Formation of the deactivation product *p*-nitrophenol was below the detectable limit and thus it was not possible to index activation:deactivation ratios. In a study examining the influence of TGP on parathion metabolism with mammalian microsomes, it was reported that TGP did not affect parathion induced inhibition of serum cholinesterase activity [32]. In spite of exposure to TGP enhancing parathion activation rates *in vitro*, there was a concomitant increase in parathion deactivation rates *in vitro*, thereby, producing an overall decrease in the activation:deactivation ratio [32].

Although it has been shown that TGP can modulate CYP1A1 expression in rainbow trout and enhance acute parathion toxicity to fathead minnows, the participation of CYP1A1 and a CYP3A-type isoform in Nile tilapia toward malathion metabolism was indirectly ruled out [33]. CYP1A1 and CYP3A induction in Nile tilapia did not alter malathion toxicity or effect AChE activity. In light of the results of the former study, induction of CYP1A1 expression in fathead minnows may not have been the mechanism by which parathion toxicity was enhanced by TGP. It is possible that increased paraoxon formation was not responsible for the enhanced acute toxicity of parathion, but rather, pre-exposure to TGP enhanced the acute toxicity of parathion to fathead minnows by increasing parathion clearance rates.

To date, only a few cytochrome P450 isoforms in species of fish have been shown to be inducible with model substrates. These cytochrome P450s include CYP1A1 in a variety of species [34], CYP2K1 and CYP2M1 in bluegill and channel catfish [35] and a CYP3A-type isoform in Nile tilapia [33]. Although only a few inducible cytochrome P450 isozymes have been characterized with teleosts, other inducible forms may exist. Therefore, it is possible that teleosts, like rats, have an inducible novel cytochrome P450 isoform(s) that is capable of parathion activation following exposure to TGP.

Summary

The first portion of this chapter described the mechanisms by which clotrimazole and propiconazole modulate *in vitro* and *in vivo* CYP1A1 expression. Analytical grade formulations of clotrimazole and propiconazole were noncompetitive inhibitors of CYP1A1 catalytic activity and did not influence CYP1A1 gene expression. However, a technical grade formulation of propiconazole was shown to produce a mixed-pattern response with respect to CYP1A1 expression. The mixed-pattern response was characterized by induction of CYP1A1 mRNA levels but inhibition of CYP1A1 catalytic activity. Induction of CYP1A1 expression by TGP was caused by unidentified AhR-active compounds within the tested TGP formulation. Although, TGP can inhibit CYP1A1 catalytic activity, CYP1A1 mRNA levels in gill tissue could potentially be used to identify exposure to TGP concentrations > 20 μg/L.

This chapter also reviewed a study which examined the potential for interaction between TGP and the OP insecticide parathion. TGP was shown to interact with parathion. The interaction between TGP and parathion may have resulted from induction of a novel cytochrome P450 isoform that was responsible for desulphuration of parathion. Although this is a possible mechanism, additional analysis is required. Future studies will address whether enhancement of acute parathion toxicity by TGP resulted either from propiconazole, from an unidentified compound(s) in the TGP formulation, from TGP influencing parathion clearance rates, or from a combination of any or all of these mechanisms. Although future studies are required to determine the mechanism by which TGP enhances the acute toxicity of parathion, the results reviewed in this chapter point out that toxic interactions can occur between these two compounds.

Acknowledgments

The authors would like to extend their thanks to the former Ciba-Geigy Corporation for supplying the technical grade formulation of propiconazole used in these studies. This research was funded by an EPA Doctoral Fellowship awarded to SLL.

References

1. Henry, M. J.; Sisler, M. D. *Pestic. Biochem. Physiol.* 1984, 2, 262-275.
2. Leslie, C.; Reidy, F.; Stacey, N. H. *Biochem. Pharmacol.* 1988, 37, 4133-4141.
3. Ronis, M. J.; Ingelman-Sundberg, J. M.; Badger, T. M. *Biochem. Pharmacol.* 1994, 44, 1953-1965.
4. Ronis, M. J.; Celander, M.; Badger, T. *Comp. Biochem. Physiol. C* 1998, 121, 221-229.
5. Levine S. L.; Oris, J. T.; Denison, M. S. *Environ. Toxicol. Chem.* 1999, 18, 275-286.
6. Hernández C.; Vargas R. In: Critical issues in Watershed Management in the 21st Century; Latin America and the Caribbean. Eds. S. E. Witter and S. Whiteford, East Lansing, USA, 1994.
7. New Scientist. Pesticides team up to damage birds, Oct. 21, 1989.
8. Manchester Guardian. Birds facing lethal cocktail peril from new pesticides, Thursday Dec. 14, 1989.
9. W. Henriques, W.; Jeffers, D.; Lacher T. E. Jr., Kendall, R. J. R. J. *Environ. Toxicol. Chem.* 1997, 16, 91-99.

10. World Resources Institute. World Resources: Oxford University Press, New York, NY, USA, 1994-1995.
11. Mortensen, S. R.; Johnson, K. A.; Weisskopf, C. P.; Hooper, M. J.; Lacher, T. E.; Kendall, R. J. *Bull. Environ. Contam. Toxicol.* 1998, 60, 562-568.
12. Castillo, L. E.; De La Cruz, E.; Ruepert, C. *Environ. Toxicol. Chem.* 1997, 16, 41-51.
13. Johnston, G.; Collet, G.;Walker, C. H.; Dawson, A.; Boyd, I.; Osborn D. *Pestic. Biochem. Physiol.* 1989, 35, 103-114.
14. Johnston, G.; Walker, C. H.; Dawson, A. *Environ. Toxicol. Chem.* 1994, 13, 615-620.
15. Kleinow K. M.; Melancon M. J.; Lech J. J. *Environ. Health Perspect.* 1987, 71, 105-119.
16. Singh S. S.; Hord N. G.; Perdew G. H.; *Arch. Biochem. Biophys.* 1996, 329, 47-55.
17. Denison M. S.; Fisher J. M.; Whitlock J. P. *Proc. Natl. Acad. Sci. USA,* 1988, 85, 2528-2532.
18. Pollenz, R. S.; Necela, B. 1998. *Aquat. Toxicol.* 1998, 41, 31-49.
19. Lech, J. J.; Vodicinik, M. J. In *Fundamentals of Aquatic Toxicology*; Editors M. Rand, B. M.; Petrocelli, S. R.; Taylor & Francis Publishing, Bristol, PA, 1985, pp 526-557.
20. Song, B. H.; Gelboin, H. V.; Park, S. S.; Tsokos, G. C.; Friedman, F. K. *Science* 1985, 228, 490-492.
21. Levine, S. L.; Oris, J. T. *Aquat. Toxicol.* 1999, 46, 279-287.
22. Jimenez, B. D.; Stegeman, J. J. *Am. Fish. Soc. Sym.* 1990, 8, 67-79.
23. Levine S.L.; Czosnyka, H; Oris, J. T. *Environ. Toxicol. Chem.* 1996, 16, 306-311.
24. Levine S.L.; Oris, J. T. *Comp. Biochem. Physiol.* C 1999, 122, 205-210.
25. Roberts, D. V. *Enzyme Kinetics*, Cambridge University Press, London, 1977; pp 48-79.
26. Tsai, J. C. Ph.D. thesis, Purdue University, West Lafayette, IN, 1977.
27. Reidy, G. F.; Rose, H. A.; Stacey, N. H. *Toxicol. Lett.* 1987 38, 193-200.
28. Wilkinson, C. F.; Murray, M.; Marcus, C. B. *Rev. Biochem. Toxicol.* 1984, 6, 27-43.
29. Kaplan, E. L.; Meier, P. *J. Am. Stat. Assoc.* 1958, 53, 457-481.
30. Ciba-Geigy Material safety data sheet for Technical Grade Propiconazole, 1996.
31. Levine S.L.; Oris, J. T. *Pest. Biochem. Physiol.* 1999, 65, 102-109.
32. Ronis, M. J.; Badger, T. M. *Toxicol. Appl. Pharmacol.* 1995, 130, 221-228.
33. Pathiratne, A.; George, S. G. *Aquat. Toxicol.* 1998, 43, 261-271.
34. Stegeman, J. J.; Brouwer, m.; Di Giulio, R. T.; Forlin, L.; Fowler, B. A.; Sanders, B. M.; Van Veld, P. In: *Biomarkers: biochemical, physiological, and histopathological markers of anthropogenic stress.* Editors, Hugget, R. J.; Kimerle, R. A.; Mehrle, P. M.; Bergman, H. L. Lewis Publishers, Chelsea, MI, 1992, pp. 235-251.
35. Haasch, M. L. *Mar. Environ. Res.* 1996, 42, 287-291.

Assessment of Risk to Non-Target Wildlife

Chapter 11

Design of a Laboratory Secondary Hazard Study

David A. Goldade, Peter J. Savarie, Jerome C. Hurley, Stanley A. Gaddis, and John J. Johnston

APHIS/WS/National Wildlife Research Center, U.S. Department of Agriculture, 4101 La Porte Avenue, Fort Collins, CO 80521–2154

Factors influencing the design and execution of a secondary hazard study are presented using the anticoagulant rodenticide, difethialone, as a model. Frequently, regulatory agencies require a determination of the potential hazard posed to non-target species by the application of pesticides. This hazard can occur through: 1) primary hazard, direct consumption of the bait or formulation containing the pesticide, or 2) secondary hazard, indirect consumption of the pesticide via ingestion of biological matrices such as animals or plants containing pesticide residues. In making this determination, the selection of appropriate test species as well as routes and level of exposure are critical. For the example study of the estimation of secondary hazards associated with difethialone, albino rats served as the representative species for the primary target animal. Ferrets and magpies served as representative mammalian and avian secondary scavenger species, respectively.

Introduction

Numerous factors should be considered in the design and implementation of a secondary hazard study including biological, chemical, and agricultural influences. Careful consideration of these factors should produce sufficient data to permit informed regulatory decisions incorporating potential hazards posed by the proposed application of the pesticide. This paper addresses the consideration of these variables by presenting a secondary hazard study performed in our laboratory for the anticoagulant rodenticide, difethialone (3-[(1RS,3RS;1RS,3SR)-3-(4'-bromobiphenyl-4-yl)-1,2,3,4-tetrahydro-1-naphthyl]-4-hydroxy-1-benzothi-in-2-one).

Definitions

In its broadest terms, a secondary hazard exists when there is a possibility of poisoning to secondary (or non-target) species from the consumption of the primary (or target) species. This is differentiated from a primary hazard which occurs from consumption or direct contact with the formulated pesticide. This is a generalized description which can cover many possible exposure scenarios from the aerial application of a liquid herbicide formulation for weed control to the use of a toxicant formulated as a pelleted bait to control populations of a pest animal. For difethialone, the potential primary hazard pertains to the consumption of formulated baits and the potential secondary hazard from consumption of the carcasses of the primary species containing pesticide residues. This is illustrated in Figure 1 where the cross-hatched arrows indicate routes of primary hazard and solid arrows indicate routes of secondary hazard.

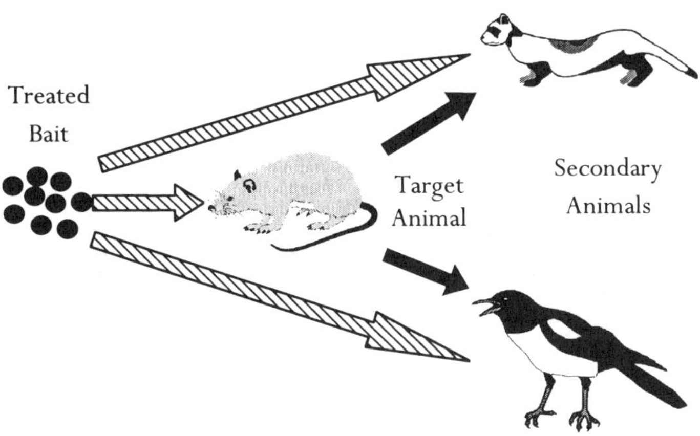

Figure 1. Primary (cross-hatched arrows) and secondary (solid arrows) hazards from a treated bait.

Overview

To approximate likely field conditions, the first step of our study was to feed the pesticide to the target species at realistic levels. Carcasses of the target species were then analyzed to determine the range of concentrations to which secondary species might be exposed. The toxicity of the pesticide to the secondary species was assessed by feeding a range of pesticide concentrations in a formulated diet to the secondary species. The formulated diet was meat-based to approximate the primary species carcasses containing pesticide residues. Secondary hazards were then estimated by comparing the toxicity of the pesticide to the secondary species with the pesticide residues in the target (primary) species.

Toxicological Considerations

The toxicological properties of the pesticide should be determined and subsequently considered in the design of a secondary hazard study. Of prime importance is the determination of the mode of action for the toxic agent. This can frequently be discerned by examining pathological and toxicological data about the pesticide. Determination of the organs affected can provide clues as to how the pesticide exerts its toxic effect and can provide direction in how best to determine residue levels. The elimination rate or half-life indicates how long the chemical is present in the organism, and therefore how long a potential hazard persists following exposure. The individual elimination rates for various tissue types need to be considered as well. If the pesticide is stored in adipose tissue, it is likely that whole body residues will need to be quantified. If the pesticide is preferentially stored in certain organs, then the residues in those organs should be determined.

An additional significant consideration is metabolism of the pesticide. If the chemical undergoes either detoxification or bioactivation, the fate and toxicity of the metabolites could prove to be very important. Thus, significant metabolites should be identified and the toxicity of these metabolites to the secondary species determined. If a detoxification process is occurring, the toxicity of the metabolite(s) to secondary species is probably lower than that of the parent compound. This would likely lessen the secondary hazards associated with residues of the pesticide. If bioactivation is occurring, it is likely that the metabolite(s) could produce toxic effects in the secondary species similar to those in the primary species (1).

Difethialone is an anti-vitamin K anticoagulant rodenticide of the chemical class hydroxy-4-benzothiopyranones. It exerts its toxic hemorrhagic effects by inhibiting the synthesis of Vitamin-K dependent blood coagulation factors. It does not cause necrosis in any specific organ, but effects the blood coagulation system throughout the body. As with most other anticoagulant rodenticides it bioaccumulates in the liver as evidenced by its long liver half-life of 108 days (2). Studies with ^{14}C-labeled difethialone indicate that it does not undergo significant metabolism (2). Therefore, the toxic effects associated with exposure to this compound are due to the parent compound. Death typically occurs from internal hemorrhage following consumption of a lethal dose (1). A lethal dose can be ingested in a single feeding (acute) or may occur following repeated feedings (sub-chronic). The toxic material is applied at a low concentration and there are no immediate toxic effects observed; therefore, there is little or no bait shyness associated with the use of this compound. This serves to increase the likelihood that a lethal dose will be ingested (1).

Mode of Delivery

The method of delivery for the pesticide being evaluated in the secondary hazard study is dependent on one important factor: how exposures are likely to occur in the field. If the chemical is sprayed onto plants, the potential exists for ingestion by herbivores. If the formulation is a pelleted or grain bait, application will likely be via broadcast. To reduce potential non-target primary hazards, application procedures may be modified such as through the use of bait stations. The degree of water solubility can be important for movement of the pesticide into ground water. If it is applied as an aerosol, potential dermal exposure to non-target species should be considered. However, for the purposes of a secondary hazard study, the most important route of exposure is oral via consumption of carcasses of the primary species containing pesticide residues.

Difethialone is prepared and marketed as a 0.0025% pelleted bait and is designed for the control of rodent populations in and around structures (EPA Registration Numbers 7173-205, 7173-206, 7173-211, and 7173-218). The pelleted bait is consumed by pest rodent species whose carcasses represent the source of the secondary hazard to scavenger species. For difethialone, the principle source of primary hazard is from consumption of the treated bait. However, since it is used for control of commensal rodents, it is not as likely to be consumed by non-target wildlife species. If the bait were applied in a broadcast fashion, the likelihood of ingestion by non-target species would probably increase. Difethialone has moderate water solubility and is applied in relatively low concentrations. Therefore, potential exposure via surface water contamination should be extremely low. The largest potential secondary hazard is from consumption of the carcasses of the primary species. The primary reasons for expecting residues to be present in the primary species are the relatively long half-life of difethialone in liver and blood (108 and 2.3 days respectively (2)) and the lack of bait-shyness for the target animals. As there is no bait-shyness associated with the formulated bait, there is an increase in the likelihood of significant residues being present in the carcass of the target species as the animal will continue to feed on the bait. Since feeding studies with the pelleted bait reveal that the time to death is typically less than seven days for the primary species, the potential exists for significant residues to be present in the carcass of the primary species at the time of its death. Therefore, the mode of delivery selected for the secondary study of difethialone was via the oral route.

For our study, the primary species were fed difethialone-treated bait. Possibilities considered for dosing the secondary species were to feed 1) carcasses of the primary species fed difethialone bait, 2) homogenized primary species fortified with difethialone, or 3) a surrogate ground substrate, such as dog food, fortified with difethialone. As difethialone residues in rat carcasses were extremely variable (c.v. = 26 %), feeding carcasses of the primary species which had fed on difethialone baits would result in extremely variable and inconsistent doses to the secondary species. This would likely produce variable toxicity data that would be of little value for subsequent risk assessment. Additionally, in order to generate toxicity data (such as LD_{50} or LC_{50} values) dosing must be performed at a sufficient number of known doses or concentration levels in order to produce a range of percent mortality results. This means that the carcasses of the primary species would have to be fortified with additional difethialone in order to produce the mortality required to generate a dose-response curve. Homogenization and subsequent fortification of the primary species to produce a homogeneous blend of fortified rat tissue was impractical as the study required 113 kg of fortified diet. The production of such a large quantity of treated diet would have involved the needless pain and suffering and subsequent deaths of nearly 600 additional rats. We chose the simpler, less expensive, more humane option of fortifying dog food with difethialone. This approach was relatively straight forward and produced a homogeneous diet for accurate delivery of the toxicant.

Selection of Test Species

The most important factor in selecting test species is to ensure they are representative of species likely to be found in a field setting. Often it is impractical to have the exact species which will be exposed, either because the species is endangered or threatened or because the species is not available from a commercial source. Likewise, it is impractical to test each and every species likely to be exposed (3). Therefore, representative species are chosen and testing conducted on them. At a minimum, the species selected should be of the same family or genus as the species found in the field. This makes extrapolation of results to other species found in the field less controversial and more reliable. The number of these animals needed for testing must be sufficient to provide statistically significant data and to provide a reasonably tight confidence interval for toxicity values (i.e. LD_{50}) of the pesticide to the secondary species. However, the degree of significance will vary from chemical to chemical and study to study. For compounds and proposed uses that generate intermediate risk assessments (see section on Risk Quotients below), a fairly precise LD_{50} or LC_{50} is desired. Another factor which must be addressed is gender. In most studies both genders should be represented, particularly if both genders are likely to be exposed in the field. This approach will also help to elucidate any gender-specific metabolic processes that may be occurring.

For the primary species, albino laboratory rats (*Rattus norvegicus*) were chosen because they were readily available, easy to handle, and are highly representative of wild rodent species likely to be found in the field. Secondary exposures are likely to occur to both mammalian and avian scavenger and predator species which feed on the carcasses of the primary species. For the laboratory study, European ferrets (*Mustela putorius furo*) and black-billed magpies (*Pica pica*) were chosen as representative secondary species. European ferrets were chosen because they were easy to obtain and highly representative of mammalian scavengers. Black-billed magpies were chosen because they were available in the region, relatively easy to obtain, and are representative of avian scavengers. They were not, however, representative of avian raptor species. An actual raptor species was not used because the use of raptor species in lethal feeding studies is less socially acceptable than is the use of scavenger species.

Level of Exposure

The selection of a level of exposure, primary or secondary, is important to provide a realistic result to the experiment. If the level of exposure is too high or too low the resulting data will be of little utility. Above all, the exposure levels should encompass those expected to be found in the field. Primary species should be exposed to levels at or slightly greater than the maximum estimated field exposure. For secondary species, toxicity values (minimally LD_{50} or LC_{50}) must be determined. Secondary animals need to be exposed to a range of concentration levels designed to result in both high and low percent mortality.

Literature values for residue levels of difethialone in the carcasses of rodents consuming difethialone baits were not available. Therefore, difethialone baits were fed *ad libitum* to rats for 3-7 days or until death occurred. Rats were frozen as soon as possible after death. The carcasses of these rodents were analyzed for residue levels of difethialone. Further, there were no literature LD_{50} values for either magpies or ferrets. Feeding studies with difethialone fortified dog food (a surrogate for rat tissues containing difethialone residues) were conducted to generate toxicity curves for the two secondary hazard species. These values were used to estimate LD_{50} values.

Analytical Chemistry Requirements

The determination of toxicant levels in the matrices chosen for the study is a critical step in the assessment of secondary hazard. To this end, sensitive and selective methods must be developed and validated for the toxicant and its metabolites as required. These methods must cover all possible matrices of interest including the treated baits, the carcasses of the primary species, and the target organs and/or carcasses of the secondary species.

Methods were developed to permit the quantification of difethialone residues in whole-body rodent carcasses, ferret and magpie liver, and dog food formulations. Briefly, the methods employed a dual retention solid phase extraction procedure with high-performance liquid chromatography separation and ultra violet detection (4,5). For each matrix a method limit of detection (MLOD) was determined. Each matrix was fortified at a range of concentrations and the percent recovery determined for each. The MLOD, validated range, and mean and standard deviation for the recoveries are given in Table I. These values demonstrate that the methods developed were sufficiently sensitive, selective, and precise to determine residue levels in all the matrices.

Assessing the Risks

The goal of the secondary hazard study is the assessment of risk associated with the secondary hazard. One commonly used approach is the EPA's risk quotient (RQ)

Table I. Method Validation Results

Matrix	*MLOD (μg/g)*	*Validated Range (μg/g)*	*Mean Recovery (%)*	*Std. Dev. (%)*
Whole-body Rodent	0.054	0.2-20	89	6.7
Dog Food	0.085	0.1-2000	89	11
Ferret Liver	0.091	0.2-200	107	17
Magpie Liver	0.16	0.2-200	107	15

method (6). Generally, the RQ is defined as the ratio of the exposure dose or concentration to the dose or concentration expected to produce lethality in 50% of the population.

$$RQ = Exposure/Effects$$

$$RQ = Dose/LD_{50} \text{ or } RQ = Concentration/LC_{50}$$

Use of this method for risk assessment provides a way to weigh known effects versus expected exposure and provide a numerical basis for decision making. In the RQ method a value greater than 1 indicates that there are significant risks associated with the proposed usc of this chemical. A RQ less than 1 indicates that the risks associated with the proposed use of this chemical may be acceptable under approved usage guidelines. This approach provides a conservative estimate of hazards as it assumes that 100% of the exposed animal's diet will consist of the pesticide formulation (primary hazard) or the tissue, organ, or carcass (secondary hazard) containing the highest concentration of residues. However, the RQ method is widely used and provides a ready framework for comparisons to other compounds. EPA further breaks down RQs of less than 1 into the following categories (7):

A RQ less than 0.1 represents a negligible risk. Values between 0.1 and 0.5 represent a moderate risk level. Finally, values greater than 0.5 represent a significant risk (7).

Difethialone Study Results

For the determination of residue levels in the carcasses of the primary species, twenty albino laboratory rats were feed a 0.0025% pelleted bait for three days (n=10) or until death occurred (n=10). The head, tail, feet, and pelt were removed from each

Table II. Risk Quotients and Associated Concerns

RQ Value	*Associated Risk*
<0.1	Use presents acceptable risk for use under approved guidelines
>0.1	Use restrictions may be imposed to protect endangered species
>0.2	Use may be restricted to certified applicators and/or mitigation techniques may be imposed
>0.5	Mitigation techniques will be imposed to protect all species of the same taxonomic order

animal and the carcass was homogenized and analyzed for difethialone residues. There was no significant difference ($p \leq 0.05$) in the mean residue levels for the two groups. The average residue level was found to be 2.0 mg/kg. (Range = 0.77-2.67 mg/kg, SD = 0.51)

The textbook definition of an LD_{50} typically involves a single, lethal dose being delivered to the test animal (1). However, as difethialone and other anticoagulant rodenticides are known to bioaccumulate, a gradual build-up to a lethal dose can occur. To better model this exposure scenario, a chronic method of determining the LD_{50} was used. For the estimation of chronic LD_{50} values ferrets and magpies were fed dog food fortified with difethialone at various concentrations. The amount of formulated dog food consumed was measured for each test animal each day and multiplied by the concentration of difethialone in the diet to give a value for the daily mass of difethialone ingested by each animal. Cumulative daily intakes were summed to yield a total dose ingested by each animal. The dose of toxicant was then divided by the individual animal weights to give a dose in mg/kg of body weight which were, in turn, used to calculate LD_{50} values. The livers of these animals were assayed for difethialone residues to confirm exposure to difethialone. The mortality of these animals were recorded and, with the mg/kg dose values, subjected to a probit analysis (8) to permit estimation of an LD_{50} value.

For magpies, the LD_{50} was estimated at approximately 4.7 mg/kg. For ferrets, the LD_{50} was estimated at approximately 760 mg/kg. The ferret LD_{50} estimate is less precise as there were only 2 ferrets per exposure level as compared to the magpie study, where 10 magpies were exposed to each difethialone concentration. For the ferret study, the variability is indicated by the fact that even though 100% mortality was observed at a difethialone concentration of 200 ppm, consistent 100% mortality was not observed for the ferrets fed higher toxicant concentrations. For both species, body weight, food consumption, difethialone concentration and mortality for each difethialone exposed animal was used to estimate the LD_{50}.

LC_{50} values could have been used for the estimation of RQ values, however the

LD_{50} value provides a more accurate assessment of toxicant exposure by using the daily food intake of each species. As food consumption rates may be quite different during a laboratory feeding study than under natural field conditions, we feel that the use of dose rather than concentration provides a more accurate assessment of toxicity and a more valid estimate of risk.

Using the daily food consumption formulas of Nagy (9) and the mean body weights for each species, the RQs for the primary hazards of difethialone were calculated according to the following formula:

$$RQ = \frac{(\text{Food Intake} \times \text{Diet Concentration})/\text{Body Weight}}{LD_{50}}$$

To demonstrate, the calculation of the RQ value for albino rodents feeding on the treated bait is calculated:

$$RQ = \frac{(15.5\ \text{g / day} \times 25\ \text{mg / kg})/300\ \text{g}}{0.29\ \text{mg / kg}} = 4.45$$

The RQ values for each species were calculated and the results are presented in Table III.

By comparing the RQ values listed in Table III with the concerns listed in Table II, the primary hazards associated with exposure to difethialone-treated baits can be estimated. As expected, the RQ value for the primary species feeding on the pelleted pesticide bait is significantly greater than 1.0 (Table III). Use of the estimated LD_{50} resulted in a RQ of 0.45 for the primary hazard of difethialone-treated baits to magpies. This suggests a moderate level of risk. However, this is surely an overestimate of the primary hazards to magpies as magpies are scavengers. It is very unlikely that magpies would directly consume difethialone baits intended for rodents. The primary hazards to magpies are likely minimal. This example illustrates the importance of considering the behavior of non-target species in addition to risk factors when estimating non-target hazards. Based on the RQ, the primary risk for

Table III. Risk Quotients for Primary Hazards

Species	*Mean Body Weight (g)*	*Food Intake (g/day)*	*RQ*
Albino Rodents	300	15.5	4.45
European Ferrets	530	40.8	<0.01
Black-billed Magpies	160	13.6	0.45

ferrets is also negligible. Even if a significant risk were generated for this species, this exposure scenario is just as unlikely for ferrets as it is for magpies. One important note is that the emphasis of the study was the estimation of secondary hazards. The fortified diet was designed to approximate secondary exposure via consumption of residue-containing carcasses. In a primary, non-target exposure scenario, the non-targets would consume the fortified bait. As the matrix may have an effect on the toxicity of a pesticide, these primary hazard estimates may be somewhat influenced by the matrix.

The RQ values for secondary hazards can be similarly determined by using the mean residue level of difethialone in the primary species as the diet for the secondary species. The values for the RQ are given in Table IV.

Table IV. Risk Quotients for Secondary Hazards

Species	*Mean Body Weight (g)*	*Food Intake (g/day)*	*RQ*
European Ferrets	530	40.8	<0.001
Black-billed Magpies	160	13.6	0.036

These results indicate that the secondary hazard for ferrets (mammalian scavengers) eating carcasses of the primary species are negligible. The secondary hazards for avian scavengers (magpies) are also negligible.

Conclusions

Design and implementation of a secondary hazard study involves many factors and considerations to ensure that valid estimates of risk are generated. These factors were presented and their consideration demonstrated through the use of a model pesticide, difethialone. If required, resulting risk estimates can be used to develop procedures to minimize undesirable non-target effects while maintaining efficacy.

Acknowledgments

Thanks go to Dennis Kohler, Tom Primus, and Doreen Griffin of the NWRC for sample analysis, method validation and quality control work on this project.

References

1. *Casarett and Doull's Toxicology: The Basic Science of Poisons-5th Edition;* Klaassen, C. D.; Amdur, M. O.; Doull, J.; Eds.; McGraw-Hill: New York, 1996.
2. Lechevin, J.C.; Poché, R.M. *Proc. Vertebr. Pest Conf.* **1988**, *13*, 59-63.
3. Suter, G. In *Ecological Risk Assessment;* Suter, G.W. II, Ed.; Lewis: Chelsea, MI, 1993; Chapter 7.
4. Goldade, D.A.; Primus, T.M.; Johnston, J.J.; Zapien, D.C. *J. Agric. Food Chem.* **1998**, *46,* 504-508.
5. Goldade, D.A. M.S. Thesis, University of Colorado at Denver, 1997.
6. Urban, D.J.; Cook, N.J. *Ecological Risk Assessment: Standard Evaluation Procedure of the Hazard Evaluation Division*; U.S. Environmental Protection Agency. Office of Pesticide Programs. U.S. Government Printing Office: Washington, DC, 1986; EPA-540/9-85-001
7. National Research Council. *Risk Assessment in Federal Government: Managing to Process*; U.S. Government Printing Office: Washington, DC; *NASNRC Publ.* **1983**.
8. SAS Operating System for Windows, Version 6.12 (1996). SAS Institute Inc. Cary, NC.
9. Nagy, K.A. *Ecol. Monogr.* **1987**, *57*, 111-128.

Chapter 12

Recognizing and Reducing Secondary and Tertiary Poisoning Risks Associated with Brodifacoum

C. T. Eason[1] and E. Murphy[2]

[1]CENTOX (Centre for Environmental Toxicology) Landcare Research, P.O. Box 69, Lincoln 8152, New Zealand (E-mail: easonc@landcare.cri.nz)
[2]Department of Conservation, Private Bag 4715, Christchurch, New Zealand

Brodifacoum has found increasing favour for rodent and possum control in New Zealand. Concurrent with the increased field use of brodifacoum, concerns have increased with regard to primary and secondary poisoning and contamination of wildlife. In this paper the results of secondary poisoning studies to determine the concentrations of brodifacoum in muscle and liver tissue from captive pigs after primary and secondary poisoning are reviewed·alongside recent data from field surveys. Following primary poisoning of pigs, with baits containing 20 mg kg^{-1} brodifacoum, liver concentrations ranged from 0.72 to 1.38 mg kg^{-1} . Pigs that had eaten the soft tissue from eight poisoned animals had brodifacoum concentrations of 0.32 to 0.80 mg kg^{-1} present in the liver. In field surveys conducted prior to September 1999, 21 of 35 wild pigs sampled from areas where rodent and/or possum control had been undertaken were contaminated with brodifacoum concentrations in the liver ranging from 0.007 to 1.78 mg kg^{-1}. Brodifacoum has also been detected in significant numbers of deer, cats, stoats, and other wildlife from these areas. In view of the potential impact on pig hunters and dogs consuming wild pig meat and offal, and, to a lesser extent, deer, the wide-scale field use of brodifacoum baits on mainland New Zealand is under review. Its use for broad-scale pest control has already been significantly curtailed. Alternative less persistent toxicants and traps are increasingly being used to minimise non-target exposure.

Introduction

Brodifacoum (3- [3-(4' - bromobiphenyl-4-yl) -1, 2, 3 4- tetrahydro-1-naphthyl] -4- hydroxycoumarin), is a potent, second-generation, anticoagulant rodenticide developed in the mid-1970s, which persists in the liver of sub-lethally poisoned animals (*1*). It has been used with success in recent rodent eradication programmes on New Zealand's offshore islands to protect populations of endangered indigenous birds (*2, 3, 4*). In addition to its worldwide use to control and eradicate rats, it is now commonly used to control brushtail possums (*Trichosurus vulpecula* Kerr) (*5*).

During the last 2 to 3 years there has been increasing field use of cereal bait containing 20 mg kg^{-1} brodifacoum (Talon® and Pestoff®) for killing rats and possums on mainland New Zealand. This wide-scale field use of brodifacoum in New Zealand is in itself unusual. In the USA and UK, the toxicant is used only for commensal rodent control (A. Buckle, *pers. comm.*), and even this and comparatively restricted use, around farm buildings, causes concern in regard to secondary poisoning of birds and mustelids (*6, 7, 8, 9*).

The toxicology and unique toxicokinetics of brodifacoum and the species at risk from primary and secondary poisoning in New Zealand, have been reviewed elsewhere (*1, 5, 10*). The acute toxicity of brodifacoum in birds varies from an LD_{50} of <1 mg kg^{-1} in pūkeko (*Porphyrio p. melanotus* Temminck) to>20 mg kg^{-1} in paradise shelduck (*Tadorna variegata* Gmelin). In several species of mammals, including rodents, pigs and possums, the LD_{50} is < 0.4 mg kg^{-1}. Because of its toxicity, all vertebrates that eat baits or poisoned prey containing brodifacoum residues are potentially at risk, including humans.

To date there are no recorded incidents of humans being poisoned after eating contaminated meat. However, a risk exists which is compounded by the unusual persistence of this toxicant and other second-generation anticoagulants in vertebrate species compared with first-generation anticoagulants (*11*). For example, brodifacoum has been shown to persist in the liver of sheep (*Ovis aries* L.) for 16 weeks (*12*) and possums for 9 months (*1*). Pigs are one of a number of species that may scavenge possum carcasses, and possums dying up to 1 year after being exposed to sub lethal amounts of brodifacoum will contain residues, particularly in the liver, which could be transferred through the food chain (*1*).

In this review paper we discuss our findings on the concentration of brodifacoum in pig serum and tissue (liver and muscle) after the primary poisoning and secondary poisoning of captive pigs. The risks of secondary poisoning to humans associated with brodifacoum use in New Zealand are discussed alongside implications from preliminary data obtained from feral pig, deer, stoat and cat samples.

Captive Animal Studies and Field Surveys

Captive Animal Studies

Eason et al. 1999 have reported results from studies conducted on captive pigs to assess the levels of brodifacoum that might occur after both primary and secondary poisoning in the field (*13*). The experimental designs and key findings from these experiments are summarised below.

Primary Poisoning of Pigs

Sixteen pigs held in captivity were divided into four groups of four each. Group 1 received non-toxic pellets; Group 2 were fed 500 g (± 31.2 g), Group 3 were fed 937 g (± 34.2 g), and Group 4 were fed 1776 g (± 81.4 g) of brodifacoum bait. The rationale for the dosages used is as follows. The published LD_{50} values for brodifacoum in pigs vary from 0.1 mg kg^{-1} (*14*) to 0.5–2.0 mg kg^{-1} (*15*). Since possum baits contain 20 ppm (or 20 mg kg^{-1}), a 20-kg pig would need to eat 1 kg of bait to ingest 1 mg kg^{-1} of brodifacoum, which would be approximately equivalent to an LD_{50} dose. Half this would be approximately an LD_{25} dose, and double this would be approximately equivalent to an LD_{90}. Furthermore, these amounts could approximate the quantity of bait wild pigs might readily obtain if they fed directly from possum bait stations. Some of these would be expected to survive, and even those eating lethal amounts of bait could be hunted and eaten prior to the toxicant taking effect.

Our analyses carried out on the pigs after post-mortem sampling (5 days after eating baits) showed that the concentration of brodifacoum in the livers and muscle of pigs that had eaten toxic bait appeared to be independent of the amount of bait eaten over the range 500 to 1776 g per pig. The lack of a dose response (in terms of increased liver concentrations) was likely to be due to saturation of absorption of brodifacoum since the baits were fed over a 2–day period. The concentration of brodifacoum in the liver was over 20 times that in the muscle. Muscle concentrations ranged from 0.02 to 0.078 mg kg^{-1} and liver concentrations ranged from 0.72 to 1.38 mg kg^{-1} (Table 1).

Secondary Poisoning Experiment

Edible soft tissue from possums poisoned with brodifacoum, including liver, kidney, heart, and muscle, and portions of the small intestine, were fed to 16 pigs over a 2-day period, and which were divided into five groups. Group 1 were fed possums that had not been dosed with brodifacoum (n=4). Pigs in Group 2 were each fed soft tissue from one poisoned possum carcass each (n=3). Pigs in Group 3 were each fed soft tissue from two poisoned carcasses (n=3), pigs in Group 4 were each fed soft tissue from four carcasses (n=3), and pigs in Group 5 were each fed soft tissue from the equivalent of eight possum carcasses each (n = 3). Five days after the pigs had first started eating the possum meat and organs, they were humanely killed for analyses. The concentration of brodifacoum in the possum tissues fed to pigs varied from 0.84 mg kg^{-1} (range 0.52 to 1.20) in the liver to 0.065 mg kg^{-1} (range 0.013 to 0.094) in the muscle.

Table I. Mean Bait Eaten with Corresponding Mean Dose of Brodifacoum and the Concentration of Brodifacoum in Pig Liver, Muscle and Serum

	Bait Eaten Per Group (g/pig ± S.E.)	*Brodifacoum Ingested (mg kg^{-1} ± S.E.)*	*Mean Concentration (mg kg-1 ± S.E).*		
			Liver[a]	*Muscle*[a]	*Serum*
1	Control non-toxic	0 negligibleb	0.04 ± 0.02*	0	0
2	Pestoff® 500 ± 31.2	0.57 ± 0.06	1.13 ± 0.07	0.05 ± 0.006	0.17 ± 0.06
3	Pestoff® 937 ± 34.2	0.96 ± 0.03	1.08 ± 1.14	0.05 ± 0.01	0.25 ± 0.04
4	Pestoff® 1776 ± 81.4	1.94 ±0.09	1.05 ± 0.06	0.05 ± 0.003	0.38 ± 0.06

[a] concentrations in muscle range from 0.02 to 0.07 and in liver from 0.72 to 1.38 mg kg^{-1} (*13*)

[b] Denotes some low level exposure in control pigs that escaped and ate some toxic bait before recapture.

The concentration of brodifacoum in the liver of pigs that ate the soft tissues from poisoned possums increased in a dose-dependent manner, with the highest concentrations present in pigs that had eaten the soft tissue and muscle from eight possums (Figure 1). The concentration of brodifacoum in serum and in pig muscle tissue was below the limit of detection except for one animal, which had a concentration of 0.01 mg kg^{-1} in muscle (and 0.8 mg kg^{-1} in the liver). The liver concentration of 0.8 mg kg^{-1} was the highest detected in this group of pigs.

Field Survey of Feral Pigs, Deer, Cats and Stoats

Samples of liver were collected from 35 feral pigs, 33 feral deer, 115 stoats and 71 cats, that were shot, or trapped, except for one feral pig found dead. All the animals were killed in areas where brodifacoum was currently in use for possum and rat control. Fourteen out of 35 pigs (approx. 40%) were not contaminated. The remaining 21 pigs, including the one found dead, were found to be contaminated with brodifacoum at concentrations ranging from 0.007 to 1.78 mg kg^{-1} (Table 2). Only five of the contaminated pigs contained liver concentrations of brodifacoum less than 0.1 mg kg^{-1}. Eleven of 33 feral deer were contaminated but the concentration did not exceed 0.03 mg kg^{-1}. Fifty-seven out of 71 cats and 97 out of 115 stoats contained residues. Concentrations in cats ranged from 0.078 to 1.84 mg kg^{-1} and in stoats from 0.008 to 1.32 mg kg.

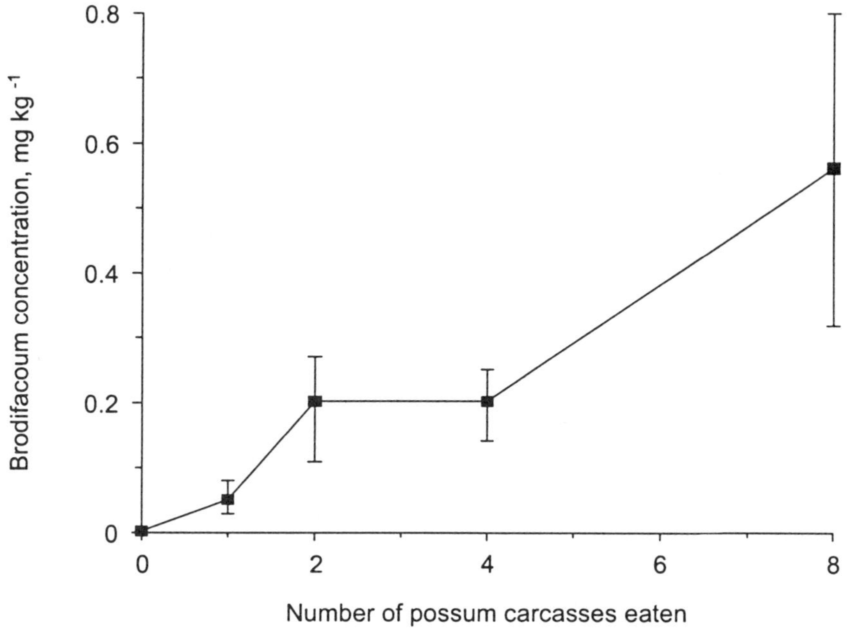

Figure 1. Brodifacoum concentrations (mg kg^{-1}) in pig liver 5 days after eating possum tissues. Range is presented as bars (in preference to S.E.) as these data do not assume a normally distributed error. (Reproduced with permission from reference (13). Copyright 1999).

Table II. Brodifacoum Concentrations in the Livers of Feral Pigs, Deer, Cats and Stoats and Number and Percent Contaminated

	Number Contaminated			
Range (0–2 mg kg^{-1})	*Pigs*	*Deer*	*Cats*	*Stoats*
0	14	22	18	14
>0–0.1	5	11	38	16
0.1–0.6	10	0	43	19
0.6–1.0	1	0	11	8
1.0–1.3	3	0	4	6
> 1.3 < 2.0	2	0	1	8
Total	35	33	115	71
% Contaminated	60	33	80	84

Discussion

It is apparent that pigs scavenging dead possums or rats are at risk of poisoning and subsequent transfer of brodifacoum residues to humans, particularly if people eat the livers of feral pigs. These risks are compounded by the unusual persistence of brodifacoum and will be heightened in people who repeatedly eat possums or pigs. Such multiple exposures of second-generation anticoagulants are most likely to lead to accumulation in the liver of any vertebrate species.

Our studies with captive animals have demonstrated the presence of residues in muscle tissue in possums and pigs, albeit at concentrations about 20 times lower than those found in the liver.

In an earlier paper (*1*) we reported mean liver concentrations of approximately 0.1 mg kg-1 in possums after administration of 0.1 mg kg^{-1} brodifacoum, which persisted in possum liver for the 9-month duration of the experiment. This concentration was associated with bleeding disorders in all animals and death in two out of 36 possums. In the feral pig survey, only five of the 21 contaminated animals had concentrations in the liver of <0.1 mg kg^{-1}. It is difficult to predict whether possums, pigs or humans would respond similarly. In risk assessment a precautionary approach is normal, hence we should assume that liver concentrations of ≥0.1 mg kg^{-1} are of considerable concern in any non-target species.

In overseas studies, liver concentrations of 1.0 to 1.9 mg kg^{-1} brodifacoum were commonly found in rats. However, concentrations between 2 and 5 mg kg^{-1} occurred with a lower frequency and, in some instances, concentrations of > 7 mg kg^{-1} of brodifacoum in rat liver were detected (*16*). In our surveys in New Zealand wildlife we did not detect concentrations above 2 mg kg^{-1} in any species. Concentrations in deer were extremely low. Nevertheless, we believe the levels of brodifacoum present in pig livers could represent a risk to humans as well as to other predators and to farm dogs; this risk is magnified by the persistence of the compound, which could lead to accumulation on repeated exposure. In this regard it is important to remember that a sublethal dose well below the LD_{50} could produce significant clotting abnormalities and some haemorrhaging (*1*). Humans that are already on antithrombotic therapy (e.g., warfarin treatment or aspirin) may be at special risk because of possible adverse drug interactions.

To date, the highest known concentration of brodifacoum in the liver of wild pigs is 1.78 mg kg^{-1}. This finding, and the detection of brodifacoum residues in native birds such as kiwi (*17*), raises serious concerns about the long-term effects of broad-scale field use of brodifacoum in New Zealand. This is confounded by the recent detection of residues in a wide range of bird species: weka, pūkeko, grey duck, harrier, black backed gull, robin, saddleback chaffinch, mynah, magpie, blackbird and mallard duck, in addition to kiwi (*18*) (Wright pers. comm.). Of far less concern was the detection of brodifacoum in cats and stoats, introduced species regarded as pests and largely responsible for the decline of native birds, such as kiwi. Nevertheless, because of the perceived indiscriminate contamination of wildlife broad-scale, the sustained field use of brodifacoum on mainland New Zealand is already being restricted. Alternative less persistent toxicants or traps are being utilised. This contrasts with 'one-off' use on New Zealand's off-shore islands, where an island sanctuary free from rodents and predators is an achievable goal, and conservationists are confident the benefit predators is an achievable goal, and conservationists are confident the benefit accrued from brodifacoum use outweighs the risks of exposure of non-target species (*19*).

Acknowledgements

Andrea Rhodes is thanked for her support and technical assistance with the captive animal studies; Geoff Wright, Les Brown and Craig Radford for analyses of brodifacoum samples referred to; Wendy Ruscoe for statistical advice and Christine Bezar for editorial comments. This research was funded by the Department of Conservation.

Literature Cited

1. Eason, C.T.; Wright, G.R.; Batcheler, D. *N.Z.J. Agric. Res.* **1996,** *39,* 397–400.
2. Taylor, R.H.; Thomas, B.W. *Biological Conservation* **1993,** *65:* 191—198.
3. Buckle, A.P.; Fenn, M.G.P. *Proceedings of the 15th Vertebrate Pest Conference,* Borreco, J.E.; Marsh, R.E. Eds. University of California, Davis, USA, **1992**, pp 36–41.
4. Robertson, H.A.; Colbourne, R.M.; Nieuwland, F. *Notornis* **1993,** *40,* 253—262.
5. Eason, C.T.; Frampton, C.M.; Henderson, R.; Thomas, M.D.; Morgan, D.R. *N.Z.J. Zool.* **1993**, *20,* 329–334.
6. Duckett, J.E. *Planter Kuala Lumpur* **1984** *60,* 3—11.
7. Hegdal, P.L.; Colvin, B.A. *Environ. Toxico. Chem.* **1988** *7,* 245–260.
8. Newton, I.; Wyllie, I.; Freestone, P. *Environmental Pollution* **1990**, *68,* 101–117.
9. Colvin, B.A.; Jackson, W.B.; Hegdal, P.L. Magallona, E.D. Ed. *Proceedings of the International Congress on Plant Protection* **1991**, *11,* 60–64.
10. Eason, C.T.; Spurr, E.B. *N.Z.J.Zool.* **1995**, *22,* 371–379.
11. Parmar G.; Bratt H.; Moore R and Batten P.L. *Human Toxicol.* **1987**, 6,: 431–432.
12. Laas, F.J.; Forss, D.A.; Godfrey, M.E.R. *N.Z.J.Agric.Res.* **1985**, *28,* 357–359.
13. Eason, C.T.; Milne, L.; Potts, M.; Morriss, G.; Wright, G.R.G.; Sutherland, O.R.W. *N.Z.J. Ecology* **1999**, *23 (2),* 219–224.
14. Godfrey, M.E.R. *Acta Zool. Fenn.* **1985**, *173,* 209—212.
15. Dublock, A.C. & Kaukeinen, D.E. *Proceedings of the 8th Vertebrate Pest Conference*, Davis, California, USA, **1978**, pp 127.
16. Brown, R.A. *Rodent Pests and their Control*, Buckle, A.P.; Smith, R.H. Eds. CAB International, University Press, Cambridge, UK. pp. 363–380.
17 Robertson, H.A.; Colbourne, R.M.; Miller, P.; Pierce, R. *N.Z..J. of Ecology.***1999.**
18. Dowding, J.E.; Murphy, E.C.' Veitch, C.R. *N.Z.J. of Ecology* **1999** *23:*
19. Empson, R.A.; Miskelly, C.A *N.Z.J. Ecology 23* **1999,** *(2),* 241–254.

Chapter 13

Chlorophacinone Residues in Rangeland Rodents: An Assessment of the Potential Risk of Secondary Toxicity to Scavengers

Thomas M. Primus, John D. Eisemann, George H. Matschke, Craig Ramey, and John J. Johnston

APHIS/WS/National Wildlife Research Center, U.S. Department of Agriculture, 4101 LaPorte Avenue, Fort Collins, CO 80521–2154

Field studies were conducted in California to assess efficacy of chlorophacinone-treated steam-rolled oats for controlling rangeland rodents. An objective of these studies was to assess the potential hazards of chlorophacinone residues in rangeland rodent carcasses and livers to mammalian and avian scavengers, especially raptors. Belding's ground squirrels, valley pocket gophers and *Microtus spp.* carcasses collected during the efficacy studies were analyzed for chlorophacinone residues. The method limit of detection (MLOD) for liver and carcass tissue samples averaged 0.036 μg/g and 0.034 μg/g, respectively. Chlorophacinone residues in Belding's ground squirrel (n=62) liver and carcass tissue ranged from <MLOD to 0.82 μg/g and <MLOD to 0.55 μg/g, respectively. Chlorophacinone residues in valley pocket gopher (n=8) liver and carcass tissue ranged from <MLOD to 0.42 μg/g and <MLOD to 1.21 μg/g, respectively. Chlorophacinone residues in whole body *Microtus* sp. (n=3) tissue ranged from 0.26 to 4.1 μg/g, respectively. Risk assessment indicated acute risk for mammalian scavengers and negligible to minimal risk for avian scavengers consuming rangeland rodents exposed to chlorophacinone baits.

INTRODUCTION

Chlorophacinone (2-[[(Chlorophenyl)phenylacetyl]-1H-indene-1,3(2H)-dione] and diphacinone [2-(Diphenylacetyl)-1H-indene-1,3(2H)-dione] are registered late-first generation or early-second generation anticoagulant rodenticides commonly used to control populations of rats and mice in urban areas. These anticoagulants are also effective in the control of other rodents such as pocket gophers (*Thomomys bottae*), Belding's ground squirrels *(Spermophilus beldingi),* and California ground squirrels (*Spermophilus beecheyi).* The acute oral toxicity (LD_{50}) for rats of both chlorophacinone and diphacinone is approximately 2 mg/kg, compared to the acute oral toxicity for other anticoagulants such as warfarin and pindone which is ~ 59 mg/kg.

Monetary damage to range grasslands attributed to pocket gophers and ground squirrels is difficult to estimate. Rangeland rodents can reduce vegetation by 20 to 40 percent, which results in less plant material for livestock grazing (*1,2*). Additionally, the combination of grazing by pocket gophers, ground squirrels, and livestock can lead to severe soil erosion. Damage to earthen irrigation ditches and dams has been observed in areas where pocket gopher and ground squirrel populations are excessive (*1,2*). Control methods for ground squirrels and pocket gophers include exclusion, shooting, trapping, flooding, use of acute toxicants including anticoagulants, and fumigants (*3*). California uses steam-rolled oat baits fortified at 0.005% (w/w) and 0.010% (w/w) chlorophacinone or diphacinone to control rangeland rodent populations.

When considering the use of pesticides to control rangeland rodents, the risk of secondary toxicity to scavengers potentially consuming target species such as Belding's ground squirrels and valley pocket gophers should be evaluated. Therefore, an objective of this study was to determine if residues of chlorophacinone were significantly high to pose a risk to scavengers. Following Environmental Protection Agency (EPA) procedures, incurred residues were compared to accumulated chlorophacinone toxicity values for various species to assess the risk of secondary toxicity for scavengers/predators. Studies reporting the toxicity of diphacinone were also cited when characterizing risk to secondary consumers, since their toxicities are generally similar.

Typically, only liver and serum are analyzed for residues of anticoagulants as anticoagulants are metabolized and accumulate in the liver. However, whole ground squirrel carcasses containing chlorophacinone residues are available to scavengers, not exclusively the liver. Based on the assumption that a majority of the chlorophacinone residue would be in the liver, the liver was removed from the carcass in this study and analyzed separately. The remaining carcass, with the appendages, head, and pelt removed, was homogenized and analyzed as an additional sample.

Several methods have been developed for analysis of indandiones in baits, formulations, and tissues. A gas chromatographic method with derivatization (*4*) is sensitive and selective, but suffers from low recoveries and is time consuming. Spectrophotometric methods (*5,6*) have been utilized for baits and formulations, but are not selective when assaying multi-residue samples. Thin-layer chromatography (*7-9*) methods are not suited for determining low levels of residues in complex matrices such as plant and animal tissues. Reversed-phase high-performance liquid chromatography (HPLC) methods (*10-14*) provide sufficient sensitivity, but often produce poor chromatographic resolution for the indandiones. Ion-pair reversed-phase HPLC (*15-20*) is sensitive and selective, but column lifetime is often short, due to adsorption of the ion-pairing reagent onto the stationary phase of the column packing material. In this sudy, reversed-phase ion-pair HPLC was used because good chromatographic resolution can be achieved and column lifetime can be extended with regular washing. Sample extraction utilized solid phase extraction (SPE) for sample cleanup with high sample throughput.

METHODS

Sample Collection

During field efficacy studies for the use of chorophacinone steam-rolled oat bait by spot baiting and with bait stations in alfalfa fields in Siskiyou County, California, carcasses of Belding's ground squirrels and *Microtus* were collected above ground *(21, 22)*. These studies were conducted in May to June of 1996. Additionally, during a field efficacy study for the use of chlorophacinone and diphacinone treated steam-rolled oat bait use by spot baiting in burrow systems in Siskiyou County, California the carcasses of valley pocket gophers were located and collected underground *(23)*. This study was conducted in October to November of 1997. Whole rodent carcasses were collected and placed in an individual plastic bags, sealed, labeled, and frozen in a portable freezer at -5 °C. The samples were stored in a freezer until shipped to our laboratory where they were stored in freezers at -20 °C until assayed. Method validation and analyses were completed under U.S. EPA Good Laboratory Practice guidelines (40 CFR 160).

Sample Preparation

Whole animal carcasses were weighed. Carcasses were weighed a second time after removal of the pelt, head, and appendages. Finally, after removing and weighing the liver, each carcass was weighed a third time. The liver was weighed separately. Individual livers and carcasses (minus head, pelt, and appendages) were frozen and homogenized with a cryogenic mill (*24*). Homogenization was completed by freezing the tissue with liquid nitrogen in a stainless steel cylinder and crushing the sample with a stainless steel piston until the tissue became a powder. The powdered, frozen liver sample was transferred to a 35-mL glass sample bottle and the powdered, frozen carcass sample was transferred to a 500-mL polyethylene bottle. The homogenized samples were stored at -20 °C and assayed within two weeks.

Reagents

Acetone, chloroform, hexane, ethyl acetate, and methanol were liquid chromatography grade reagents (Fischer Scientific, Denver, CO). Deionized water was purified using a Milli-Q water purification system (Millipore, Bedford, MA). Concentrated phosphoric acid (Fischer Scientific, Denver, CO) was used to make 4 N phosphoric acid in water. Concentrated formic acid (Fischer Scientific, Denver, CO) was used to prepare 1% formic acid in 1:1 acetone:chloroform extraction solution. Anhydrous sodium sulfate (Fischer Scientific, Denver, CO) was mixed with tissue samples to remove water.

Chlorophacinone (98.9%) was obtained from LiphaTech (Milwaukee, WI). Concentrated stock standards of chlorophacinone were prepared by first drying the technical grade compound for 4 hours at 110 °C, then dissolving 10.000 mg in 10.0 mL of ethyl acetate. Working standards, ranging in concentration from 0.030 μg/mL to 10.0 μg/mL, were prepared by dilution of stock solutions with mobile phase. All standard solutions were stored at 5 °C.

Tetrabutylammonium dihydrogen phosphate (97%) was purchased from Aldrich (Milwaukee, WI) and was used to prepare a 5 mM solution in methanol. An aqueous solution of 5mM tetrabutylammonium dihydrogen phosphate with 50 mM potassium dihydrogen phosphate buffer (Alltech, Inc.; Deerfield, IL) was prepared.

Liver and Carcass Tissue Sample Extraction and Cleanup

Sample Extraction

Homogenized tissue samples were weighed (1.0 - 1.1 g) into a mortar and 10.0 g of anhydrous sodium sulfate was added. The tissue and sodium sulfate were ground together with a pestle for five minutes. The solid mixture was transferred to a 50-mL tube with a powder funnel. The mortar was rinsed with three 5 mL aliquots of extraction solution and transferred to the 50 mL tube. Sample tubes were vortex mixed thoroughly and shaken horizontally on a mechanical shaker at high speed for 20 minutes. Sample tubes were centrifuged at approximately 2500 rpm for 5 minutes.

The extract was transferred to a 50-mL glass tube. The extraction was repeated twice following two subsequent 10 mL additions of extraction solution. Extract solvent was removed by placing the tubes in a warm water bath ($\leq$ 60 °C) and allowing nitrogen gas to flow over the surface of the extract until no solvent remained.

The residue was reconstituted with 5.0 mL of hexane, gently vortex mixed, and sonicated for 10 minutes.

Analyte Concentration

Each silica SPE (2 g) column was conditioned with approximately 5 mL of hexane. The packing material was not allowed to dry. The reconstituted sample extract was added to the SPE column with a Pasteur pipet. The entire solution was passed through the column at 1 to 2 mL/min (vacuum was typically not necessary). The eluate was collected in a 25-mL glass tube. Each SPE column was rinsed with hexane by adding 5 x 2.5 mL aliquots (12.5 mL total volume) to the 50-mL tube and transferring the solution to the SPE columns. This eluate was discarded. Each SPE column was rinsed with 20 mL (8 x 2.5 mL) of 1:1 ethyl ether:hexane and this eluate discarded.

Liver Sample Analyte Elution

A clean 15-mL screw top centrifuge tube was placed under each SPE column in the manifold. The analyte was eluted from each SPE column by adding 15 mL (6 x 2.5 mL) of 12% (v/v) methanol in ethyl ether. After the last 2.5 mL aliquot of eluant passed through the SPE column, vacuum was used to collect eluant that remained in the SPE packing material.

Carcass Sample Analyte Elution

A clean 15-mL screw top centrifuge tube was placed under each SPE column in the manifold. The analyte was eluted from each SPE column by adding 20 mL (8 x 2.5 mL) of 15% (v/v) methanol in ethyl ether. After the last 2.5 mL aliquot of eluant passed through the SPE column, the vacuum was applied to collect eluant remaining in the SPE packing.

Sample Reconstitution

The volume of eluate was reduced by placing tubes in a warm water bath and blowing a stream of nitrogen over the solution until the solvent was removed (early in the solvent removal procedure the tube was kept out of the water bath). The residue was redissolved with 1.0 mL of 75:25 methanol:water (with 5 mM tetrabutyammonium phosphate), vortex mixed and sonicated for 5 minutes. The reconstituted samples were filtered through a 0.45 μm Teflon syringe filter into a vial and capped before HPLC analysis.

High Performance Liquid Chromatography

The HPLC system consisted of a Hewlett-Packard 1090 liquid chromatograph (Palo Alto, CA) and a Hewlett-Packard 1050 variable wavelength detector. The mobile phase was prepared by mixing aqueous and methanolic solutions of 5 mM tetrabutylammonium dihydrogen phosphate (32:68 v/v) and adjusting the pH to 8.0 with 4 N phosphoric acid. The mobile phase was degassed by sparging with helium. At the end of each set of analyses, the column was washed with a mixture of 1:1 (v/v) methanol:water for 40 minutes. Each tissue sample was analyzed in duplicate. The HPLC parameters are listed in Table I.

Quality Control Samples and Fortification of Controls

Belding's ground squirrels and valley pocket gophers were trapped and euthanized by California Department of Food and Agriculture representatives at two sites in Siskiyou County, California prior to any baiting operations (*25*). These animal carcasses and livers were processed and screened for chlorophacinone and diphacinone prior to combining control samples into a composite. Control liver and carcass tissue samples were fortified at 0.10, 1.0, and 10 ppm chlorophacinone with aliquots of fortification standards of chlorophacinone in ethyl acetate. The quality control samples were then assayed with the appropriate method described previously.

Table I. HPLC parameters for the analysis of liver and carcass extracts

Parameter	*Conditions*
Mobile Phase:	Combine the aqueous IPC solution and methanolic IPC solution in the ratio 68:32 (Methanol:Water)
Column Conditioner:	1:1 Methanol:Water
Flow Rate:	1.0 mL/min
Injection Volume:	100 μL
Column:	Keystone ODS/H (C18), 5 μm, 250 mm x 4.6 mm i.d. or equivalent (use guard column containing identical HPLC packing)
Column Temp.:	35 °C
Detector:	UV @ 285 nm and 325 nm
Run Time	Standard: 20 minutes Samples: 35 minutes

Chlorophacinone concentrations were determined by comparing the area of the chlorophacinone peak in the sample extract to a working standard. The retention time of chlorophacinone over the dates of analyses (2/7/97 to 3/27/97) varied from 15.2 to 17.5 min, respectively.

Microtus Analysis

During the collection of animal carcasses for the field portion of one of the studies, several *Microtus spp.* carcasses were found and collected. These samples were handled and stored under the same conditions as the ground squirrels and pocket gophers. Each whole animal was homogenized and assayed as described previously.

RESULTS AND DISCUSSION

Analytical Methods

Mean recoveries of chlorophacinone of liver (n = 24) and carcass (n = 28) quality control samples were 80.4 ± 17.2% and 75.5 ± 10.0% (Table II). Two lots of the silica solid phase extraction columns were used to complete the analyses, with no difference in recoveries observed between the two lots.

Table II. Analytical recoveries of chlorophacinone in Belding's ground squirrel and valley pocket gopher tissues for quality control samples

Fortification Levels (ppm)	*Tissue*	*Range (%)*	*Mean (%)*	*Std. Dev. (%)*	*CV (%)*
Belding's Ground Squirrel					
0.010 - 10	Carcass (n = 17)	60 - 134	83	17	21
0.010 - 10	Liver (n = 17)	55 - 89	74	10	14
Valley Pocket Gopher					
0.010 - 1.0	Carcass (n = 7)	70 - 87	76	5.4	7.1
0.010 - 2.5	Liver (n = 11)	62 - 98	79	11	14

Response Linearity

Two sets of six calibration standard solutions were prepared ranging in concentration from 0.030 to 10 μg/mL. Each standard solution was injected two times and a linear regression performed on the data set. The relation between chromatographic response and concentration was linear. The response was directly proportional to concentration over the range of interest.

Method Limit of Detection

The method limit of detection (MLOD) was calculated as the concentration of chlorophacinone required in the sample to generate a signal equal to 3 times the baseline noise (peak to peak) observed in the chromatogram of the control extract. The MLOD was estimated from the chromatographic response in height of a control tissue extract and an extract from a control tissue sample fortified at 0.10 μg/g. The MLOD for all liver and carcass tissue samples averaged 0.036 μg/g and 0.034 μg/g, respectively. Chromatograms of carcass sample extracts with positive and negative chlorophacinone responses are shown in Figure 1.

Carcass and Liver Residues

Chlorophacinone residues in Belding's ground squirrel (n = 62) liver and carcass tissue ranged from <MLOD to 0.82 μg/g and <MLOD to 0.55 μg/g, respectively (Table III). Chlorophacinone residues in valley pocket gopher (n = 8) liver and carcass tissue ranged from <MLOD to 0.42 μg/g and <MLOD to 1.21 μg/g, respectively. Total residue was calculated by multiplying the analyte concentration in the liver or carcass by the liver or carcass weight and summing the quantities, respectively. For samples containing <MLOD, the MLOD was used to calculate mean residue and total residue.

The primary wavelength for quantitative analysis was 285 nm, though absorption at 325 nm was also determined. The ratio of absorbance at 285 and 325 nm was used to qualitatively confirm presence of the analyte. The molar absorptivity of chlorophacinone at 285 nm is twice the molar absorptivity at 325 nm.

Microtus Residues

Chlorophacinone residues in 3 whole animal *Microtus* sp. were 0.26, 0.36 and 4.1 μg/g, respectively.

Secondary Toxicity Assessment

Many factors determine whether rodenticide residues in poisoned animals pose a secondary hazard to non-target (scavenger) species. These include the chemical and toxicological properties of the active ingredient, composition of the formulated bait and how it is applied, behavior of the non-target species at risk, local environmental factors, and the variability of residue concentrations among carcasses (*26*). For example, ground squirrels which are diurnal may not be preyed upon by nocturnal predators, however, their carcasses may be available to either nocturnal or diurnal scavengers. A study of the anticoagulant brodifacoum on plains pocket gophers

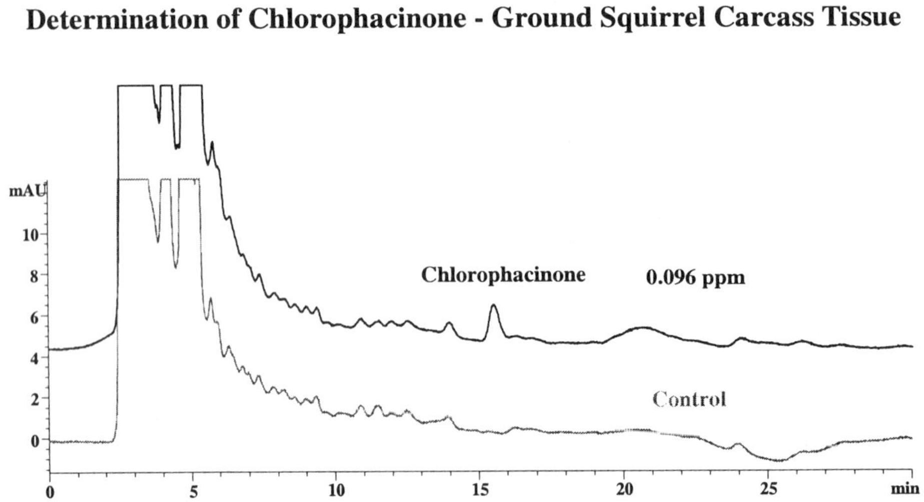

Figure 1. Chromatograms of a control blank and a 0.096 μg/g chlorophacinone-fortified control carcass tissue samples with ultraviolet detection at 285 nm.

showed that 90% of radio-equipped animals expired underground (*27*), which minimizes the risk of secondary poisoning. Environmental factors play a role in determining how long a carcass is available to scavengers. During hot and dry weather, carcasses above ground are often dessicated and consumed by insects in less than two days (*28*).

Table III. Residues of chlorophacinone in Belding's ground squirrel and valley pocket gopher tissues

Sample Type (# analyzed)	*Range of Residues (μg/g)*	*Mean Residue* (μg/g)*	*Range of Total Residue (μg)*	*Mean Total Residue (μg)*
Ground Squirrel				
Livers (62)	<MLOD - 0.82	0.133	0.10 - 11	1.8
Carcass (62)	<MLOD - 0.55	0.131	1.1 – 123	20
Pocket Gopher				
Liver (8)	<MLOD - 0.42	0.161	0.060 - 2.4	0.92
Carcass (8)	<MLOD - 1.21	0.357	1.0 – 126	23
Microtus spp.				
Carcass (3)	0.26 - 4.1	1.58	2.1 – 57	21

*To calculate the mean residue for samples reported as <MLOD, the MLOD was used as the value for these samples.

A commonly used approach for evaluating non-target hazards is the calculation of a risk quotient (RQ) (*29*). Acute risk quotients predicting the potential for lethal exposure are routinely calculated using the median lethal dietary dose (LC50) of the most sensitive species in relation to the expected pesticide concentration in the diet. This method divides the expected environmental concentration (EEC) (in the case of secondary toxicity this would be the maximum observed tissue concentration) by the LC50 of the most sensitive species.

$$RQ = EEC / LC50$$

Acute dietary toxicity to mammals is normally not known. In conducting a screening level assessment for mammals, the LC50 is determined by dividing the median lethal acute oral dose (LD50) value (usually the rat LD50) by the animal's daily dietary intake in relation to the animal's body weight. A risk quotient is then determined as above, by dividing the EEC by the derived LC50 value.

$$RQ = EEC / [LD50 * \% \text{ Body Wt. Consumed per day}]$$

Significant risk to non-target avian and mammalian species is predicted if the RQ is greater than 1. However, a value between 0.5 and 0.1 usually requires that the product be used under specified restrictions. The following risk assessment focuses on direct mortality resulting from the secondary exposure to chlorophacinone contaminated carcasses. Sub-lethal effects leading to indirect mortality will not be addressed.

The residues found in ground squirrel livers collected in this study were used to represent the maximum expected environmental concentration for conducting a secondary hazard assessment for predators and scavengers. Although higher residues were found in the pocket gopher, these carcasses had to be dug out of the ground and obviously presented little hazard to scavenging species. Acute toxicity data for chlorophacinone and diphacinone were obtained from the EPA (*30, 31*) and RTECs (*32*) databases (Table IV). Risk quotients calculated using the maximum residues found in Belding's ground squirrel livers indicate little risk for avian species (Table V). However, acute risk is predicted for all three weight classes of mammals. If these (Table V) calculations were made for more typical feeding situations where the entire carcass is consumed and the maximum residue for ground squirrel carcasses being 0.55 μg/g, risk is lower but still indicated. The highest chlorophacinone residue found in a *Microtus* carcass was 4.1 μg/g. Risk quotients calculated on a EEC of 4.1 μg/g are 5 times higher than shown in Table V. Even at the higher EEC no risk for avian species is predicted. However, risk quotients for mammals would be increased by 5 and indicate that all weight classes of mammals and the coyote are potentially at risk from acute exposure if only *Microtus* were consumed. The most realistic approach may be to use the mean residue value for the Belding's ground squirrel carcasses as the EEC. Most scavengers will not specifically consume the liver and very few carcasses will be at the maximum concentration (Table V).

It is possible a scavenging or predatory mammal's diet may consist solely of contaminated *Microtus* for a period long enough to obtain a lethal dose, but it is more likely *Microtus* would only be scavenged occasionally. In this study treated colonies were exhaustively searched to locate carcasses. *Microtus* are approximately 7 times smaller than *Belding's* ground squirrels. Because of the small size of a *Microtus*, carcasses might have been overlooked. This may explain why 20 times more ground squirrel carcasses were found than *Microtus* carcasses despite the fact ground squirrels are fossorial and many may die underground. The dessication rate of a smaller carcass will be much quicker than that of the ground squirrel, reducing the length of time it will be attractive to vertebrate scavengers. Because *Microtus* live above ground, the potential for scavenging a carcass is high. In a treated ground squirrel colony with large numbers of squirrel carcasses above ground, there may be a tendency for scavengers to focus on the abundant food source and overlook small *Microtus* carcasses.

Table IV. Toxicity data for Chlorophacinone and Diphacinone

Species	*LD50 (mg/kg)* Chlorophacinone	Diphacinone
Rat *(Rattus spp.)*	2.1 ‡	1.5 ‡
Mouse *(Mus. Spp.)*	1.1 ‡	28 ‡
Rabbit (*Oryctolagus spp.*)	50 ‡	35 ‡
Mallard*(Anas platyrhynchos)*	100 ‡	3160 ‡
Northern bobwhite *(Colinus virginianus)*	260 †, 430 ‡	1200 †
Dog (*Canis domesticus*)	---	3.0 ‡
Cat (*Felis catus*)	---	15 ‡
Pig *(Sus spp.)*	---	150 ‡
Coyote (*Canis latrans*)	---	0.6 (*33*)

Species	*LC50 (ppm)* Chlorophacinone	Diphacinone
Coyote (*Canis latrans*)	0.95 (*36*)	---
Mallard *(Anas platyrhynchos)*	170 †, 430 †	910 †
Northern bobwhite† *(Colinus virginianus)*	56 †, 240 †	4500 †
Golden eagle (*Aquila chrysaetos*)	---	2.7* NOEL (*33*)
Barn owl (*Tyto alba*)	1.3^* NOEL (*34*)	—
American kestrel (*Falco sparverius*)	5.7^* NOEL (*35*)	—

* no mortality ^ extrapolated (reference) † EPA Data (*30, 31*) ‡ RTECS (*32*)
NOEL = (No Effect Level)

Studies (*33-35*) to assess the secondary hazards of indandione anticoagulants on raptors can be evaluated and compared to the RQs values calculated with our data. One study investigated the effects of secondary exposure of golden eagles to sheep muscle laced with 2.7 ppm diphacinone (*33*). Of seven golden eagles fed this diet, none expired (four eagles were fed for 5 days and three were fed for 10 days).

Hematocrit and prothrombin levels indicated subacute symptoms of toxicity and two of seven eagles treated demonstrated sublethal symptoms. Assuming that chlorophacinone and diphacinone toxicities are similar (Table IV), the total quantity of residue consumed by the eagles was approximately three times higher than found in one *Belding's* ground squirrel analyzed. The highest residue detected in Belding's ground squirrel tissue was 50% lower than that fed to the eagles and no mortality occurred. Therefore, risk appears to be even lower than estimated by previous studies *(33)*.

Table V. Risk quotients for avian and mammalian species using maximum observed residue in liver tissue

			Liver		*Carcass*	
Birds	*LC50*		*Max Residue (μg/g)*	*RQ*	*Mean Residue (μg/g)*	*RQ*
Northern bobwhite	56		0.82 ^	0.015	0.131	0.0023
Mallard	170		0.82	0.005	0.131	0.0008
Mammals	*% BW Consumed*	*Estimated LC50 #*	*Max. Residue (μg/g)*	*RQ*	*Mean Residue (μg/g)*	*RQ*
15 g	95	1.15	0.82	0.71	0.131	0.11
30 g	66	1.67	0.82	0.49	0.131	0.078
1000 g	15	7.33	0.82	0.11	0.131	0.018
Coyote	-	0.95*	0.82	0.86	0.131	0.14

Estimated LD50 is based upon the mouse LD50 of 1.1 mg/kg

^ Belding's Ground Squirrel maximum liver residue and mean carcass residue observed

* *(36)*

Mendenhall and Pank assessed anticoagulant rodenticides (6 compounds including chlorophacinone) hazards to owls (*34*). For ten days, two barn owls were fed tissue from rats that had been poisoned with a mean consumption of 12.6 mg of chlorophacinone in treated bait. Mortality of rats occurred within 6 days on average. Neither bird expired and no symptoms of toxicity were observed. No residue analysis was completed on the rat tissue. If a "worst case" scenario is adapted and no excretion or metabolism occurred for the poisoned rats, one barn owl consumed 712 g of rat tissue containing 9.2 mg of chlorophacinone, the concentration of chlorophacinone in the rat tissue would have been 13 μg/g. Metabolism studies with domestic rats administered 1.4 mg of chlorophacinone have shown that after two days, 90% of chlorophacinone was metabolized or excreted (*37*). If it is then assumed that 10% of the chlorophacinone was retained by the rats, the concentration of chlorophacinone in the rat carcass would have been 1.3 μg/g. This no effect level is greater than the highest residue concentration of chlorophacinone in poisoned Belding's ground squirrels or pocket gophers.

A third publication reports the effect of chlorophacinone poisoned *Microtus* on American kestrels (*35*). For 21 days, four American kestrels were fed *Microtus* that were poisoned with an average of 1.14 mg chlorophacinone in a treated bait. On average, *Microtus* mortality occurred within 6 days. None of the birds expired and no external toxic symptoms were observed, but several birds subsequently euthanized and examined internally showed evidence of hematomas. Unfortunately, no residue analysis was completed on the *Microtus*. Based on the previous assumption that after two days 90% of chlorophacinone is metabolized or excreted, the concentration of chlorophacinone in the *Microtus* tissue with an average body mass of 40 g would have been approximately 5.7 μg/g. This is almost 500% greater than the maximum concentration observed for chlorophacinone in pocket gophers and 39% greater than the maximum residue observed in the three *Microtus* carcasses analyzed. These studies indicate that secondary hazards to birds consuming chlorophacinone burdened rodent carcasses are minimal.

Two references (*33, 36*) yielded information on indanedione hazards to mammals. In one study *(33)*, acute toxicity was noted when pairs of coyotes were orally gavaged with diphacinone at seven levels from 0.16 to 10.0 mg/kg. This study yielded an LD_{50} of approximately 0.6 mg/kg. An experiment investigating the secondary chlorophacinone exposure of coyotes (*36*) was conducted with California ground squirrels exposed to chlorophacinone fortified bait. Coyotes were fed one California ground squirrel a day for five consecutive days. Three of the seven exposed coyotes died. Residues in the ground squirrels ranged from 0.16 to 2.8 μg/g with a mean value of 0.95 μg/g. The mean residue concentration reported by Marsh and Howard *(36)* was seven times higher than the mean residue observed in this study for Belding's ground squirrel tissues. The maximum residue concentration was three times higher than the maximum residue reported for Belding's ground squirrel tissues in this study. The RQ calculated using an LC50 of 0.95 μg/g and the maximum residue observed in Belding's ground squirrel livers in our study (0.82 μg/g) is 0.86. If the mean carcass residue is used as the EEC, (0.131 μg/g) the RQ is lowered to 0.14, indicating the actual field risk of chlorophacinone exposure may be lower than that simulated in the previous study *(36)*. These results indicate coyotes consuming a diet of only contaminated Belding's ground squirrels are at risk for acute affects.

CONCLUSION

The methodology developed for analyzing chlorophacinone liver and whole body tissue proved to be reliable, efficient and simple. The same method was used to analyze tissue from three different species. Chlorophacinone residues in Belding's ground squirrel (n=62) carcass and liver tissues ranged from <MLOD to 0.55 μg/g and <MLOD to 0.82 μg/g. Chlorophacinone residues in valley pocket gopher (n=8) carcass and liver tissues ranged from <MLOD to 1.21 μg/g and <MLOD to 0.42 μg/g. Chlorophacinone residues in whole body *Microtus* spp. (n=3) tissue were 0.26, 0.36,

and 4.1 $\mu g/g$, respectively. In estimating potential secondary hazards for proposed use of indandione rodenticides, every likely scenario cannot be investigated. Three studies combined with the residue data from this work reinforce the avian risk quotients and suggest that the potential chlorophacinone secondary hazards to avian scavengers are minimum to negligible. However, the implications from the two studies with coyotes and indandione secondary toxicity are reinforced by the residue data from this work and the associated mammalian risk quotients. These studies suggest potential secondary hazards for chlorophacinone to some mammalian scavengers.

ACKNOWLEDGMENTS

Funding for these studies was provided by the California Vertebrate Pest Control Research Advisory Committee through a cooperative agreement with the California Department of Food and Agriculture. The authors are grateful to HACCO, Inc. and LiphaTech, Inc. for supplying diphacinone and chlorophacinone. Mention of commercial products is for identification only and does not constitute endorsement by the U.S. Department of Agriculture.

REFERENCES

1. *Wild Mammals of North America*; Chapman, J. A.; Feldhammer, G. A., Eds.; The Johns Hopkins University Press, Baltimore and London, 1982, pp. 251-252.
2. *Wild Mammals of North America*; Chapman, J. A.; Feldhammer, G. A., Eds.; The Johns Hopkins University Press, Baltimore and London, 1982, pp. 202-203.
3. Salmon, T. P.; Schmidt, R. H. *Proc. Vert. Pest. Conf.* **1984,** 11, 32-37.
4. Bullard, R. W.; Thompson, R. D.; Holguin, G. *J. Agric. Food Chem.* **1976,** 24, 261-263.
5. Kawano, Y.; Chang, W. *J. Assoc. Offic. Anal. Chem.* **1980,** 63, 996-998.
6. Caswell, R. L. Report of Rodenticides *J. Assoc. Offic. Anal. Chem.* **1959,** 42, 104-106.
7. Mallet, V.; Surette, D.; Brun, G. L. *J. Chromatogr.* **1973,** 79, 217-222.
8. Owen, P.; Pendlebury, A.; Moffat, A. C. *J. Chromatogr.* **1978,** 161, 187-193.
9. Opong-Mensah, K.; Porter, W. R. *J. Chromatogr.* **1988,** 455, 439-443.
10. Addison, J. B. *J. Assoc. Off. Anal. Chem.* **1982,** 65, 1299-1301.
11. Houglum, J. E.; Larson, R. D.; Neal R. M. *J. Chromatogr.* **1989,** 481, 458-460.
12. Reynolds, J. D. *Proc. Am. Assoc. Vet. Lab. Diagn.* **1980,** 23, 187-194.
13. Bennett, B. R.; Grimes, G. S. *J. Assoc. Off. Anal. Chem.* **1982,** 65, 927-929.

14. Mura, P.; Piriou, A.; Papet, Y.; Lochon, D.; Reiss, D. *J. Anal. Toxicol.* **1992,** 16, 179-181.
15. Vigh, G.;Varga-Puchony, E.; Papp-Hites, E.; Hlavay, J.;Balogh, S. *J. Chromatogr.* **1981,** 214, 335-341.
16. Hunter, K. *J. Chromatogr.* **1984**, 299, 405-414.
17. Hunter, K. *J. Chromatogr.* **1985,** 321, 255-272.
18. Hunter, K.; Sharp, E. A. *J. Chromatogr.* **1988,** 437, 301-305.
19. Chalermchaikit, T.; Felice, L. J.; Murphy, M. J. *J. Anal. Toxicol.* **1993,** 17, 56-61.
20. Primus T. M.; Goldade D. A.; Petty E. E.; Johnston. J. J. *J. Chrom. Sci.* **1996,** 34, 389-393.
21. Ramey, C.A.; Matschke, G. H.; McCann, G.R. 1999. Chlorophacinone/spot baiting – Belding's ground squirrel field study. Unpublished report, QA-475, National Wildlife Research Center, Fort Collins, CO 176 p.
22. Matschke, G. H.; Ramey, C.A.; McCann, G.R. 1999. Chlorophacinone/bait station – Belding's ground squirrel field study. Unpublished report, QA-474, National Wildlife Research Center, Fort Collins, CO 170 p.
23. Matschke, G. H.; Ramey, C.A.; McCann, G.R.; Bourassa, J.; Tope, K; Brooks, J.; Stewart, W.B. 1999. Chlorophacinone and diphacinone: field study of hand baiting to control valley pocket gophers *(Thomomys bottae).* Unpublished report, QA-478, National Wildlife Research Center, Fort Collins, CO 160 p.
24. Sterner, R. T.; Mauldin, R. E. *Arch. Environ. Contam. Toxicol.* **1995**, 28, 519-523.
25. Matschke, G. H. Personal Communication, Product Registration Section, Denver Wildlife Research Center, **1996**.
26. Record C.R.; Marsh. R.E. *Proc. Vertebr. Pest Conf.* **1988,** 13, 163-168.
27. Matschke, G. H.; Hegdal, P. L.; Stockwell, C. A. *The 6th International Biodeterioration. Symposium,* **1984**, 60-65.
28. Sullivan, D. *The Proceedings of the Vert. Pest Conf.* **1992**, 60-65.
29. Urban, D. J.; Cook, N. J. Ecological Risk Assessment: *Standard Evaluation Procedure of the Hazard Evaluation Division,* U.S. Environmental Protection Agency. Office of Pesticide Programs. U.S.Government Printing Office: Washington, D.C.; *NASNRC Publ.* **1986;** EPA-540/9-85-001.
30. U.S. Environmental Protection Agency. Office of Pesticide Program Ecotoxicity Data Base.
31. U.S Environmental Protection Agency. Reregistration Eligibility Decision - Rodenticide Cluster. EPA 738-R-98-007. **1998,** Unpublished studies.
32. RTECS Ecotoxicity Data Base.
33. Savarie, P. J.; Hayes, D. J.; McBride, R. T.; Roberts, J. D. In *Avian and Mammalian Wildlife Toxicology.* **1979,** p. 69-79. Am. Soc. for Testing and Materials, Spec. Tech. Publ. 693.
34. Mendenhall, V. M.; Pank, L. F. *Wildl. Soc. Bull.* **1980,** 8(4), 311-315.

35. Radvanyi, A.; Weaver, P.; Massari, C.; Bird, D.; Broughton, E. *Bull. Environ. Contam. Toxicol.* **1988,** 41, 441-448.
36. Marsh, R. E.; Howard, W. E. Unpublished report. **1986,** 78 p.
37. Belleville, M. Absorption, Distribution, Metabolism and Excretion Studies in the Rat Using Carbon-14 Labeled Chlorophacinone: Lipha Report No. 86-001. Unpuplished study prepared by Lipha Research Center. **1981**, 16 p.

Chapter 14

Wildlife Primary and Secondary Toxicity Studies with Warfarin

Richard M. Poché and Jeff J. Mach

Genesis Laboratories, Inc., P.O. Box 1195, Wellington, CO 80549

Through the results of several toxicity studies, warfarin was evaluated as a potential field rodenticide. Although warfarin is generally accepted as being a safer compound compared to the second generation anticoagulants, little information is available on the potential for secondary poisoning to non-target wildlife that may consume rodents killed by the bait. Studies were completed to determine primary and secondary laboratory efficacy and field efficacy of warfarin baits. The laboratory efficacy study had 5 male and 5 female black-tailed prairie dogs (*Cynomys ludovicianus*) in each of six treatment groups (0, 50, 100, 250, 500, and 1000 ppm warfarin). The resulting efficacy after a 15-day exposure and 5-day post-test period was 0, 30, 50, 60, 100, and 80%, respectively. Also, 16 montane voles (*Microtus montanus*) and 4 meadow voles (*Microtus pennsylvanicus*) were housed by species in large stock tanks and were exposed to 250 ppm warfarin bait, apples, and alfalfa cubes. All voles died by day 11 of the test. A field efficacy test with a 250 ppm warfarin bait on black-tailed prairie dogs proved to be 86.4% and 78.6% effective according to 2 census techniques after 3 spot-baiting applications. European ferrets (*Mustela putorius furo*) and black-billed magpies (*Pica pica*) were secondarily fed black-tailed prairie dogs and Norway rats (*Rattus norvegicus*), respectively, that had consumed 500 ppm warfarin bait. After the 5-day secondary exposure and 21-day post-test period, the ferrets and magpies displayed no signs of toxicosis. Our research demonstrates a broad spectrum for potential warfarin use.

INTRODUCTION

The anticoagulant rodenticide warfarin was registered by the United States Environmental Protection Agency (USEPA) in 1952 and quickly took hold as an effective means of rodent control. Warfarin gained an advantage in the market place over the acute toxicants (1). Warfarin baits have been used and proved effective on numerous rodent species including: nutria (*Myocaster coypus*) (2), rats (*Rattus spp.*) (3), and the house mouse (*Mus musculus*) (4).

Warfarin should be considered as a field rodenticide because it is a cost-effective rodenticide, relatively nontoxic to birds (3, 5), degrades quickly in the gastrointestinal tract (half-life of 42 hours) (6), is rapidly excreted from the body (7), often "bait shyness" is not developed (8), and is relatively safe with respect to secondary poisonings (9). Little research had been conducted on warfarin as a field rodenticide because it was perceived that more toxic material would be required and render the product less economical.

Our goal was to identify a concentration of warfarin that would be most effective in a field bait. As a result, several tests were designed to determine if a warfarin bait would have potential for managing black-tailed prairie dogs and various vole species. A primary objective was to determine the mortality rates of various concentrations and then develop a final bait to be tested for secondary toxicity and field efficacy. If results suggested benefits over alternative registered compounds, the product should be considered for federal registration.

This paper describes studies that were performed with warfarin as a primary and secondary toxicant. The three experimental types, listed by section title, are: Laboratory Primary Toxicity Studies, Field Toxicity, and Laboratory Secondary Toxicity Studies. The results indicate warfarin may be a practical toxicant for controlling an array of rodent species.

Laboratory Primary Toxicity Studies

Black-tailed Prairie Dog

The black-tailed prairie dog (*Cynomys ludovicianus*) is distributed over the Great Plains. At the turn of this century an estimated five billion prairie dogs inhabited the range (10). Prairie dogs decreased plant biomass planted by early settlers and were considered a pest to ranchers and farmers (11, 12, 13, 14). Also, they serve as reservoirs for diseases that affect humans and other wildlife (15). For these reasons, prairie dog control remains a common practice, utilizing toxicants and firearms. The result has contributed to lower population densities and widely scattered towns (16, 17).

Prairie dog populations have declined considerably over this century because of widespread agricultural tillage and control programs (10). Farmers and ranchers continue to use control techniques on prairie dogs in attempts to reduce economic damage. Registered toxicants have exhibited limited effectiveness and pose considerable hazards to the environment (18).

To assess the potential of warfarin use as a field rodenticide, a laboratory bioassay study using black-tailed prairie dogs was conducted with five dietary concentrations of warfarin baits (0.005%, 0.01%, 0.025%, 0.05%, and 0.1%) to test for efficacy. In a previous study, it was observed that there was sufficient amounts of vitamin K in alfalfa pellets to provide an antidote for warfarin toxicosis (19). In this study, alfalfa was excluded to minimize consumption of dietary vitamin K.

Meadow and Montane Vole

Voles of various species are a common problem throughout much of the U.S (20). Economic damage is in the range of $100 million annually. The small rodents often girdle fruit trees over the winter, resulting in decreased fruit production or tree mortality (21, 22). First generation anticoagulants were developed for use in controlling agricultural pests such as ground squirrels, pocket gophers, and voles. Chlorophacinone and diphacinone are commonly used. We wished to examine the potential for using the less-toxic anticoagulant, warfarin, for control of voles. As a result, a pilot laboratory-choice test was conducted on wild voles to asses the feasibility.

Field Toxicity Study

Black-tailed Prairie Dog

Because of the success of the laboratory efficacy tests with the warfarin baits with the black-tailed prairie dog, we chose to proceed with a field efficacy study (23). This study was designed to generate product performance data for 250 ppm warfarin bait against the target species, the black-tailed prairie dog (*Cynomys ludovicianus*). The 250 ppm bait was chosen for use in the field efficacy study because of the available toxicity data for product registration for 250 ppm warfarin baits, and because it was effectively used as a commensal rodenticide for many years. A 500 ppm warfarin product would require complete toxicity testing, a costly venture. A secondary objective of the study was to determine the potential hazard to non-target primary and secondary animals posed by spot-baiting applications.

Laboratory Secondary Toxicity Studies

European Ferret

To assess the mammalian secondary toxicity of warfarin, a laboratory study was conducted with European ferrets (*Mustela putorious furo*) (24). The European ferret served as a model for wild mustelids which could be present in rodent colonies where a rodenticide might be applied. Several species of weasels, and as a worst case scenario, the black-footed ferret (*Mustela nigripes*), may be found in areas of high prairie dog activity. The mustelids have been documented as being susceptible to anticoagulants. Bromadiolone was tested which proved to be effective against the primary target, but was also secondarily toxic to domestic ferrets (*Mustela putorius furo*) (25). The least weasel was reported (*Mustela nivalis*) as being secondarily susceptible to warfarin (26).

The study was designed to determine what potential risk warfarin may have secondarily to domestic ferrets when fed warfarin-treated black-tailed prairie dogs. Black-tailed prairie dogs were fed warfarin bait for the secondary toxicity test because they represent a typical model of a target vertebrate pest in which warfarin baits could be used to manage their population numbers.

Black-billed Magpie

To assess the avian secondary toxicity of warfarin as a field rodenticide, a laboratory study was conducted with the black-billed magpie (*Pica pica*) (27). The species is a common scavenger and feeds readily on road-kill animal carcasses. The objective of the study was to determine if exposure to warfarin-poisoned animals has the potential to cause toxicosis to black-billed magpies.

METHODS

Laboratory Primary Toxicity Studies

Black-tailed Prairie Dog

Sixty-six (66) wild black-tailed prairie dogs were trapped with Tomahawk® live traps from private land in Larimer County, Colorado. Traps were checked at least twice daily. The prairie dogs were transported to the lab and dusted with an EPA registered flea powder containing pyrethrin to control ectoparasites. Animals were randomized to treatment groups. The prairie dogs were held in individual metal screen bottom cages having a surface area of at least 1400 cm^2 and a minimum height of 36 cm (28). The test animals received a basal diet of laboratory pellets (Manna Pro Lab Cubes), rolled barley, and water, *ad libitum*. Bedding changes and cleaning of water bottles was performed weekly, and the water and feed were checked daily. Exposure to human activities was held to a minimum because handling of animals could cause lesions, bruises, or injury that could bias mortality estimates. Frequent handling can cause higher mortality rates in treatment groups (29).

Prairie dogs were monitored daily to determine health status. Observations included physical and behavioral signs which could indicate sickness, or during the test, warfarin toxicosis. The 10-day acclimation period involved an observation once daily. During the 15-day exposure and 10-day post-test periods, the prairie dogs were observed twice daily.

Five warfarin formulations were prepared: 50, 100, 250, 500, and 1000 ppm with a large commercial Hobart mixer. A blank control bait was mixed in the same manner. The formulations were offered to 5 males and 5 female in each treatment group.

All bait formulations and the control were presented daily during this time until death or the end of the test period. The baits were presented in stainless steel feed cups. Seventy grams of feed were offered daily to the prairie dogs. Bait consumption was measured daily, and the consumed portion of bait was refilled with the respective fresh bait. After the exposure period, the prairie dogs were observed for 10 days for possible signs of toxicosis. During this period, the rodents were fed the basal laboratory diet.

Body weights were taken to assess maturity. It was determined that 675 g and 775 g for females and males, respectively, would be sufficient weights to label individuals as "adults" (30). Underweight individuals on the day of dosing (day 0) were arbitrarily replaced with "extra" animals of sufficient weight that had also been acclimated. Body weights were taken at test termination or death to assess weight loss or gain.

The baits were analyzed for freezer storage and animal room stability, and homogeneity. All animals that died during the test were necropsied for internal signs of anticoagulant poisoning. A sample of tissue was analyzed after the completion of the bioassay.

Meadow and Montane Vole

Four meadow voles (*Microtus pennsylvanicus*) and 16 montane voles (*Microtus montanus*) were live-trapped from Larimer County, Colorado and Whitman County, Washington, respectively. They were shipped to Genesis Laboratories, Inc. and housed according to species in large stock tanks with a floor surface area of approximately 2 m^2. The voles' maintenance diet consisted of alfalfa pellets, sliced apples, and Harlan Teklad 8664 laboratory rodent diet.

An experimental non-grain warfarin bait (250 ppm) was formulated and offered to the voles along with a choice of the maintenance diet for a 11-day exposure. When bait or portions of the maintenance had been entirely consumed, additional bait or feed was offered. Daily observations were made for signs of toxicosis once daily. Water was offered with a stainless steel ball-type waterer, *ad libitum*. The grass and wood shaving-type bedding was changed once per week.

Field Toxicity Study

Black-tailed Prairie Dog

Three study sites were located in Larimer County, Colorado. Sites were selected based on an estimated population of approximately 20 prairie dogs as determined by preliminary visual counts. Census plots and buffer zones were marked with colored flagging. Buffer zones were established as needed to control reinvasion of prairie dogs at distances of approximately 85 meters from the edge of the census plot. Two treatment plots and one control plot were established.

Two activity indices were used: visual counts and closed burrow counts. Both are indices used to measure changes in activity rather than actual population numbers. During pre-treatment and post-treatment phases, three counts were made on each of three consecutive days during peak activity periods. Counts were made at approximately the same times each day, from 11:00 am to 3:00 pm. From the nine counts of the pre- and post-treatment, the highest single count was used as the population index (31).

Closed burrow counts were conducted as soon as possible after visual counts were completed. Technicians moved through the plots in a systematic pattern using shovels to close prairie dog holes with sod and/or soil. Snow and cold weather during the pre-treatment counts prevented technicians from counting open burrows until seven days after closure. The same time period was used during the post-treatment census.

The warfarin bait (250 ppm) used for the field test was formulated at Genesis Laboratories, Inc. Baiting began the same day that the pre-treatment closed burrow census was completed. Bait was first applied on November 20, 1997. Bait was spot-baited at a rate of approximately 72 g per active burrow on the census plot and the buffer zone of treatment plots by technicians walking a systematic pattern. Bait was spread over approximately 5 ft^2 instead of piling the bait. Bait was re-applied in the same manner every third day for three applications. Technicians replenished bait placements only as needed to maintain a continuous supply.

Laboratory Secondary Toxicity Studies

European Ferret

Prairie dogs were presented with a 500 ppm warfarin-treated bait in a no-choice presentation for five days. The animals were euthanized at the end of the exposure period, if not already found dead. These carcasses were frozen and later fed to European ferrets. The total amount of bait consumed by the prairie dogs was recorded and used to calculate the amount of warfarin potentially ingested by the ferrets given certain assumptions.

Twelve (6 males and 6 females) ferrets, were obtained from Marshall Farms. The ferrets were 12-15 weeks old at the time of receipt. The ferrets were housed individually in plastic-coated wire pens (61 x 76 x 46 cm) suspended over metal collection pans during the exposure period and post-test observation period. The surface area of each pen was 4,636 cm^2. Bedding changes were performed at least once weekly using wood shavings as absorbent material in the collection pans. The ferrets were provided with Purina Ferret Diet and water *ad libitum* during the acclimation and post-test observation periods.

The treatment group of ferrets (5 males & 5 females) were exposed to treated carcasses for seven consecutive days. Two control ferrets were presented with untreated prairie dog carcasses. On day 0 of the test, each ferret in the treatment group was provided a thawed, pre-cut (abdominal and thoracic region exposed) warfarin-fed prairie dog in a no-choice situation.

Carcass consumption was monitored daily for seven days by physical description of consumed portions of the carcass. Only one carcass was presented to each ferret. If the ferrets consumed the carcass before the end of the 7-day test period, they were placed on a maintenance diet. Remaining portions of uneaten carcasses were weighed at the end of the 7-day exposure period to determine consumption. If the hide was the only part of the carcass remaining, it was weighed and removed from the cage before the end of the 7-day exposure period to prevent further desiccation.

All ferrets were observed at least once daily for signs of toxicosis due to warfarin poisoning during the 7-day exposure and 21-day post-test periods. The body weight of each ferret was taken at day 0 of the test, and at the termination of the post-test observation period (day 28).

Black-billed Magpie

Black-billed magpies (*Pica pica*) were trapped from locations near Wellington, Colorado. Sixteen magpies (7 males and 9 females) were housed individually in cages (61 x 76 x 46 cm) suspended over metal pans. Wood shaving were used as absorbent material in collecting droppings and feeding residues. Commercial dog food, soaked in water, was provided *ad libitum.* Lighting was maintained on a 12 hour-light cycle per day.

Laboratory rats were fed 500 ppm warfarin bait until death occurred (within 7 days). Bait consumption was recorded daily for each rat. Rat body weights were documented before the introduction of warfarin bait and upon completion of the exposure and observation periods. All fifteen rats died during the 7-day exposure period to 500 ppm warfarin bait. Control rats were fed cubes and remained healthy. The control rats were euthanized and presented to the control magpies. Warfarin-fed and control rats were frozen until offered to the magpies.

Magpies (7 male and 7 female) were exposed to treated rats for five consecutive days. One control group (2 female magpies) received untreated rat carcasses. The weight of each rat carcass was taken before presentation to a magpie. Carcass consumption was monitored daily by physical description of consumed portions of the carcass. A new rat carcass was presented only if a magpie consumed virtually all of the carcass. On day 5, the remaining rat carcasses were collected and weighed to determine consumption by each magpie.

The body weight of each magpie was taken at day 0 of the exposure period and at termination of the post-test observation. All magpies were observed at least once daily for signs of toxicosis during the 5-day exposure and 22-day post-test periods. The birds' gender were determined at necropsy.

RESULTS AND DISCUSSION

Laboratory Primary Toxicity Studies

Black-tailed Prairie Dog

The five warfarin dietary concentrations resulted in dose-response curves based on mortality and mean warfarin consumption (Table I).

Table I. Mortality and Mean Warfarin Consumption

Concentration (ppm)	*Mortality (%)*	*Mean Warfarin Consumption-mg (Range)*
0	0	0.0 (0.0)
50	30	13.7 (6.3-21.5)
100	50	29.9 (16.3-43.3)
250	60	55.3 (16.0-107.7)
500	100	161.4 (65.1-208.2)
1000	80	239.9 (128.6-352.4)

Note: Concentration units are parts per million (ppm). Mortality units are percent. Mean warfarin consumption is in milligrams of technical warfarin during the test period. Source: Reproduced from Reference 19.

There was a high correlation (R=0.917) between bait concentration and warfarin (technical) consumption for groups 50 through 1000 ppm, while total bait consumption remained similar between the same groups. This demonstrates that palatability of the baits may not have been a factor in this study.

Bait consumption in the control group was significantly higher than any of the treatment groups. Consumption could have been affected by the presence of warfarin, causing decreased palatability, but most likely, the treatment groups were reducing intake because of illness associated with warfarin toxicosis. Signs of warfarin poisoning appeared at a similar time (days 5-11) and then consumption decreased throughout the treatment groups because of illness (days 10-15). The control prairie dogs continued to eat because they were healthy or showing no signs of poisoning from warfarin. Warfarin consumption was highly correlated with treatment level. Palatability likely did not affect bait consumption because ingestion of warfarin did not decrease at the higher treatment levels. Higher warfarin concentrations (1,000 to 2,000 ppm) are reported as being unpalatable (36), but in anticipation of this problem, we used a high purity of warfarin and disguised the taste with feed additives, which resulted in no palatability problems. Even at the highest dose (1,000 ppm), neophobic response was not evident.

The necropsies revealed considerable variation in hemorrhaging among individuals. Hemorrhaging was observed in the stomach, liver, intestines, kidneys, heart, lungs, brain, and subcutaneous membrane. Extensive fat reserves were observed in the abdominal and thoracic region of all prairie dogs.

The body weights of individuals in the treatment groups decreased, whereas, the control animals gained weight (Table II).

Table II. Body Weight Changes.

Treatment Group	*Mean Treatment Group Loss/Gain*	*Mean Male Loss/Gain*	*Mean Female Loss/Gain*
0	31.2	51.4	11.0
50	-77.0	-111.2	-42.8
100	-80.7	-110.8	-50.6
250	-112.7	-118.2	-107.2
500	-94.7	-128.2	-61.2
1000	-142.2	-170.2	-114.2

Note: Treatment group units are parts per million. Mean loss/gain units are grams.
Source: Reproduced from Reference 19. Copyright 1998 Proceedings Vertebrate Pest Conference.

The mean treatment group weight change correlation coefficient (R=0.707) demonstrated that when the warfarin concentration increased, the amount of weight loss correspondingly increased. If a higher warfarin concentration was offered to prairie dogs, a higher amount of weight loss would be probable.

The computer program "Toxstat, version 3.4" produced an LC_{50} value of 97 mg/kg body weight for black-tailed prairie dogs and an LC_{90} for black-tailed prairie dogs of 831 mg/kg body weight. A break in the mortality pattern occurred at the 1000 ppm treatment group, perhaps due to small sample size. Therefore, other analyses were performed using dummy variables in this same group.

An LC_{50} value was calculated for the actual results and then 2 alternative analyses were performed. Little change was noticed in the LC_{50} of all analyses, but a noticeable difference was evident with the LC_{90} values. The two animals that survived the test in the 1,000 ppm treatment caused the actual LC_{90} value to more than double, compared to the dummy analyses. The expected result would be 100% mortality in the 1,000 ppm treatment, if the 500 ppm group had 100% mortality. Efficacy should increase as dosage increases. If 100% mortality had been achieved in the 1000 ppm group, the LC_{90} would be one-half of the actual analysis.

In a previous pilot study, prairie dog mortality did not exceed 20% in identical warfarin treatment groups (19). Reduced efficacy was attributed to the use of dry alfalfa cubes. Dry alfalfa is reported as containing as much as 14.2 mg of fat-soluble vitamin K per kg (32). Vitamin K is required by the liver to produce prothrombin (Factor II), a major component of the blood clotting mechanism (33, 34). This mechanism has positive feedbacks that continue to amplify the reaction intensity. When this mechanism is inhibited by warfarin, the result is a failure in blood clotting (35).

Preliminary analysis of data indicated that prairie dogs may be more tolerant of warfarin when given a supply of dietary vitamin K (19). Removal of alfalfa cubes in the current study, however, demonstrated that the species is susceptible to the anticoagulant. Sufficient mortality (100%) was achieved in the 500 ppm treatment group demonstrating the absence of dietary vitamin K is an important factor in determining warfarin efficacy. EPA requires a minimum 70% decrease in population from a field rodenticide before federal registration. This is the criteria used to determine what formulation is effective. Dietary vitamin K_3 has been shown to have antidotal effects in anticoagulants studies with rats and mice (37).

Vitamin K is fat soluble and is able to be stored within fat along with vitamins A, D, and E (35, 38). This allows consumption of vitamin deficient diets over a longer period of time compared to water-soluble vitamins, before deficiency signs appear. In this study necropsies performed on test animals revealed that prairie dogs had extensive fat reserves, even though weight loss was common. The large amounts of fat were present in the abdominal region and within the thoracic cavity around the heart and lungs.

This could be a reason for marginal recoveries and the ability of many prairie dogs to not succumb to warfarin intoxication. When sufficient vitamin K is stored within the fat reserves of the prairie dogs, the vitamin K could be used to reverse the effects of warfarin.

Gut bacteria also contributes to vitamin K availability (39). Bacterial synthesize vitamin K that is then absorbed in the distal section of the intestinal tract, where the bacterial population is greatest (38). Specifically, vitamin K is absorbed in the large intestine of mammals sufficiently to prevent deficiency symptoms (40). The extent of bacteria synthesized vitamin K inhibiting the action of warfarin is not known, but it is expected to contribute a portion to the antidotal effect.

In the absence of dietary vitamin K, it would appear that the fat reserves and the bacteria of the gut could furnish a sufficient amount of vitamin K to produce some degree of antidotal effects. The two extremes of warfarin consumption, 6.30 and 352.40 mg warfarin/kg body weight resulted in mortality. This exhibits the wide range of physiological responses to warfarin. It appears that some prairie dogs are able to acquire vitamin K from the fat or bacteria more efficiently than others. Less susceptible individuals are able to metabolize the warfarin more efficiently (41).

Bait applied during a time when the prairie dogs have a leaner body weight, as in early spring, will increase control of population numbers. When fat reserves are low, the absence of vitamin K may allow full activity of the warfarin. It would be less effective to bait when prairie dogs are storing fat nutrients for winter survival.

To avoid the problem of vitamin K accumulation in prairie dogs, one must apply the bait during early spring months when the prairie dogs are active, yet vegetative growth remains dormant. The initiation of spring vegetative growth must be avoided because of its high nutritional quality, including vitamin K. Baiting at this time would be before the animals had time to build up fat reserves.

Meadow and Montane Vole

A 250 ppm warfarin bait provided to voles of two species in a choice test resulted in 95% mortality in the test animals. The first dead vole was found on day 4 and by day 11, 19 of the 20 voles succumbed to warfarin. Three (3) of 4 montane voles and 16 of 16 meadow voles died during the 11-day exposure period.

These data identified two vole species as being susceptible to a warfarin bait. With the possibility of vole damage to orchards and forests, it is very important to control the vole populations to protect such a valuable investment. Voles present in alfalfa fields (high in vitamin K) must be baited in early spring when the alfalfa is dormant. However, it is not understood why the voles succumbed to the bait when alfalfa cubes were offered as an alternate choice during the exposure period. Perhaps the voles are very sensitive to warfarin or they did not find the alfalfa palatable in the dry, pelleted form.

Field Toxicity Study

Black-tailed Prairie Dog

Census plots were the following sizes (not including buffer zones): plot 1–0.53 ha (treatment), plot 2–0.77 ha (control), plot 3–1.59 ha (treatment). The bait was applied to a total area of 2.12 ha in the last half of November, 1997.

The two treatment plots received three applications of the 250 ppm warfarin bait. Applications were made every third day. Consumption of the bait began almost immediately. Alternative food sources were available in the form of dried prairie grasses. Little bait remained on the plots when post-treatment censusing initiated. Treatment plot 1 received 6.3 kg/ha and plot 3 received 5.3 kg/ha.

The activity indices showed similar results in decreased prairie dog counts: 86.4% and 78.6% on the treatment plots (Table III). The mean of the two activity indices is 82.5%. The control plot population had a decrease of 5.9% (visual counts) and an increase of 7.0% (closed burrow counts). Visual count efficacy calculations were adjusted to incorporate the decrease in the control plot. Pre-treatment and post-treatment censuses were conducted 37 days apart.

Table III. Baiting Efficacy Based on Two Census Methods

Plot #	*Plot Type*	*Visual Counts*	*Burrow Counts*
1	Treatment	83.4	78.0
3	Treatment	89.4	79.2
	Mean Efficacy	86.4	78.6

Note: The control plot population index changed according to the visual and burrow counts by 5.9% (decrease) and 7.0% (increase), respectively. Census method units are percent.

Source: Reproduced from Reference 23. Copyright 1997 Unpublished Genesis Laboratories, Inc. Report.

A dead long-tailed weasel (*Mustela frenata*) was found on a treatment plot. A necropsy revealed blood in the abdomen and hemorrhaging in the lungs. Search of the stomach contents uncovered *Peromyscus* spp. hair. We concluded that the weasel may have died from warfarin, and most likely by ingesting a warfarin-intoxicated field mouse. Least weasels (*Mustela nivalis*) were susceptible to warfarin at levels of 200 ppm (26). Mink (*Mustela vison*), in contrast, were fed warfarin-treated rabbits for 28 days and no mortality occurred (9).

This study demonstrates the effective use of a warfarin bait to control populations of the black-tailed prairie dog. Its primary use was effective, but the non-target hazards suggest secondary susceptibility with the long-tailed weasel. The next portion of the text will explain methods and results of two secondary hazard studies that suggest warfarin could still be used as an effective rodenticide with some degree of safety secondarily.

Laboratory Secondary Toxicity Studies

European Ferret

All prairie dogs died during the 7-day exposure period to 500 ppm warfarin bait and none of the control prairie showed signs of intoxication. The control prairie dogs and were euthanized before presentation to the respective treatment group of ferrets. The ferrets showed no signs of toxicosis or mortality during the 7-day exposure period or the 21-day post-test period.

The mean ferret body weights taken at test initiation for the control and treatment groups were 1,269.0 g (n=2) and 1,244.1 g (n=10, s.d.=427.5), respectively. The mean body weights taken at test termination increased to 1,393.5 g and 1,377.0 g (s.d.=510.5) for the control and treatment groups, respectively. Both sexes and treatment groups exhibited an increase of weight that suggests normal health.

Carcass consumption by each ferret was monitored daily during the 7-day exposure period. Generally, ferrets started eating the anterior portion and consuming tissue to the posterior of the prairie dog. The means for the carcass consumption by the control and treated groups were 728.5 g (n=2) and 522.6 g (s.d.=171.4), respectively. The total amount of warfarin ingested ranged from 31.7 to 160.4 mg warfarin/kg body weight (mean = 81.3 mg warfarin/kg body weight). This was based upon the total amount of bait consumed by the prairie dog, assuming no degradation over time and the equal distribution of the warfarin in the carcass.

Warfarin is known to be concentrated in the liver, but can also be amounts in the adrenal glands, lungs, bone marrow, kidneys, and lymph nodes (38). Ferrets consumed major organs, which have the ability to accumulate warfarin, yet they showed no signs of intoxication.

Based upon the mortality and lack of toxicosis sign data, the 28-day secondary hazard study using 500 ppm warfarin-fed black-tailed prairie dogs fed to European ferrets, confirmed that there was little potential secondary toxicity of warfarin.

Black-billed Magpie

The mean body weight of magpies at the experimental start in the control and treated groups was 153 g (n=2) and 168 g (n=10, s.d.=18), respectively. The mean body weight of magpies at the experimental termination in the control and treated groups was 150 g (n=2) and 162 g (s.d.=18), respectively. The mean body weight throughout the test of the control and treated groups was -2.5% and -3.5%, respectively. It is believed that the small loss of weight could be due to the stress of being in a laboratory environment because warfarin is reported as being relatively non-toxic to birds (3, 5).

The mean carcass consumption by the control and treated groups was 205 g (n=2) and 180 g (s.d.=36), respectively. The average maximum potential ingestion of warfarin by the treatment group was 354 mg/kg body weight. This level of warfarin consumption was more than four times the amount the domestic ferrets consumed and yet the birds showed no signs of intoxication.

The actual amount of warfarin intake by the magpies is unknown since more than half of the warfarin consumed by a rat is excreted and metabolized. Rodent mortality increases if warfarin is fed over several days, therefore the rats received warfarin bait for several consecutive days. The amount of warfarin intake by the magpies estimated based on the total amount of warfarin consumed by the rats and the amount of rat carcasses consumed by each magpie. We assumed a worse case and calculated the warfarin ingested by assuming there was zero degradation and an equal distribution of the drug in the rat carcass.

The magpies showed no signs of toxicosis to the secondary feedings of the warfarin-treated rats. All treated and control magpies survived the test and post-test observation periods. The gross necropsy showed no signs of toxicosis in any of the birds at the end of the 27-day study.

The 27-day secondary hazard study using 500 ppm warfarin-fed Norway rats presented to black-billed magpies, confirmed that there was little potential for secondary toxicity of warfarin to magpies.

CONCLUSIONS

Information presented herein indicated that there may be a few problems to non-target wildlife (long-tailed weasel) as a result of using warfarin baits, but the laboratory secondary hazards studies suggest there is a degree of safety since no signs of toxicosis were observed in the laboratory. We do not fully understand why the warfarin bait caused a non-target death during the field study when the laboratory secondary toxicity study suggested that warfarin was safe. Speculation into the difference leads to possibilities of: vitamin K in the diet of the ferrets, lack of activity of the laboratory ferrets delaying naturally-induced injury from daily activity, and the ferret not being a fair representation of a wild weasel.

We conclude that warfarin, when formulated as a 250 ppm bait, may be an efficacious and safe field rodenticide. Based on these studies, warfarin appears to have potential for control of many rodent species in agriculture. Further research is recommended before large-scale use of such products is implemented.

Literature Cited

1. Prier, R. F.; Derse, P. H. Evaluation of the hazard of secondary poisoning by warfarin-poisoned rodents. *JAVMA*. 1962, 140, 351-354.
2. Evans, J.; Ward, A. L. Secondary poisoning associated with anticoagulant-killed nutria. *JAVMA*. 1967, 151, 856-861.
3. Hagan, E. C.; Radomski, J. L. The toxicity of 3-(acetonylbenzyl)-4-hydroxycoumarin (warfarin) to laboratory animals. *J. Amer. Pharm. Assoc.* 1953, 42, 379-382.
4. Bentley, E. W.; Larthe ,Y. The comparative rodenticidal efficiency of five anti-coagulants. *J. Hyg.* 1959, 57, 135-149.
5. Christopher, M. J.; Balasubramanyam, M.; Purushotham, K. R. Toxicity of three anticoagulant rodenticides to male hybrid leghorns. *Zoology*. 1984, 71, 275-281.
6. Ford, M. In *Handbook of Medical Toxicology*; Viccellio, P., Ed.; Rodenticides; Little, Brown, and Co., Boston, MA. 1993; pp 322-327.
7. Wong, L. T.; Solomonraj, G. Biliary and urinary excretion of [^{14}C] warfarin in rabbits. *Xenobiotica*. 1980, 10, 201-210.
8. Meister, R. T.; 1995 Farm chemicals handbook pesticide dictionary; Meister Publishing Company, Salem, MA, 1995; pp 445.
9. Aulerich, R. J.; Ringer, R. K.; Safronoff, J. Primary and secondary toxicity of warfarin, sodium monoflouroacetate, and methyl parathion in mink. *Arch. Env. Cont. Tox.* 1987, 16, 357-366.
10. Merriam, C. H. In *Yearbook of the U.S. Department of Agriculture, 1901;* Washington, D.C. 1902; pp 257-270.
11. Taylor, W. P.; Loftfield, J. V. G. *Damage to range grasses by the Zuni prairie dog.* U. S. Department of Agriculture Bulletin. 1924; pp 1227.
12. Hansen, R. M.; Gold, I. K. Black-tailed prairie dogs, desert cottontails and cattle trophic relations on shortgrass range. *J. Range Manage.* 1977, 30, 210-214.
13. O'Meilia, M. E.; Knopf, F. L.; Lewis, J. C. Some consequences of competition between prairie dogs and beef cattle. *J. Range Manage.* 1982, 35, 580-585.
14. Knowles, C. J. Some relationships of black-tailed prairie dogs to livestock grazing. *Great Basin Naturalist.* 1986, 46, 198-203.
15. Barnes, A. M. Surveillance and control of bubonic plague in the United States. *Symp. Zool. Soc. (London).* 1982, 50, 237-270.
16. Clark, T. W.; Campbell, T. M. III; Socha, D. C.; Casey, D. E. Prairie dog colony attributes and associated vertebrate species. *Great Basin Naturalist.* 1982, 8, 46-49.
17. Foster, N. S.; Hygnstrom, S. E. *Prairie Dogs and Their Ecosystem*; University of Nebraska-Lincoln, Lincoln, NE, 1990; pp 8.
18. Hygnstrom, S. E.; Virchow, D. R. *In Prevention and Control of Wildlife Damage;* S. E. Hygnstrom; Timm, R. M.; Larsen, G. E., eds. Prairie Dogs;

Nebraska Cooperative Extension Service, University of Nebraska-Lincoln, 1994; pp B85-B96.

19. Mach, J. J. Laboratory efficacy study with a warfarin bait to control the black-tailed prairie dog. (R. O. Baker and C. Crabb, Eds.) *Proceed. Vert. Pest Conf.* 1998, 18, 184-190.
20. Poché, R.M. Vole control in the eastern United States. *Proc. East. Wildl. Damage Conf.* 1985, 2, 138-144.
21. Jackson, J. J. Controlling vertebrate animal damage in southern pines (L. R. Lewis and R. E. Marsh, Eds.) *Proceed. Vert. Pest Conf.* 1990, 14, 199-202.
22. Sullivan, T. P.; Sullivan, D. S.; Crump, D. R.; Weiser, H.; Dixon, E. A. Predator odors and their potential role in managing pest rodents and rabbits (Crabb, A. C. and Marsh, R. E., Eds.) *Proceed. Vert. Pest Conf.* 1988, 13, 145-150.
23. Fiedler, D. A.; Mach, J. J. Field efficacy study using prairie dog bait containing 0.025% warfarin to control the black-tailed prairie dog (*Cynomys ludovicianus*) in Colorado. *Unpublished.* 1997.
24. Carlet, L.; Mach, J. J. Secondary hazard study using warfarin-killed black-tailed prairie dogs (*Cynomys ludovicianus*) fed to domestic ferrets (*Mustela putorius furo*). *Unpublished.* 1997.
25. Fisher, D. D.; Timm, R. M.; Poche, R. M.; Hygnstrom, S. E. Laboratory study on bromadiolone: effectiveness on prairie dogs and secondary hazards to domestic ferrets. (Hygnstrom, S. E.; Case, R. M.; Johnson, R. J.; Eds.) *Proceed. Great Plains Wildlife Damage Conf.* 1991, 10, 70-72.
26. Townsend, M. G.; Bunyan, P. J.; Odam, E. M.; Stanley, P.I.; Wardall, H. P. Assessment of secondary poisoning hazard of warfarin to least weasels. *J. Wldlf Manage.* 1984, 48, 628-632.
27. March, K. L.; Mach, J. J. Secondary hazard study using warfarin-killed laboratory rats fed to black-billed magpies (*Pica pica*) *Unpublished.* 1997.
28. National Research Council. *Guide for the Care and Use of Laboratory Animals.* National Academy Press: Washington, D. C., 1992; pp 125.
29. Penumarthy, L.; Oehme, F. W. Treatment and prothrombin responses during warfarin tòxicosis in rats and mice. *Toxicology.* 1978, 10, 377-401.
30. Hoogland, J. L. *The Black-tailed Prairie Dog: Social Life of a Burrowing Mammal.* The University of Chicago Press: Chicago, IL, 1995; pp 557.
31. Fagerstone, K. A. In *Vertebrate Pest Control and Management Materials* (Kaukeinen, D. E., Ed). An Evaluation of Visual Counts for Censusing Ground Squirrels; Symposium ASTM Special Technical Publication: Philadelphia, PA, 1983, 14, 239-246.
32. Anonymous. *United States-Canadian Tables of Feeds Composition, 3rd Ed.* National Acadamy Press: Washington, D.C. 1982; pp 866.
33. Donoco, E.; Haft, J. I. *Thrombosis, platelets, anticoagulation, and acetylsalicylic acid,.* Stratton International Medical Book Corporation: New York, NY, 1976; Vol. II, pp 344.
34. Seegars, W. H.; Walz, D. A. *Prothrombin and Other Vitamin K Proteins.* CRC Press, Inc.: Boca Raton, FL, 1986; Vol. I, pp 256.

35. Church, D. C.; Pond, W. G. *Basic Animal Nutrition and Feeding.* John Wiley & Sons: New York, NY, 1988; pp 472.
36. Meehan, A. P. *Rats and Mice: Their Biology and Control.* Rentokil Limited: Grinstead, West Sussex, 1973; pp 383.
37. MacNicoll, A. D.; Gill, J. E. Vitamin K_3 in feedstuffs: antidotal effects in captive anticoagulant-resistant rats and mice. *J. Wldlf. Manage.* 1993, 57, 835-841.
38. Machlin, L. J. *Handbook of Vitamins: Nutritional, Biochemical, and Clinical Aspects.* Marcel Dekker, Inc.: New York, NY, 1984; pp 464.
39. Hadler, M. R.; Buckle, A. P. Forty-five years of anticoagulant rodenticides-past, present and future trends. (J. E. Borrecco & R. E. Marsh, eds.) *Proceed. Vert. Pest Conf.* 1992, 15, 149-155.
40. Hollander, D.; Gitnick, G. L. *Principles and Practices of Gastroenterology and Hepatology.* Elesevier: New York, NY, 1988, pp 1622.
41. MacNicoll, A. D. A review of biochemical mechanisms of warfarin resistance in the house mouse. *Pest. Science.* 1995, 43, 57-59.

Chapter 15

Non-Target Hazard Assessment of Using DRC-1339 Avicide to Manage Blackbirds in Sunflower

John D. Eisemann[1,3], George M. Linz[2], and John J. Johnston[1]

[1]APHIS/WS/National Wildlife Research Center, U.S. Department of Agriculture, 4101 LaPorte Avenue, Fort Collins, CO 80521–2154
[2]National Wildlife Research Center, 2110 Miriam Circle, Suite B, Bismarck, ND 58501

Terrestrial hazard assessments were conducted for the spring blackbird baiting program to protect sunflower crops. Risk Assessment methodology proposed by the Ecological Committee on FIFRA Risk Assessment Methods (ECOFRAM) and the method currently used by the U.S. Environmental Protection Agency ($LD_{50}s/ft^2$) were compared for their predictive strengths and for the ease of adapting the assessment to site specific conditions. While the ECOFRAM and $LD_{50}s/ft^2$ methods identified the same groups of organisms as being at risk, the flexibility of the ECOFRAM methodology allowed more latitude in adapting the assessment to unique behaviors of individual species. These risk assessment approaches indicate that blackbird baiting with DRC-1339 presents acute hazards to select nontarget birds like western meadowlarks and mourning doves but few hazards to most mammals or small granivorous birds like sparrows and finches. However, field experiments indicate that the mitigation measures currently employed in the baiting program, minimize the nontarget hazards.

Sunflower production in the United States is centered in North Dakota and South Dakota. In 1997, the U.S. sunflower harvest was 1.5 million metric tons (1 million hectares planted) with North Dakota and South Dakota accounting for 57% (526,000 hectares planted) and 28% (283,000 hectares planted) of the total harvest, respectively[1].

[3]USDA, National Agricultural Statistics Service.
http://www.nass.usda.gov/nd/cesuna98.htm, search date 9/9/99
http://www.nass.usda.gov/sd/cesuna98.htm, search date 9/9/99

Ripening sunflower seeds provide a highly desirable forage source for a variety of pests. The primary vertebrate pests to sunflower include the red-winged blackbird (*Agelaius phoeniceus*), the common grackle (*Quiscalus quiscula*), and to a lesser extent the yellow-headed blackbird (*Xanthocephalus xanthocephalus*) (*1, 2*). In recent years, damage attributed to these species was estimated at over $5 million per year (*3*). A variety of techniques have been used to manage avian depredation in sunflower fields, including the use of chemical toxicants. DRC-1339 (3-chloro-p-toluidine hydrochloride) is the only lethal toxicant currently registered in the U.S. for managing blackbird damage in sunflower during both the spring and fall migrations. Blackbird damage occurs in late-summer when the sunflower crop is ripening. Because it is difficult to lure blackbirds away from ripening sunflower heads to rice baits on the ground, the efficacy of late-summer baiting is likely to be limited (*4*). This paper focuses on the nontarget hazards associated with the spring blackbird management program when naturally occurring food sources may be limited.

Concern has been raised about the impacts of DRC-1339 to other vertebrates, particularly nontarget birds likely to forage on treated bait sites. The purpose of this paper is to conduct nontarget hazard assessments for DRC-1339 using methodology currently employed by the U.S. Environmental Protection Agency (EPA) and new methodology developed by the EPA-sponsored Ecological Committee on FIFRA Risk Assessment Methods (ECOFRAM) (*5*). We compare the hazard assessments for environmental relevance, and potential for customizing the assessment for site-specific environmental and nontarget species information. Finally, nontarget hazards identified through these assessments are characterized for select species to provide a picture of the potential risk presented by this program.

Environmental Fate and Toxicology of DRC-1339

The environmental fate and toxicology of DRC-1339 have been throughly reviewed (*6, 7*). Reported half-lives of DRC-1339 range from 1-3 days and are highly dependent upon climatic conditions. The half-life in soil under aerobic conditions is approximately 25 hours. The aquatic photolysis half-life is between 6.5 and 41 hours. DRC-1339 is highly soluble in water but does not hydrolyze. High affinity to soil organic matter explains the low soil mobility of DRC-1339.

The acute toxicity database for DRC-1339 is noteworthy, with 46 North American mammals and 8 African bird species tested for acute oral toxicity. As illustrated in Figure 1, laboratory studies, while not definitive, support the possibility that DRC-1339 exhibits a differential toxicity and mode of action among taxonomic families *(8, 9, 10, 11, 12, 13, 14*). Target species such as blackbirds, grackles, starlings and corvids are highly sensitive, with LD_{50}s in the range of 1 to 10 mg/kg. Doves, galliformes, and some passerine species are also acutely sensitive to DRC-1339 (LD_{50}s < 20 mg/kg). Additionally, the only species of owl tested, the common barn owl, (*Tyto alba)*, was found to be sensitive. One species DRC-1339 is nephrotoxic to sensitive species in that it destroys proximal convoluted tubules, resulting in uremia or increased levels of uric acid in the blood. Metabolism studies have shown that as much as 90% of a dose

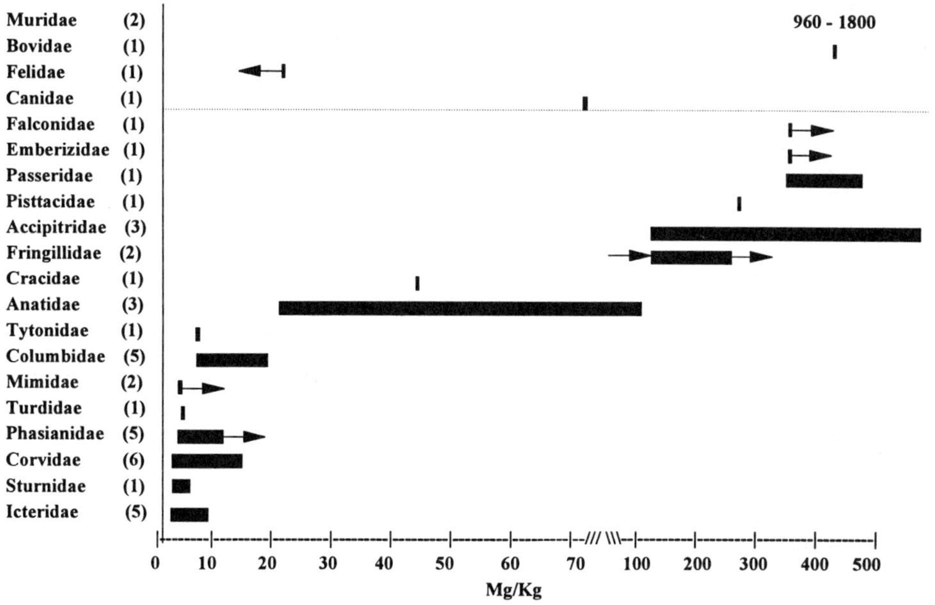

Figure 1. Acute toxicity data for 45 species of North American vertebrates, representing 20 taxonomic families. The range of LD_{50}s for the family Muridae is off the scale at 960 to 180 mg/kg. Arrows represent greater than (→) and less than (←). Numbers in parenthesis are the numbers of species tested in each taxonomic group. The dotted line divides the data for mammals and birds.

administered to birds is excreted in the form of parent com pound or metabolite within 30 minutes (*15, 16, 17, 18*). Sparrows, finches, raptors and most mammals appear to be relatively insensitive to DRC-1339, with LD_{50}s greater than 100 mg/kg. Mammals and possibly non-sensitive avian species do not exhibit kidney necrosis, excrete acetylated metabolites in the urine, or show increased levels of methemoglobin in the blood. Non-sensitive animals probably succumb to CNS depression and respiratory failure (*19, 20, 21, 22*).

DRC-1339 Use Practices

DRC-1339- treated brown rice is formulated at a concentration of 2% and diluted with untreated rice the day of application at a ratio of 1:25. This mixture is broadcast in swathes 6.5 to 17 m wide with a seed-spreader mounted on an all-terrain vehicle at a rate of 12 to 23kg/ha. Treated plots are about 0.8 ha and are located near roads under roost-to-field flight paths. Plots are pre-baited with untreated rice for a period sufficient to habituate foraging blackbirds to bait sites and monitor nontarget activity. Up to 4 subsequent applications can be made after 75% of the previous application has been consumed or 10mm of precipitation has fallen (EPA Registration Numbers. 56228-30, and SD-980005). Decoy birds, housed at the bait site in large cages, are used to draw larger numbers of blackbirds to bait sites.

Methods

Current EPA Methodology

The first assessment method follows the current standard used by the U.S. EPA to assess the risk of granular and bait pesticide products, $LD_{50}s/ft^2$ (*23*). This method relates the amount of pesticide in a given area of field to the LD_{50} of the most sensitive species or species of interest. The resulting risk quotient (RQ) is compared to the following established Levels of Concern (LOCs):

If the Acute RQ is:	LOC	Presumption
$\geq$	0.5	Acute high risk to all species
$\geq$	0.2	Risk that may be mitigated through restricted use
$\geq$	0.1	Acute risk to endangered species

The RQ ($LD_{50}s/ft^{2)}$ is calculated by the following equation where mg ai/ft^2 = (lb ai/acre (453,590 mg/lb) ÷ 43560 ft^2/acre.

$$RQ = mg\ ai/ft^2 \div (LD_{50})(Body\ weight\ in\ kg)$$

ECOFRAM Methodology

Screening Level Assessment

Hazard assessment procedures also were conducted following methods outlined by the ECOFRAM workgroup. In the ECOFRAM draft document, exposure assessment focuses on a dietary dose equation modified from that proposed by Pastorok et al. (*24*). The proposed exposure equation allows the determination of total dose by providing a means for summing the total pesticide intake for each contaminated food item and allowing for behaviors unique to individual species (Table I). In the screening assessment, we assumed that 100% of an animal's daily diet consists of brown rice obtained at the bait site and the entire daily food requirement was consumed at one time. Additionally, Avoidance (AV), Percent of Time in the treated field (PT) and Fresh/Dry Ratio (FDR) are assumed to have no impact and have been eliminated from the equation.

The basic screening assessment using the dietary dose equation is generic in that standard avian and mammalian body weights and food intake rates are used in place of species-specific data. The toxicity reference value, which serves as the denominator of the risk quotient calculation, is the concentration considered hazardous to only five

Table I. Dietary dose equation as proposed by the U.S. EPA ECOFRAM workgroup

	$\mathbf{DD = (FIR * C_i * PD_i} * AV_{Ci} * PT_i * FDR_i\mathbf{) / BW}$
DD	Daily Dose (mg DRC-1339 / kg BW / day)
FIR	Food intake rate (g / day)
C_i	Concentration of DRC-1339 in food type i (mg / kg wet weight)
PD_i	Proportion of food type i in the diet
AV_{Ci}	*Avoidance Factor of food type i at DRC-1339 concentration C*
PT_i	*Proportion of food type i obtained in the treated field*
FDR_i	*Fresh to dry weight ratio for food type i*
BW	Body weight (g)

Note: *AV*, *PT* and *FDR* are italicized because they are assumed to equal 1.

percent of the species utilizing the affected environment. This consists of the 5th percentile of the log-normal distribution of the LD_{50} values. Additionally, to account for the uncertainty around this estimate, ECOFRAM has recommended using the one-sided 95% confidence around the 5th percentile. The ECOFRAM document proposed two methods for determining the 5th percentile when only a few toxicity values are available. Both methods incorporate extrapolation factors. Because there are 40 avian median lethal dose estimates for DRC-1339, this assessment calculated the 5th percentile directly instead of using an extrapolation factor. Additionally, no extrapolation factor was used to calculate the 5th percentile for mammals.

When calculating the 5th percentile of species sensitivity, we assumed the toxicity data for both birds and mammals have a log normal distribution. We also assumed a mean LD_{50} value when more than one LD_{50} was reported for a species, and LD_{50} estimates reported as > or < were eliminated from the calculation. DRC-1339 concentrations were assumed to be homogenous at a concentration of 769 mg/kg rice. No allowance was given for the probability that animals could select between treated vs. untreated rice grains. An animal's food intake rate (FIR) was calculated for passerine birds, non-passerine birds and rodents using Nagy's weight-based allometric equations (*25*). Finally, because the dietary dose equation accounts for some of the uncertainty in the assessment, the RQ will be compared to a LOC of 1.0

First Level of Refinement

The first level of refinement utilized the same basic data used in the screening level assessment. However, actual body weights obtained from Dunning (*26*) and acute toxicity information for species found in the sunflower growing region were substituted for generic body weights and the 5th percentile. Mammalian body weights were obtained

Table II. U. S. EPA standard risk assessment results for granular bait

Species	LD_{50} *(mg/kg)*	*Body Wt. (g)*	$LD_{50}s / ft^2$ *(RQ)*
Red-winged blackbird	2.4	53	3.15 ***
Northern bobwhite	2.6	178	0.87 ***
Mallard	100	1082	< 0.01
Lab rat	326	300	< 0.01
Lab mouse	960	30	0.01

*** Exceeds the Acute High Risk LOC

from the DRC-1339 Reregistration Eligibility Decision (*7*). Because the dietary dose equation accounts for some of the uncertainty inherent in the assessment, the RQs generated by this method were compared to a LOC of 1.0.

Second Level of Refinement

The second level of refinement moved the risk assessment from deterministic risk quotients to probabilistic quotients by expanding parameter estimates from point estimates to distributions by performing Monte Carlo sampling of equation parameters for which distributions of data were available. This was conducted using the risk analysis software @Risk (*27*). Food intake rates (FIR) were still based upon Nagy's allometric equations, however, body weight estimates were randomly selected from either truncated normal or normal distributions. Given that an average rice grain weighs 20 mg, the total number of grains per day was determined given FIR for a species. Percent of the diet that is treated (PD) was set as the ratio of 1 treated to 25 untreated rice grains or 0.038. To determine the total number of treated rice grains consumed during a day, PD was multiplied by values drawn randomly from a binomial distribution based upon the total number of grains consumed per day and the probability of selecting a treated grain. Since the rice bait is formulated to yield 0.4 mg per treated rice grain, the Daily Dose (DD) was determined by multiplying the total number of treated grains per day by 0.4 mg. Risk quotients were then calculated by dividing DD by the LD_{50} which had been normalized for the body weight of any given animal (mg DRC-1339/animal).

Results

Results of the current U. S. EPA LD_{50}/ft^2 screening assessment, indicate that there is concern for some species of birds with respect to the DRC-1339 blackbird baiting program (Table II). High risk is predicted for the northern bobwhite (*Colinus virginianus*), and the red-winged blackbird (*Agelaius phoeniceus)*, the target species. However, the risk to the mallard (*Anas platyrhynchos*) and mammals is so low the LOC for endangered species (0.1) is not triggered.

Table III. Quotient based screening assessment using generic species data and the median concentration hazardous to 5% of species (5th percentile).

Species	*FIR (g/day)*	*DD (mg/day)*	*5th percentile (mg/kg)*	*Risk Quotient*
1000g Bird	53.9	41.5	0.62	67 *
150g Bird	12.9	66.5	0.62	107 *
30g Bird	7.1	183.8	0.62	296 *
300g Mammal	15.4	39.7	68.6	0.58
30g Mammal	4.2	108.4	68.6	1.56 *

Note: DRC-1339 concentration on rice is homogenous at 769 mg/kg.
* Exceeds the Acute Risk LOC of 1.0.

The screening assessment based upon the ECOFRAM dietary dose equation indicate significant risk to all classes of animals, except larger mammals. (Table III). The 5th percentile for birds and mammals is 0.62 and 68.6 mg/kg, respectively. Clearly, the results of this screen indicate significant risk to all classes and weights of animals at the 5th percentile of species protection level. These results indicate further refinement of the assessment is necessary to determine what environmental impacts are most likely and what types of mitigation measures should be considered to reduce those impacts to acceptable levels.

The second level of refinement in this assessment uses point estimate data for model inputs but incorporats site-specific information to represent species most likely to be found at the application site (Table IV). The surrogate species used in Table IV are not necessarily representative of similar sensitivity to DRC-1339, but are more representative of body weight and dietary parameters. For example, the red-winged blackbird is used as a surrogate for other Icterids such as the western meadowlark (*Sturnella neglecta*). It is not necessarily representative of the toxicological sensitivity of all Icterids. The sensitivity of the meadowlark is unknown and it cannot be assumed to be the same as the red-winged blackbird. However, given the large data set for the acute toxicity of DRC-1339, and the narrow range of sensitivities among the five Icterids tested, it in not unreasonable to assume the red-winged blackbird is representative of all Icterids. This is not the case for the mallard which is used to represent the body size and foraging patterns of other waterfowl. Because, the range of sensitivity to DRC-1339 for waterfowl (Family Anatidae) ranges from 20 to 100 mg/kg, the uncertainty is to large to say the mallard is representative of the sensitivity of all waterfowl. The house sparrow (*Passer domesticus*) is considered a surrogate for other small granivorous species such as sparrows and finches. While the magnitude of the predicted hazard is significantly lower, the basic trend is the same, with acute high risk

Table IV. Refined deterministic assessment using species specific data. DRC-1339 concentration on rice is considered homogenous at 769 mg/kg rice

Species	*Weight (g)*	*FIR (g/day)*	*DD (mg/day)*	*LD_{50} (mg/kg)*	*Risk Quotient*
Red-winged blackbird	53	11.6	168.7	2.4	70.3 *
Northern bobwhite	178	14.8	63.7	2.6	24.5 *
Ring-necked pheasant	1135	59.2	40.1	10	4.0 *
Mourning dove	119	10.9	70.4	5.4	13.0 *
House sparrow	28	6.8	185.7	316	0.59
Mallard	1082	54	40.6	103	0.39
Lab rat	300	15.5	39.7	915	0.04
Lab mouse	30	4.2	108.4	960	0.11
Deer mouse	20	3.4	129.4	1800	0.07

* Exceeds the acute risk LOC of 1.0

anticipated for most species of birds and lower hazard to mammals. Compared to the RQs calculated by $LD_{50}s/ft^2$ methodology, hazard predictions for the birds and mammals have increased and now raise concern for restricted use and endangered species, respectively.

The results of the second level of refinement which employed probabilistic methods (Table V) yield roughly the same risk quotient patterns as those shown in the previous table. Significant hazard still is indicated for the red-winged blackbird, northern bobwhite, ring-necked pheasant (*Phasianus colchicus*) and mourning dove (*Zenaida macroura*). Little hazard is indicated for the house sparrow, mallard and mammals. The advantage of conducting a probabilistic assessment is that a distribution of hazard quotients is generated and the probability of a risk quotient exceeding the LOC can be determined by inspecting this distribution. For example, the house sparrow risk quotient is 0.6 with a range from 0 to 1.3. The probability a risk quotient exceeds 1.0 for the house sparrow is <5% (95 percentile RQ = 0.85).

Discussion

Model Comparison

We conducted hazard assessments using two methods: the current methodology used by the EPA based upon $LD_{50}s/ft^2$ and methodology proposed by the EPA ECOFRAM workgroup. Both assessments focused on the primary hazards associated with birds and

Table V. Refined risk assessment using species specific data and Monte Carlo sampling of parameter distributions

Species	*BW (g)* $\bar{x} \pm SD$ *(range)*	LD_{50} *(mg/kg)*	*Rice Grains / Day* $\bar{x} \pm SD$ *(range)*	*DD (mg/day)* $\bar{x} \pm SD$ *(range)*	*RQ* $\bar{x} \pm SD$ *(range)*
Red-winged blackbird	53 ± 4.4 (29, 81)	2.4	581 ± 40.8 (430, 710)	8.8 ± 2.00 (2.4, 16)	69 ± 15.0 (23, 119)
Northern bobwhite	178 ± 19.3	2.6 [d]	735 ± 59.9 (540, 925)	11.2 ± 2.26 (4, 19)	24 ± 4.5 (9, 43)
Mourning dove	119 ± 1.9	5.4	544 ± 6.6 (521, 565)	8.3 ± 1.78 (2.4, 15.2)	13 ± 2.8 (4, 23)
Ring-necked pheasant	1135 ± 136 [a] (953, 1861)	10	3015 ± 228.7 (2601, 4019)	45.9 ± 5.38 (29, 65)	4 ± 0.4 (3, 6)
House sparrow	28 ± 2.2 (20, 35)	316	338 ± 28.8 (256, 401)	5.2 ± 1.48 (0, 10.4)	0.6 ± 0.16 (0, 1.3)
Mallard	1082 ± 129 (720, 1580)	100 [e]	2854 ± 2498.6 (2121, 3653)	43.4 ± 5.52 (24, 65)	0.4 ± 0.04 (0.27, 0.59)
Lab rat	200 ± 7.8 [b] (183, 220)	915	616 ± 12.6 (586, 650)	9.3 ± 1.94 (2.4, 16)	0.14 ± 0.030 (0.02, 0.24)
Lab mouse	21 ± 8 [b] (16, 26)	960	172 ± 13.3 (148, 195)	2.6 ± 1.04 (0, 6.8)	0.13 ± 0.052 (0, 0.04)
Deer mouse	20 ± 3.47 [c]	1800	167 ± 16.9 (100, 219)	2.6 ± 1.02 (0, 6.8)	0.07 ± 0.029 (0, 0.20)

[a] Assumed to be the same mean to standard deviation ratio as mallards [b] *(28)*; [c] *(29)*; [d] *(10)*; [e] *(11)*

Table VI. Birds observed on or in the vicinity of DRC-1339 bait sites in North Dakota and South Dakota

Species	*Frequency of observation (%)*
American tree sparrow, western meadowlark;	21-25
American coot, killdeer, Lapland longspur	16-20
Canada Goose, mallard, vesper sparrow, song sparrow	6-10
American robin, ring-necked pheasant, horned lark mourning dove, rock dove	1-5

Note: Percentages are representative of the total number of counting periods during which a species was observed

mammals consuming DRC-1339-treated rice. The assessments assumed acute exposure resulting from an animal's entire daily dietary requirement consisting of brown rice obtained from the treated bait plots and did not consider other routes of exposure.

Both assessment methods identified the same basic patterns among species: hazard to mammals and mallards were low, hazards to other bird species were significant but variable. Both methods correctly identified the red-winged blackbird, a target species, as a species at high risk. While simple to use, the $LD_{50}s/ft^2$ method is extremely limited in the amount of refinement that can be done with the model. On the other hand, the dietary dose equation provides a single equation for exposure that can be easily refined to include site- and species-specific information, thus allowing the assessment to be easily adapted to address specific objectives. The primary limitation with the dietary dose equation is the availability of data.

Neither assessment method adequately predicts potential hazards for compounds like DRC-1339 that exhibit two modes of action, do not bioaccumulate, and are rapidly metabolized or eliminated from the body. To address these issues, factors for depuration and elimination rates would need to be included in the model. The hazard associated with single feeding (gorge feeding) bouts can be estimated using the dietary equation which is useful for fast acting compounds like organophosphates or carbamates. Without specific information on dietary intake, the assessment is conservative in that the total daily intake is related to the LD_{50} and no allowance is made for subacute exposure extended throughout a day. The assessment can be used for more than a screen if specific foraging information is input to the equation for FIR.

The conservative nature of this assessment can be demonstrated by substituting estimates of the daily food intake rate other than that recommended by Nagy. Kendeigh (*30*) calculated existence metabolism requirements (kcal/bird-day) regression equations for both passerine and non-passerine birds. The majority of the 13 passerine species were sparrow sized granivores. The five non-passerine species included 3 pheasant species, the Canada goose and mallard. Using Kendeigh's allometric equations and a value of 360 cal/100g uncooked brown rice (*31*) daily food intake estimates in terms of number of rice grains per day are approximately 50% lower than those estimated by Nagy's method and shown in Table V. Risk quotients based upon this estimate would be approximately 50% lower and still show the same trend; no concern for house sparrows and mallards, while RQs for the red-winged blackbird, northern bobwhite, mourning dove, and ring-necked pheasant still exceed the LOC of 1.0.

Field data from Cummings et al. (*32*) also can be used to illustrate the conservative nature of the assessment by illustrating the effect FIR has on the outcome. Cummings collected 118 red-winged blackbirds as they left bait plots in Louisiana. He reported that the median number of rice grains in the GI tract of these birds was 28 and ranged from 0 to 83. Considering the red-winged blackbirds contained a maximum of 83 and a median of 20 grains of rice, the RQ for a single feeding bout would be reduced to approximately 10 or 3.3 respectively. Single foraging bout data for other species of concern would undoubtedly lower their respective risk quotients.

Risk Characterization for Nontarget Species

Concern has been expressed for all nontarget species in the treatment area. However, the major emphasis has been on the western meadowlark and ring-necked pheasant, and sparrows. To identify the species most at risk in the northern Great Plains, Knutsen (*33*) conducted avian censuses during 2 consecutive years in cornfields during spring blackbird baiting. Employing point-count and video monitoring census methodology, Knutsen observed 774 individual birds, representing 31 species, either at or in the area of bait plots. Twenty-six of these species were nontarget species. Fourteen species observed during at least one percent of the observation periods (Table VI).

Two years prior to Knutsen, Kenyon (*34*) conducted avian surveys around bait sites in South Dakota. Fifty-seven, 15-minute point-count surveys were done at sites baited with untreated brown rice. No nontarget birds were observed during (52%) of the point-counts. A total of 476 nontarget birds (20 species) were observed in the remaining (48%) of the observation periods. Ring-necked pheasants, western meadowlarks, and waterfowl (mallard, Canada goose, and green-winged teal) were observed in less than 10% of the point-counts. Insectivorous birds (American robin, northern flicker, downy woodpecker, and killdeer) as a group were observed during nearly 30% of the point-counts. Granivores (American tree sparrow, song sparrow, clay-colored sparrow, dark-eyed junco, horned lark, and mourning dove) as a group were observed in approximately (21%) of the point-counts.

Comparison of the species list derived from these two studies to the refined risk assessment results identifies species of highest concern. While not conclusive , it appears that there is some basis for assuming a taxonomic-based grouping for species sensitivity to DRC-1339. The western meadowlark belongs to the family Icteridae. All species tested in this family are highly sensitive to DRC-1339 (LD_{50} $\leq$10 mg/kg). Because they are roughly the same size as a blackbird and are assumed to be equally sensitive to DRC-1339, risk quotients and the probability of survival would likely be similar. During the spring, the meadowlark's diet is 80% invertebrates and 20% seeds gleaned from the ground (*35*). However, when invertebrates are scarce, plant material can comprise a substantial portion of their diet.

The mean RQ for the ring-necked pheasant is 4, therefore, acute risk is indicated. In the dietary dose model, pheasants are predicted to consume more than 3000 rice grains (115 treated grains) per day. Because of the sheer volume of seeds a large bird can eat, it is likely it will eat more treated rice grains than smaller birds. Given the relatively high sensitivity of the ring-necked pheasant to DRC-1339 (10 mg/kg) an adult bird need eat only 28 treated grains, or 750 total grains, at one feeding to ingest a dose equal to the LD_{50}. At 20 mg per grain of rice, 750 grains of rice equals 3.75 g of rice. While reasonable to assume a ring-necked pheasant can consume 3.75 g of food during one feeding bout, field observations show that pheasants move rapidly through baited plots and are unlikely to take a sufficient rice grains to cause acute effects (*33*). Further, Avery et al. (*36*) speculated that there may be some learned aversion to treated bait sites by demonstrating that in large enclosure trials at least some pheasants do not return to food sites where they previously ate DRC-1339-treated rice. Additionally, they found female pheasants preferred cracked corn and sorghum over brown rice.

The northern bobwhite, with a LD_{50} of 2.4 mg/kg, is sensitive to DRC-1339. The risk quotients for this species is approximately 24, indicating a high potential for acute effects if they consume treated rice. The northern bobwhite is not found in the northern Great Plains, however, the results for the bobwhite could be considered an indicator for other Phasianids for which there is no acute toxicity data available. During the four years Kuntsen (*33*) and Kenyon (*34*) conducted observations, excluding the ring-necked pheasant, only one other Phasianid, a single gray partridge (*Perdix perdix*), was observed in the baited sites.

Sparrows and longspurs are frequently observed at or around bait sites. Despite this, the risk to these species appears to be low, given the relative insensitivity of the Emberizidae and Fringillidae (LD_{50}s 100 to 400 mg/kg). Cummings et al. (*32*) examined the risk to sparrows using three methods. Thirteen savannah sparrows were collected as they left bait sites baited with untreated brown rice. Upon inspection of the gastrointestinal tract, only one sparrow contained rice and only one grain was found in this bird. In a separate effort, 20 sparrows (9 savanna, 8 white-crowned, and 3 field sparrows) were live trapped at bait sites baited with DRC-1339-treated brown rice. Only two savannah sparrows died during the 10-day holding period. These deaths were attributed to capture related injuries. Finally, 72 sparrows (54 savanna, 9 white-crowned, 3 song, and 3 field sparrows) were live trapped in untreated fields. After acclimating to cages, they were denied food overnight (12 hours) and provided DRC-1339 treated brown rice for a 12-hour period. No sparrows died during the 5-day post-treatment observation period.

The studies reported by Cummings et al. (*32*) were conducted during spring baiting in Louisiana. The availability of alternative foods and other environmental factors are undoubtedly different from those found during spring baiting in North Dakota and South Dakota. These differences could certainly affect the dietary preferences of sparrows. However, the results do provide evidence that brown rice treated with DRC-1339 is not a preferred food for sparrows, even under the severe conditions of a single choice test following a starvation period. These results support the dietary dose equations of low risk and the probability that risk to sparrows under operational use is low.

As reported by Knutsen (*33*), mourning doves (*Zenaida macroura*) and rock doves (*Columba liva*) were found on 1-5% of all observed bait sites. As indicated in Figure 1, the LD_{50}s of the 5 species tested in the family Columbidae ranged from 8 mg/kg to 20 mg/kg. Under the conditions of the dietary dose equation, significant risk is predicted for the mourning dove and Columbids. The food intake model predicts an average daily intake of 544 grains of rice, with an average hazard quotient of 13. Consumption of 42 rice grains (2 treated grains) during a single feeding bout would be approximately equivalent to the mourning dove LD_{50}. Thus, doves feeding on the treated rice bait are at high risk of acute exposure.

The susceptibility of the horned lark (*Eremophila alpestris*), a predominantly granivorous species (*35*), is unknown. No toxicity data are available for this species or other members of their taxonomic family. If one is conservative and assumes these birds are as sensitive as the most sensitive species or even as sensitive as the calculated 5th percentile, significant risk is indicated. Although the American robin (*Turdus migratorius*) is sensitive to DRC-1339 (LD_{50} = 3.2 mg/kg), the risk to this species may be low because they feed primarily on fruit and invertebrates during the spring and they were present on less than 5% of the bait sites observed by Knutsen. The killdeer (*Charadrius vociferus*) was found in 15% to 20% of Knutsen's observation periods. The sensitivity of the killdeer to DRC-1339 is unknown, however, the risk to this species may be low because it's diet is predominantly animal material (*35*) and the probability it will consume rice is low.

Finally, both risk assessment methods identified mammals as at little risk from the baiting program. With the exception of cats, the mammals tested for acute toxicity are not sensitive to the effects of DRC-1339. Cats are unlikely to eat uncooked rice so the risk of primary exposure is low.

The spring blackbird baiting program presents a potential hazard to some nontarget species, if they eat the DRC-1339-treated rice bait. However, to reduce nontarget hazards, mitigation measures have been incorporated into the baiting operations. First, the spring baiting program is conducted prior to the arrival of most spring migrants *(Linz, G.M. USDA National Wildlife Research Center, Bismarck, ND, pers. comm. 1999*). Prebaiting is conducted to habituate target birds to feeding at a particular site and to feeding on brown rice. Prebaiting also provides an opportunity to observe nontarget activity on the bait site and to change bait locations if necessary. Brown rice is used exclusively as the bait material. Blackbirds readily accept brown rice while some nontarget species prefer other more familiar foods (*37*). The bait is always diluted at a ratio of at least 1 part treated rice to 25 parts untreated rice, which significantly reduces the probability a bird will pick up a treated grain. This is especially important for small birds which consume only a few grains during a foraging bout. Finally, caged decoy blackbirds often attract large flocks of blackbirds which may deter nontarget species from feeding on bait sites *(Allen, A.E. USDA Wildlife Services, Crowley, LA, pers. comm. 1999*).

Acknowledgments

We would like to thank David Fischer for his help with developing the probabilistic model, and Michael Avery, John Cummings, Robert Troup, Ed Schafer, Kathy Fagerstone, Mark Tobin, Alain Baril and Brett Petersen for their technical review.

Literature Cited

1. Homan, H.J.; G.M. Linz; W.J. Bleier, *Am. Mid. Nat.* 1994 131:381-385.
2. Twedt, D.J., W.J. Bleier, and G.M. Linz, 1991, *Condor.* 93:975-986.
3. Lilleboe, D. *The Sunflower.* 1996, 22:10-13.
4. Linz, G.M.; Bergman, D. L, *Crop Protection.* 1996, 15:307-310.
5. U.S. Environmental Protection Agency. Ecological Committee on FIFRA Risk Assessment Methods Terrestrial Draft Report. 1999, May 10, 1999.
6. U.S. Department of Agriculture, Animal Damage Control: Final Environmental Impact Statement. 1994. Vol. 3, pp194-202.
7. U.S. Environmental Protection Agency. Reregistration Eligibility Decision (RED): Starlicide (3-chloro-p-toluidine hydrochloride). 1995, EPA-738-R-96003.
8. Schafer, E.W.; W.A. Bowles; J. Hurlbut, 1983, *Arch. Environ. Contam. Toxicol.* 12:355-382.
9. DeCino, T.J.; D.J. Cunningham,; E.W. Schafer, *J. Wildl. Manage.* 1966. 30:249-253.
10. Fletcher, D.; C. Pedersen, Final Report. Lab Project No. BLAL # 89 DD 71; 1990, Bio-Life Associates, Ltd., Neillsville, WI.
11. Fletcher, D.; C. Pedersen, Final Report. Lab Project No. BLAL # 89 DD 135; 1990, Bio-Life Associates, Ltd., Neillsville, WI.
12. Peoples, S.A. Progress Report on Starling Control to the Denver Wildlife Research Center, Univ. Calif.-Davis, CA, 1965.
13. Schafer, E.W.; W.A. Bowles. *Arch. Environ. Contam. Toxicol.* 1985, 14:111-129.
14. Schafer, E.W. *Toxicol. Appl. Pharm.* 1972, 21:315-330.
15. Apostolou, A. *PhD. Dissertation.* University of California at Davis, 1969.
16. Apostolou, A.; Peoples, S. A. *Toxicol. Appl. Pharmacol.* 1971, 18:517-521.
17. Mull, R.L. *Ph.D. Dissertation.* University of California at Davis, 1971.
18. Giri, S.N.; D.E. Gribble; S.A. Peoples. 1976, *Fed. Proc.* 1976. 35:328.
19. Borison, H.L.; S.R. Snow; D.S. Longnecker; R.P. Smith. *1975. Toxicol. Appl. Pharma.* 31:403-412.
20. Felenstein, W.C.; R.P. Smith; R.E. Gossein, 1974. *Toxicol. Appl. Pharm.* 28:110-125.
21. Palmore, W.P. *Florida Veterinary Journal.* 1978.
22. Giri, S.N.; D.M. Siegel; S.A. Peoples. 1978, *Toxicol.* 11:153-165.
23. Urban, D.J.; Cook, N.J. Standard Evaluation Procedure for Ecological Risk Assessment. 1986, EPA/540/09-86/167.

24. Pastorok, R.A.; M.K. Butcher; R.D. Nelson. Human and Ecological Risk Assessment. 1996, 2:444-480.
25. Nagy, K.A. 1987, 57:111-128
26. Dunning, J.B. CRC Handbook of Avian Body Mass; CRC Press, Boca Raton, FL, 1993.
27. Palisade Corp. @RISK: Advanced risk analysis for spreadsheets. Version 3.5.2. 1997, Newfield, NY.
28. McCann, G.M.; G.H. Matschke. Final Report, QA-506, National Wildlife Research Center, Fort Collins, CO.
29. U.S. Environmental Protection Agency. EPA/600/R-93/187a, 1993.
30. Kendeigh, S.C. *The Condor*. 72:60-65.
31. Altman, P.L. and D.S. Dittmer. Metabolism; Fed. of Amer. Soc. for Exper. Biol. Bethesda, MD, 1968.
32. Cummings, J.L.; J.F. Glahn; E.A. Wilson; J.E. Davis; D.L. Bergman; G.A. Harper. Final Report. Bird Section Research Report No. 481, Denver Wildlife Research Center, Denver, CO, 1992.
33. Knutsen, G.A. M.S. Thesis. North Dakota State Univ. Fargo, ND, 1998.
34. Kenyon, M.J. MS Thesis. North Dakota State Univ. Fargo, ND, 1996.
35. Martin, A.C.; H.S. Zim; A.L. Nelson, American Wildlife and Plants: A guide to wildlife food habits. Dover Pub., New York, 1951.
36. Avery, M. L.; M. J. Kenyon; G. M. Linz; D. L. Bergman; D. G. Decker; J. S. Humphrey. *J. Wildl. Manage.* 1998, 62:388-394.
37. Linz, G.M.; H.J. Homan; R.L. Wimberly; G.A. Knutsen. In Pesticides & Wildlife, Johnston, J.J. Ed.; ACS Symposium Series, American Chemical Society: Washington, DC, in press.

Chapter 16

Ecotoxicological Risks of Potential Toxicants for Brown Tree Snake Control on Guam

John J. Johnston, Richard E. Mauldin, Pete J. Savarie, Joseph E. Brooks, and Thomas M. Primus

APHIS/WS/National Wildlife Research Center, U.S. Department of Agriculture, 4101 LaPorte Avenue, Fort Collins, CO 80521–2154

INTRODUCTION

The brown tree snake (Boiga irregularis) is a nocturnal, arboreal, rear-fanged, mildly venomous, colubrid snake which can reach lengths of up to 2.3 m and weigh as much as 2 kg(1). Originally, the species' range included the northern and eastern coasts of Australia, Papua New Guinea and nearby islands (2). It is believed that sometime in the 1950's, that snakes were inadvertently transported from New Guinea to Guam, where they proliferated (3). By the mid-1960's, marked decreases in Guam's bird life were observed. By the mid-1980's, snake densities were estimated at 50 to 100/hectare (13,000 to 26,000/sq mile), higher densities than those recorded for any other snake (3,4).

Brown tree snakes are dietary generalists, being observed to eat chicken bones, cooked spare ribs, lizards, birds, rodents, domestic fowl hatchlings, puppies, piglets, rabbits (in hutches), and pet birds (in cages inside homes) (1,5). Human infants have also been attacked, resulting in very serious bites (6,7). Snake predation has resulted in the extirpation or severe reduction in the populations of virtually all Guam's avifauna and has essentially resulted in the extinction of four endemic species/subspecies: 1) Bridled white-eye *(Zosterops conspicillatus conspicillatus)*, 2) Guam flycatcher *(Myiagra freycineti)*, 3) Micronesian kingfisher *(Halcyon cinnamomina cinnamomina)*, and 4) Rufous fantail *(Rhipidura rufifrons uraniae) (8,9)*. The Mariana crow *(Corvus kubaryi)* has also been severely reduced, with 8 birds remaining on Guam, and an additional 300 to 600 remaining on the nearby island of Rota (10). The crow is listed as an endangered species and, as a scavenger that might consume lethally-dosed snake carcasses resulting from chemical toxicant control operations, plays a significant role in secondary hazard assessments of the use of such toxicants.

In addition to the ecological and agricultural damage, snakes crawl along power lines in search of prey. This activity frequently results in short circuits leading to extensive damage to power transmission equipment, subsequent power blackouts to human population centers, and millions of dollars in economic losses (7).

The large military presence on Guam and shipment of associated cargo coupled with the high snake densities increase the likelihood of dispersal of the snake to other locations where the whole damage scenario might be repeated. Individual snakes have been observed on other islands in the Marianas (Saipan, Tinian) and other islands such as Kwajalein, Wake, Diego Garcia, and Hawaii (11,7). One individual was found in a cargo container in Corpus Christi, TX which had been shipped from Guam some six months earlier (12). The United States Department of Agriculture Wildlife Services personnel on Guam utilize a variety of measures to prevent snake accidental snake relocations, but the only long term solution is the reduction or eradication of the brown tree snake population on Guam. As part of a multiagency snake control program funded by the U.S. Department of Defense's Legacy Program scientists from the NWRC were asked to evaluate traps, lures and chemical toxicants. Several candidate compounds with demonstrated toxicity to poikilothermous vertebrates were screened for effectiveness. Among the most effective were: 1) pyrethrum, an extract of *Chrysanthemum* flowers containing a family of six pyrethrins that is registered with the U.S. EPA for insecticidal use, 2) rotenone, a natural product extracted primarily from roots of the tropical plant genus *Derris*, registered as both a piscicide and insecticide, and 3) propoxur, a carbamate insecticide.

EXPERIMENTAL

Toxicity Testing

On Guam, the acute toxicity of toxicants to brown tree snakes was evaluated by oral gavage, oral dosing in bait, and dermal application. For oral gavage, the toxicants were dissolved in propylene glycol or ethanol and introduced directly into the entrance of the snake's esophagus by means of a ball tipped feeding needle (13). As future wide scale snake population reduction might utilize a baiting program, snake preferences for various potential bait matrices were evaluated (14). Of 21 bait matrices tested, geckos and mice, processed meat (SPAM) and juvenile quail were well accepted. These latter matrices were subsequently combined with each toxicant at varying concentrations and offered to snakes. Lastly, toxicants dissolved in ethanol were applied to the dorsal surface of restrained snakes using a syringe fitted with a ball tipped needle. In all procedures, five snakes were used to test each toxicant concentration. During experimentation, snakes were housed in plastic cages in racks kept outdoors under shadecloth. Snakes which received non-lethal toxicant doses were euthanized using halothane (14). All snakes were wrapped in aluminum foil and frozen for subsequent residue determination.

Analytical Chemistry

Chemical analyses were required to generate the data required for risk assessments. Residues of the toxicants in snakes following dosing were determined. To determine the potential secondary hazards associated with the use of these potential snake toxicants, residue methods were developed to quantify rotenone, pyrethrins and propoxur in whole body brown tree snakes. For all three methods, frozen (-20 °C) snakes were cut into 2 inch pieces and placed into a cylindrical stainless steel container containing liquid nitrogen. The frozen snake was then shattered into a homogeneous powder with a steel bar (15).

To quantify propoxur residues, a silica gel matrix solid phase dispersion method was developed to clean up and concentrate the residues in 2 g portions of homogenized tissue. Extracts were analyzed by reversed phase high performance liquid chromatography (HPLC) with fluorescence detection (excitation = 225 nm, emission =305 nm). The mean recovery and standard deviation (std dev) were 86.7 and 7.8 percent, respectively. The method limit of detection (MLOD) was 9 parts per billion (ppb) (16).

To quantify pyrethrins, a liquid extraction followed by C8 solid phase extraction clean up was developed to clean up and concentrate pyrethrins in 6 g portions of homogenized tissue. Extracts were analyzed by gas chromatography (GC)/electron capture detection (ECD). Mean recovery was 70.8% with a std dev of 5.7%. MLOD was 6.5 ng/g (17)

To quantify rotenone residues, a silica/florisil solid phase extraction method was developed to clean up and concentrate rotenone residues in 2 g portions of homogenized tissue. Rotenone residues in the extracts were separated by HPLC and quantified by ultraviolet detection at 295 nm. Mean recovery was 84.7% with a std dev of 7.4%. MLOD was 0.012 μg/g (18).

RESULTS

Toxicity testing

Pyrethrins

Snakes were gavaged with pyrethrum solutions in ethanol and propylene glycol (Table I). Doses ranged from 5 to 40 mg/kg. Oral gavage with pyrethrins yielded 100% mortality only at the highest tested dose of 40 mg/kg. This dose is equivalent to a dose of approximately 0.25 to 8 mg active ingredients per snake for average snakes ranging in weight from 50 to 200 g. No mortality was found in controls given ethanol or propylene glycol only.

Table I. Mortality Following Gavage with Pyrethrum

	Dose (mg/kg)			
Carrier	*5*	*10*	*20*	*40*
Ethanol	ND*	1/5**	4/5	5/5
Propylene glycol	0/5	4/5	3/5	5/5

*Not determined
** #Dead/ #tested
Source: Reference 13.

Incorporation of pyrethrum into SPAM and quail chick bait matrices greatly reduced toxicity (Table II). For example, when given in a treated bait, only 50% mortality was achieved at the highest dose of 40 mg/bait (40 mg/snake). This is about 10 to 20 times greater dose than the highest dose administered by oral gavage (which produced 100% mortality). Obviously, combination with a bait severely attenuated the effectiveness of the pyrethrum.

Table II. Acute toxicity of pyrethrum fortified baits

Dose (mg/bait)	*Number of Snakes Consuming Bait*	*Percent Mortality*
20	5	20
40	4	50

Source: Reference 14.

The whole body pyrethrin residues in snakes given 40 mg baits ranged from 4.1 to 501 μg/g (Table III). The higher residues were found in fatally dosed snakes. This suggests that snakes surviving the initial pyrethrin dose will rapidly metabolize/excrete the pyrethrins. Residue concentrations in surviving snakes were less than in fatally dosed snakes. Quantification of these residue levels was necessary to estimate the potential secondary hazards to predators and/or scavengers potentially feeding on pyrethrin- containing brown tree snake carcasses. For secondary hazard estimates, the highest residue concentrations for each toxicant was used. This conservative approach generally results in "worst case" risk assessment calculations. For pyrethrins, the concentration of 501 μg/g was used.

Table III. Pyrethrin dose and residues

Bodyweight (grams)	*Dose (mg/kg)*	*Tissue Residue (μg/g)*	*Fate*
121	331	113	Died
83	482	501	Died
206	194	29	Survived
45	889	4.1	Survived

Source: References 19,20.

Rotenone

Snakes were also orally gavaged with varying doses of rotenone (Table IV). By this route of administration, rotenone appears to be more toxic than pyrethrins as the lowest dose that achieved 100% mortality was 2.5 mg/kg (0.125 mg - 0.50 mg/snake) compared to 40 mg/kg for pyrethrins. All concentrations higher than 2.5 mg/kg also produce 100% mortality. Again, no mortality was noted in control snakes gavaged only with carrier.

Table IV. Acute Toxicity following gavage with rotenone

Dose (mg/kg)	0.61	1.25	2.5	5	10	20	40
Mortality (dead/treated)	0/5	1/5	5/5	5/5	5/5	5/5	5/5

Source: Reference 13.

However, when rotenone was incorporated into SPAM and quail chick baits, no acute toxicity was observed at any concentrations tested (Table V). The highest concentration, 10 mg/bait, was 40 – 80 times the 2.5 mg/kg dose which yielded 100% mortality in the orally gavaged snakes. Incorporation of rotenone into baits decreased the toxicity to even a greater extent than was observed for pyrethrins.

Table V. Acute Toxicity of Rotenone Fortified Baits

Concentration (mg/bait)	*Number of Snakes Consuming Bait*	*Percent Mortality*
2.5	3	0
5	5	0
10	5	0

Source: Reference 14.

Whole body rotenone residues were determined in the snakes fed baits containing 10 mg rotenone (Table VI). As all the snakes survived, the magnitude of residues were similar to those observed for the surviving pyrethrins dosed snakes and less than the fatally pyrethrins dosed snakes. The highest observed level was 61 μg/g.

Table VI. Rotenone Dose and Residues

Bodyweight (grams)	*Dose (mg/kg)*	*Tissue Residue (μg/g)*	*Fate*
111	90	61	Survived
131	76	0.67	Survived
185	54	12.4	Survived
98	102	48.4	Survived

Source: Reference 21.

Propoxur

Brown tree snakes were orally gavaged with solutions containing varying concentrations of propoxur. The lowest concentration to yield 100% mortality was 40 mg/kg which is equivalent to a dose of 2 - 8 mg propoxur per snake.

Table VII. Acute Toxicity Following Gavage with Propoxur

Dose (mg/kg)	5	10	20	40
Mortality (dead/treated)	0/5	2/5	3/5	5/5

Source: Reference 13.

Propoxur baits were prepared by fortifying quail chicks and SPAM at 20 mg/bait which delivered a dose ranging from 146 – 220 mg/kg (7 - 43 mg/snake). While this dose is 4 to 5 times the 40 mg/kg oral gavage dose that resulted in 100% mortality, mortality was only 75%. Again, incorporating the toxicant into a biological matrix reduced toxicity (Table VIII).

Table VIII. Acute Toxicity of Propoxur Fortified Baits

Concentration (mg/bait)	*Number of Snakes Consuming Bait*	*Percent Mortality*
20	4	75

Source: Reference 14.

Propoxur residues in snakes consuming baits containing 20 mg propoxur were similar for both surviving and fatally dosed snakes (Table IX). The highest observed residue was 141μg/g.

Table IX. Propoxur Residues in Snakes Fed Propoxur Baits

Bodyweight (grams)	*Dose (mg/kg)*	*Tissue Residue (μg/g)*	*Fate*
131	153	106	Died
124	161	134	Died
137	146	116	Survived
91	220	141	Died

Source: Reference 22,23.

Dermal application was evaluated as a potential mean of applying toxicants to snakes. By far the most effective compound tested was rotenone, yielding 100% mortality at 10 mg/kg, or 0.5 to 2 mg/snake (Table X). This level of toxicity was half that observed for administration via oral gavage.

Table X. Acute toxicity* Following Dermal Dosing

	Dose (mg/kg)						
Toxicant	0	2.5	5	10	20	40	80
Pyrethrins	-**	-	-	-	1/5	2/5	-
Rotenone	-	0/5	2/5	5/5	5/5	5/5	5/5
Propoxur	-	-	-	-	0/5	3/5	2/5

Note: * # dead/#tested, ** Not determined
Source: Reference 13.

Table XI. Residues and Acute Toxicity Following Dermal Application of Rotenone

	Dose (mg/kg)					
	2.5	*5*	*10*	*20*	*40*	*80*
Residue (μg/g)	0.221 (s)	0.390 (s)	4.07 (d)	6.84 (d)	11.1 (d)	35.2 (d)
	0.183 (s)	0.579 (s)	3.04 (d)	7.76 (d)	14.4 (d)	17.2 (d)
	0.112 (s)	1.70 (d)	4.74 (d)	8.94 (d)	13.5 (d)	23.3 (d)
	-	1.72 (d)	-	-	18.2 (d)	-
	-	-	-	-	16.0 (d)	-
Mean residue	0.172	1.1	3.95	7.85	14.6	25.2
Std. dev.	0.05	0.71	0.86	1.05	2.7	9.2

Note: (s) = survived (d) = died
Source: Reference 22.

As rotenone was the only toxicant that appeared to be promising with respect to dermal application, the residue and toxicity data in Table XI is limited to snakes dermally dosed with rotenone. Using the minimum 100% lethal dosage of 10 mg/kg, the highest tissue concentration found was 4.74 μg/g.

Secondary hazard assessment

When evaluating the use of chemical toxicants to control snake populations, consideration must be given to those non-target species which could accidentally ingest toxicant by scavenging or preying on dead or dying snakes. On Guam such scavengers include feral cats, wild pigs, feral dogs, monitor lizards *(Varanus indicus)*, and the Mariana crow. Obviously, the endangered crow elicits the greatest concern from a secondary hazard standpoint, while the other scavenger species are introduced and may be considered pest species themselves. Ideally, secondary hazard considerations should not be limited to non-target species found on Guam. Toxicants developed for brown tree snakes may be required to control future introduced brown tree snake populations at other locations. These locations will likely contain a wider variety of potential non-target species than are currently found on Guam. Also, the brown tree snake population on Guam may be suitably reduced to permit the reintroduction of other species such as the Micronesian kingfisher, which may have preyed on small snakes (6,24). In this scenario, where regular chemical control may be required to keep snake populations minimized, the reintroduced species represent potential non-target species.

A widely used, straight forward approach for estimating non target hazards is the risk quotient (RQ) method (25). The RQ is the expected dose or dietary concentration divided by the dose or concentration expected to produce lethality in 50% of the population, respectively (dose/LD_{50} or concentration/LC_{50}). RQs provide a numerical basis for decision making. A RQ greater than 1 indicates that there are appreciable non target risks associated with use of this chemical. A RQ less than 1 indicates that the non target risks from use of this chemical may be acceptable under approved usage guidelines. To provide a "worst case" estimate of non target hazards, we assumed that 100% of the exposed animal's diet would consist of the pesticide formulation (primary hazard) or the tissue, organ, or carcass (secondary hazard) containing the residue highest concentration. EPA further breaks-down RQ values less than 1 into the following categories (26):

For the potential brown tree snake toxicants, RQs were calculated for the crow, dog, pig and cat. Crows, feral dogs and feral cats are potential consumers of brown tree snake carcasses on Guam. For pyrethrin bait-dosed snakes, the highest tissue residue concentration of 501 μg/g was used for all calculations. For a worst case exposure estimate, this concentration was multiplied by the average food consumption for crows, 0.076 g food/g bodyweight/day (27). To estimate the acute toxicity of pyrethrins to crows, we relied on the literature value of 7070 μg/g, the LD_{50} for Japanese quail. The resulting RQ for crows consuming brown tree snakes killed by

Table XII. Risk Quotient Values and Associated Concerns

RQ Value	*Associated Risk*
<0.1	Use presents acceptable risk for use under approved guidelines
>0.1	Use restrictions may be imposed to protect endangered species
>0.2	Use may be restricted to certified applicators and/or mitigation techniques may be imposed
>0.5	Mitigation techniques will be imposed to protect all species of the same taxonomic order

Table XIII. Pyrethrum Risk Quotients

Animal	*Calculation*		*RQ*
Crow	$\frac{501\ \mu g/g \times 0.076\ g/g}{7070\ \mu g/g}$	=	0.002
Dog	$\frac{501\ \mu g/g \times 0.006\ g/g}{200\ \mu g/g}$	=	0.15
Pig	$\frac{501\ \mu g/g \times 0.04\ g/g}{200\ \mu g/g}$	=	0.1
Cat	$\frac{501\ \mu g/g \times 0.07\ g/g}{200\ \mu g/g}$	=	0.18

pyrethrins is 0.002, well below the level of concern for endangered species (Table XIII). RQs were similarly calculated for dog, cat, and pig using the oral LD_{50} values for the rat, 200 mg/kg (28) and literature referenced consumption rates for dog, cat, and pig (29). The resulting RQs ranged from 0.1 to 0.18 indicating that the potential secondary hazards for these species are minimal.

The same procedure was used to calculate RQs for rotenone and propoxur. The resulting RQs are summarized in Table XIV. For rotenone, the highest snake residue concentration of 61μg/g was utilized. For crow, acute toxicity was estimated with the LC_{50} (1608 μg/g)from ring-necked pheasant (29) to give an RQ of 0.003. For dog, pig, and cat, no rotenone LD50s were available, so the LD50 for the rat (60 mg/kg) was used. The resulting RQs for the dog, pig and cat are 0.06, 0.04, and 0.07. These risk quotients suggest that the secondary hazards associated with the use of rotenone to control brown tree snakes on Guam are minimal.

To calculate the RQs associated with the use of propoxur, the highest tissue residue concentration of 141μg/g was used. For crow, LC_{50} for the house finch, 3.55 mg/kg was used (EPA data base). For dog, pig, and cat, the rat oral LD_{50} of 41 mg/kg was used (RTECS)]. The resulting propoxur RQs ranged from 0.14 to 0.24 for mammals and 3.0 for the crow. The RQ of 3.0 for the crow triggers significant concern for secondary hazards, especially when an endangered species is potentially exposed.

Table XIV. Risk Quotients for Oral Dosing

Toxicant	*Crow*	*Pig*	*Dog*	*Cat*
Pyrethrin	0.002	0.1	0.15	0.18
Rotenone	0.003	0.04	0.06	0.07
Propoxur	3.0	0.14	0.21	0.24

Risk Quotients were also calculated using the highest rotenone concentration found in snakes dermally dosed at 10 and 20 mg/kg (Table XV). The resulting risk quotients were quite low for all species of concern, especially the crow. These data suggest that secondary hazards associated with dermal rotenone dosing to control brown tree snakes on Guam are minimal.

Table XV. Rotenone Dermal Risk Quotients

	RQ	
Animal	*10 mg/kg*	*20 mg/kg*
Crow	0.0002	0.0004
Dog	0.005	0.01
Pig	0.05	0.1
Cat	0.005	0.01

Conclusions

When administered orally in solutions, the acute toxicity of the potential brown tree snake toxicants evaluated was rotenone > propoxur = pyrethrins. Incorporation of the pesticides into biological matrices (SPAM or quail chicks) reduced the acute toxicity of all the pesticides. The greatest reduction was noted for rotenone. When administered in fortified baits, the toxicity was propoxur > pyrethrins > rotenone. With respect to secondary hazards, the most favorable (least risk, lowest RQ) compound appears to be rotenone followed by pyrethrum. Secondary hazards associated with propoxur appear to be manageable for mammalian scavengers, but suggest high risk for birds.

Acute toxicity for the pesticides when administered in solutions via dermal application was rotenone > pyrethrins = propoxur. Acute toxicity of dermally applied pesticides was about half to one quarter of that observed for gavage. However, pesticide residues and associated secondary hazard risk quotients were significantly less for dermal application.

From a secondary hazard perspective, dermal application of rotenone appears to be a promising technique for the control of brown tree snakes on Guam. However, the development of an efficient and selective dermal application procedure for the brown tree snake is not available. Based on our findings with these toxicants, oral dosing in biological based baits appears to be a less promising approach due to decreased toxicity noted when the toxicant was combined with the bait matrix. However, in the absence of the biological matrix, oral application of rotenone is highly toxic to brown tree snakes. We are currently attempting to capitalize on these observations by developing a synthetic lure which will combine a brown tree snake attractant into a synthetic matrix that will not decrease the toxicity of the pesticides. If successful, a toxicant such as rotenone could be combined with the synthetic matrix to produce an efficacious oral bait for reducing brown tree snake populations. At this point, subsequent reevaluation of residues and risk quotients may be needed. Further work is being conducted to identify additional compounds with high toxicity to brown tree snakes, minimal secondary hazards to non-target species, and adequate efficacy when incorporated into biological matrices.

References

1. Rodda G.H.; Sawai, Y; Cjozaar. D., Tanaka, H. Problem Snake Management, The Habu and the Brown Treesnake. Comstock Publishing Associates: Ithica, N.Y. 1999; pp 44.
2. Kinghorn, J.R. The Snakes of Australia. Revised Edition. Halstead Press: Sydney, Australia. 1964, 197 pp.
3. Fritts, T.H. Elapio. 1997, 47, 17-18.Rodda G.H.; Sawai, Y; Cjozaar. D., Tanaka, H. Problem Snake Management, The Habu and the Brown Treesnake. Comstock Publishing Associates: Ithica, N.Y. 1999; pp 236.
4. Rodda, G.H., Fritts, T.H., Conry, P.J. Pacific Science 1992, 46, 46-57.
5. Savidge, J.A. J. Herpetol. 1988 22, 275-282.
6. Fritts, T.H. and McCoid, M.J. The Snake. 1991, 23, 75-80.
7. Fritts, T.H.; Scott, N.J.; Savidge, J.A. The Snake. 1987, 19, 51-58.
8. Savidge, J.A. Ecology. 1987, 660-668.
9. National Research Council. The Scientific Basis for Preservation of the Mariana Crow National Academy Press: Washington, D.C. 1997. p. 18.
10. Aguan, C. Division of Aquatic and Wildlife Resources, Guam Department of Agriculture. personal communication, 1999.
11. McCoid and Stinson, 1991.
12. McCoid, M.J., Fritts, T.H., Campbell, E.W. The Texas Journal of Science 1994, 46, 365-368.
13. Brooks, J.E., Savarie, P.J. and Johnston, J.J. Wildlife Research. 1998, 25, 427-435. Brooks, J.E.; Savarie, P.J.; Johnston, J.J.. Laboratory and Field Evaluation of Oral Toxicants for Brown Tree Snake Control (QA 484), Unpublished progress report, USDA/APHIS/WS/NWRC, 1996, 9 pp.
14. Sterner, R.T. and Mauldin, R.E. Arch. Environ. Contam. Toxicol. 1995, 28, 516-523.
15. Primus, T.M.; Tawara, J.N.; Goodall, M.J.; Brooks, J.E.; Savarie P.J.; Johnston, J.J. J. Agric. Food Chem. 1998, 46, 2647-2650.
16. Johnston, J.J.; Furcolow, C.A.; Volz, S.A.; Mauldin, R.E.; Primus, T.M.; Savarie, P.J.; Brooks, J.E. J. Chromatogr. Sci. 1999, 37, 5-10.
17. Mauldin, R.E.; Johnston, J.J. J. Agric. Food Chem. (In Press).
18. Furcolow, C.E. Pyrethrin Residues in Whole Body Brown Tree Snakes (QA 490) 97-014. Unpublished analytical services report. USDA/APHIS/WS/NWRC, 1997, 5pp.
19. Furcolow, C.E. Pyrethrin Residues in Whole Body Brown Tree Snakes (QA 484) 97-018. Unpublished analytical services report. USDA/APHIS/WS/NWRC, 1997, 5pp.

20. Furcolow, C.E.; Mauldin, R.E. Whole Body Residues in Brown Tree Snakes (QA 484, 490) 97-123. Unpublished analytical services report. USDA/APHIS/WS/NWRC, 1998, 6pp.
21. Goodall, M.J. Whole Body Propoxur Residues in Brown Tree Snakes (QA 490) 96-065. Unpublished analytical services report. USDA/APHIS/WS/NWRC, 1996, 2pp.
22. Goodall, M.J.; Whole Body Propoxur Residues in Brown Tree Snakes (QA 490) 96-066. Unpublished analytical services report. USDA/APHIS/WS/NWRC, 1996, 4pp.
23. Baker, R.H. The avifauna of Micronesia, its origin, evolution, and distribution. University of Kansas Publications: Lawrence, KS, 1951; p. 359.
24. Urban, D.J.; Cook, N.J. Ecological Risk Assessment: Standard Evaluation Procedure of the Hazard Evaluation Division; U.S. Environmental Protection Agency. Office of Pesticide Programs. U.S. Government Printing Office: Washington, DC, 1986; EPA-540/9-85-001.
25. National Research Council. Risk Assessment in Federal Government: Managing to Process; U.S. Government Printing Office: Washington, DC; NASNRC Publ. 1983.
26. U.S. Environmental Protection Agency, Office of Research and Development. Wildlife Exposure Handbook. 1993; pp 3-5.
27. National Institute of Occupational Safety and Health. Registry of Toxic Effects of Chemical Substances (RTECS) Database. July, 1999.
28. Guide to the Care and Use of Experimental Animals. Canadian Council on Animal Care: Ottawa, Canada. Volume 1, 1984.
29. U.S. Environmental Protection Agency. Office of Pesticide Program Ecotoxicity Data Base.

Acknowledgements

The authors wish to acknowledge the critical support of the United States Department of Defense Legacy Program for funding NWRC brown tree snake research and Dr. Richard Bruggers, Assistant Director, NWRC for coordination of the NWRC brown tree snake research. Mention of specific products is for identification purposes only and does not constitute endorsement by USDA.

Chapter 17

History and Risk Assessment of Triazine Herbicides in the Lower Mississippi River

W. R. Hartley[1], L. E. White[1], J. E. Bollinger[2], A. Thiyagarajah[1], J. M. Mendler[2], and W. J. George[2]

[1]School of Public Health and Tropical Medicine, Environmental Diseases Prevention Research Center and [2]Division of Toxicology/Pharmacology Department, Tulane University, 1430 Tulane Avenue, New Orleans, LA 70112

All available historical water quality data for triazine herbicides in the Mississippi River south of Memphis, TN were characterized for temporal trends and ecological and human health risks. The most abundant of the triazine compounds detected was atrazine, which also represented the largest contribution to the dataset. Temporal analysis of the data indicated no distinct increase in median atrazine concentrations from 1975 to 1997; the median atrazine concentration was 0.47 ug/L (ppb) in the river water. In the 1990-1994 LA Department of Environmental Quality (LADEQ) survey of Mississippi River fish, atrazine was detected in 1.3% of fish examined, with a maximum concentration of 0.058 mg atrazine/kg fillet (ppm). Based on the safe lifetime dose for atrazine, protection for equivocal evidence of carcinogenicity, the Maximum Contaminant Level (MCL), and ecological criteria, there were no significant health risks due to atrazine from use of the river water as a drinking water source, nor were there significant risks to aquatic organisms.

INTRODUCTION

The purpose of this paper is to characterize the water quality data on triazine herbicides in the lower Mississippi River south of Memphis, TN, and conduct health and environmental risk assessment based on temporal trends in the data. We focused on potential health and ecological risks associated with atrazine, cyanazine, and simazine based upon occurrence in the study area and known usage patterns in the

Mississippi River Basin. The triazines are a group of chemically similar herbicides including simazine, propazine, prometryn, prometon, cyanazine, ametryn and atrazine. The USEPA (*1*) in a 1988 evaluation of the STORET database found atrazine in 4,123 of 10,942 surface water samples analyzed, representing 1,659 surface water locations in 31 states. The 85th percentile value of all non-zero samples was 2.3 ppb. Cyanazine was found in 1,708 of 5,297 surface water samples analyzed, representing 392 surface water locations in seven states. The 85th percentile value for all non-zero samples was 4.11 ppb. Atrazine is one of the most broadly used pesticides in the U.S. based upon pounds of active ingredient applied per year. Primary concerns over the use of triazines are surface and groundwater contamination and subsequent human and ecological health issues. Atrazine is primarily used for general weed control on industrial and other non-agricultural land and selective weed control in crops including corn, sorghum, sugar cane, pineapple, wheat, macadamia nuts, and Christmas trees. Atrazine is also used on turf and lawns and in conifer restoration (*2,3*). With regard to regulatory status, atrazine is classified as a Restricted Use Pesticide (RUP) due to potential for groundwater contamination. RUPs may only be purchased and used by trained and certified pesticide applicators. In November 1994, the USEPA initiated a Special Review of atrazine including use and health effects aspects (*4*). In June 1996, the USEPA published a proposed rule prohibiting use of three triazines (atrazine, simazine, and cyanazine) unless USEPA approved State and Tribal management plans were developed. In 1994, 21-34 million pounds of cyanazine were applied to control weeds on cornfields. The use of cyanazine is being gradually phased out and will not be used after December 2002. Five to seven million pounds of simazine and 200,000 to 400,000 pounds of propazine (under emergency exemptions) per year are used (*3*).

Triazine herbicides have been found in Mississippi River water reflecting usage during the past 30 years. DeLeon et al. (*5*) reported atrazine levels of ND (source), 0.640 ppb (Cario, IL), 1.10 ppb (Memphis, TN), and 0.410 ppb (New Orleans). Larson et al. (*6*) studied atrazine flux in the Mississippi River from 5/1/91 to 3/1/92 and reported concentrations were highest during May and June, immediately following pesticide application and spring rains and dropped below detection limits in late summer. The authors reported maximum concentrations of 3.6 ppb atrazine in June, 1991, in the Mississippi River near Baton Rouge, LA. During an intense rainstorm event (May 15-17,1990) in Iowa, Illinois, Indiana, and Ohio, the peak concentrations of triazine herbicides in the tributaries of the Upper Mississippi River reached 36 ppb and resulted in an upriver gradient of 0.2 ppb per 100 km (*7*). During the 1993 flood, Goolsby et al. (*8*) reported that median (range) atrazine and cyanazine concentrations at six stations on the Mississippi River including Baton Rouge, LA were 2.2 ppb (1.27-3.31) and 1.18 ppb (0.45-1.91) respectively.

Pereira and Hostettler (*9*) evaluated primary sources of atrazine in the Mississippi River. They concluded that inputs of atrazine, cyanazine, and simazine to the Mississippi River (study periods: July-August 1991, October-November 1991, and April-May 1992) are mainly from the Minnesota, Des Moines, Missouri, and Ohio rivers. Their study suggested that during base-flow conditions, there is significant groundwater and surface water interactions in the Mississippi River. Conservative estimates of annual mass transport of atrazine and cyanazine to the Gulf of Mexico were 160 tons and 71 tons respectively. Individual rivers such as the Minnesota River which drains the agricultural region of southern Minnesota annually contributes up to 1-2 tons of atrazine to the Mississippi River (*10*). In a study by the USGS of contaminants in the Mississippi River from 1987-1992 (*11*), levels of atrazine were unexpectedly elevated with the highest concentrations near St. Louis, MO, because of inputs from the Missouri and Illinois rivers, associated with drainage from corn fields. However, maximum concentrations of atrazine only exceeded the drinking water Maximum Contaminant Level (MCL) of 3.0 ppb under extreme localized conditions. Concentrations of atrazine dropped rapidly with distance upriver of the Gulf of Mexico. Transport of atrazine in to the Gulf of Mexico was further documented by McMillin and Means (*12*) in which atrazine was ubiquitous over the entire northwestern Gulf of Mexico coastal shelf area in the spring, summer, and fall.

MATERIALS AND METHODS

Water Quality

As a central aspect of the Tulane University Mississippi River Database Project, Tulane researchers have developed a database for use as a repository for results from the analyses of water samples collected from the Lower Mississippi River, from just south of Memphis to the receiving waters of the Gulf of Mexico. To date, roughly three million records of over 800 water-quality parameters have been accumulated. These data, collected from analysis conducted by academia, government, industry and municipal water works, were integrated into a relational information system and GIS, representing the most comprehensive source of water-quality information available on the lower Mississippi River as a single database (*13*).

Triazine herbicide data presented in this report were generated by the Louisiana Department of Environmental Quality (LADEQ) and the US Geological Survey (USGS) through the US Environmental Protection Agencies' Storage and Retrieval System (STORET), as well as through reports in the literature (*14,15,16*). Additional data were obtained directly from municipal water works (Jefferson Water Works (JWW) and Orleans Water Works (OWW)). Atrazine data were also obtained from the surface water monitoring program conducted by Novartis (formerly Ciba-Geigy Corporation) (*17*) and other published results from academic sources (*5*). All data were subjected to a validation process, which removed duplicate records and replaced non-detects with the reported detection limit value for each analytical method used. Visual inspection of the triazine data indicated no apparent outlying points.

Variations in triazine concentrations due to sampling location were minimal throughout the study area. This is attributed to the fact that the Lower Mississippi

River receives very little discharge south of the Arkansas River confluence (*18, 19, 20, 21*). The relative abundance of atrazine data allowed for statistical comparisons between source agencies and between filtered and unfiltered water fractions. However, a lack of sufficient overlap between data sets precluded similar comparisons for the other triazine compounds studied. Visual inspection of the various agencies' data with respect to these other triazine compounds revealed no obvious differences in their reported results, with the exception of higher concentrations in the unfiltered samples of simazine and cyanazine. Therefore, where statistical comparisons of data sets were not possible, data from different agencies were assumed comparable and were grouped for ease of display. Filtered and unfiltered fractions for these sets were indicated separately on summary plots. Risk assessment was conducted using the unfiltered samples for simazine and cyanazine as a worst case scenario.

Nonparametric statistical comparisons were conducted on weekly means of atrazine data, grouped by source agency or water fraction, to determine whether these data could be integrated into a single large set. Paired sets of data from coinciding weeks were analyzed using the Kruskel-Wallis test (using analysis of variance on rank-transformed data), with a p= 0.05 level of significance (*22*). Using these statistical methods, data from JWW prior to 1990 were found to be inconsistent with other records available from the same period. These data were therefore not included in continuing analyses. Because no other significant differences were found with respect to source of data (where temporal overlap existed) or between filtered and unfiltered data, the remaining atrazine data were combined for the purposes of temporal characterization and risk analysis.

Health and Ecological Assessment

Health and ecological assessment included determination of the potential impact of triazine herbicides on use of the Lower Mississippi River as a drinking water source and on the consumption of fish. The impact of the triazine herbicides in the Lower Mississippi River on aquatic organisms was also considered. With regard to drinking water, two health assessment methods were utilized (*1*). In the first approach, Margins of Exposure (MOEs) were calculated based upon some of the exposure assumptions of the Safe Drinking Water Act (SDWA) in which the reference dose (RFD) is used as the safe lifetime daily dose and the exposure assumptions are 2 L/day water consumption, 70 kg body weight person and 100% gastrointestinal absorption. The MOE does not, in this case, consider other potential sources of the contaminant. In the second approach, the ratio of the median herbicide concentration to the Maximum Contaminant Level (MCL) is used. The MCL-value is a legal standard and considers both risk assessment and risk management including a relative source contribution of 20% from drinking water, potential carcinogenicity, technical feasibility and economics. Fish consumption (*23*) exposure assumptions included one fish meal per week (30g fish/person/day), 70kg body weight person, 100% gastrointestinal absorption, and RFD as the safe lifetime daily dose. Cooking method was not assumed to result in any significant contaminant removal. Canadian Government aquatic-life guidelines were used to assess aquatic toxicity. There are currently no US Government guidelines for aquatic toxicity of triazine herbicides.

RESULTS AND DISCUSSION

There were approximately 5,368 records located on triazine herbicides compiled from six existing databases (Figure 1). Most of the records came from JWW, Novartis, and LADEQ. The number of available records available by year was skewed. Prior to 1990, only two hundred (200) or fewer records were available per year whereas from 1991-1997 more than 500 records per year available (Figure 2).

Atrazine, cyanazine, and simazine were found in measurable concentrations while, when detected, propazine, prometryn, prometon, and ametryn occurred at trace concentrations (Figure 3). Based on previously cited application patterns, it is expected that atrazine, cyanazine, and simazine would be found in ranges to allow characterization whereas propazine, prometryn and prometon were not. Based on these results atrazine, cyanazine and simazine were further characterized in water quality and health and environmental risk assessments.

From 1975-93, the atrazine records are primarily from filtered water samples and from 1994- 1998 are primarily from unfiltered water samples (Figure 4). Based on the distribution of the atrazine data on unfiltered and filtered water, this represents primarily a shift in analytical methods. Also considering the solubility of atrazine, the atrazine measured in whole and filtered water are not statistically significantly different to warrant separate analyses of the data sets.

The annual median atrazine levels range from 0.2 to 0.7 ppb from 1975-1997. In 1990 and 1993 the annual median concentrations rise to 1.0ppb. In both of these years, peak rainfall and/or flood events occurred in the upper Mississippi River (Figure 5). This observation correlates well with previous observations associating usage patterns and rainfall events.

A seasonal analysis of the atrazine levels by month (Figure 6) shows expected higher atrazine concentrations in June, July and August which is associated with peak triazine pesticide usage on crops, particularly in the upper Mississippi River. The lower Mississippi River is confined by levees and it is unlikely that substantial loads of atrazine would enter from local agriculture sources except in the event of extreme flooding that has not occurred.

Cyanazine in filtered and unfiltered water samples from the Mississippi River (Figure 7) has remained relatively stable based on annual averages throughout the study period with the exception of 1993 for which there was a notable increase, probably due to flooding of the upper Mississippi River.

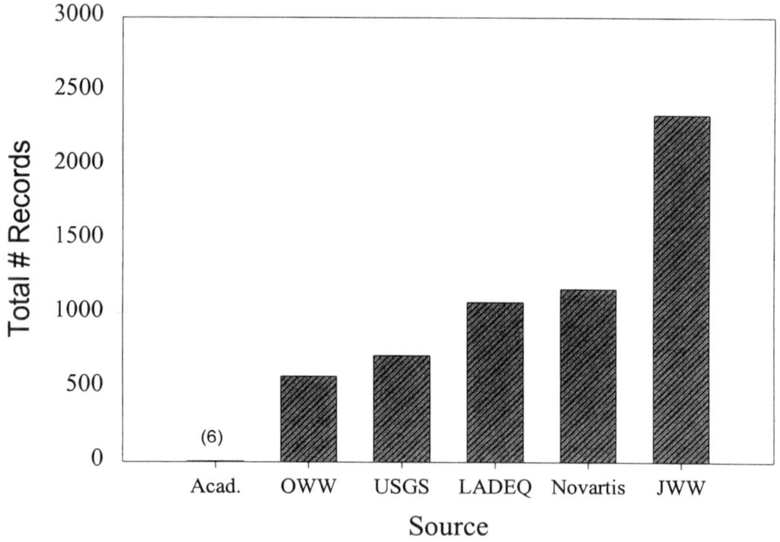

Figure 1. Total number of triazine records from 1975-1997 by source agency, including all JWW records.

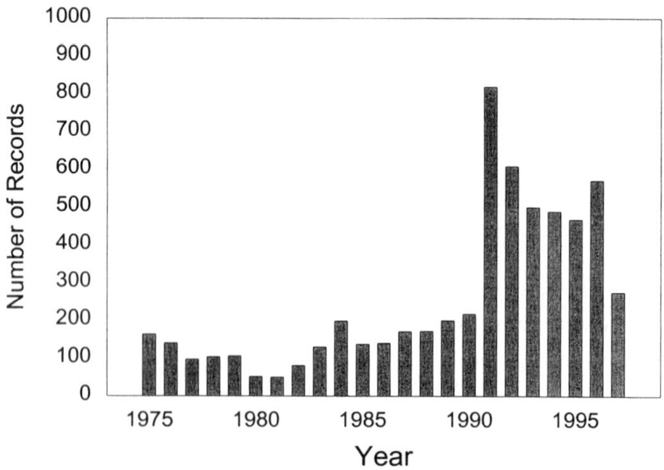

Figure 2. Total number of triazine records available by year including all JWW records.

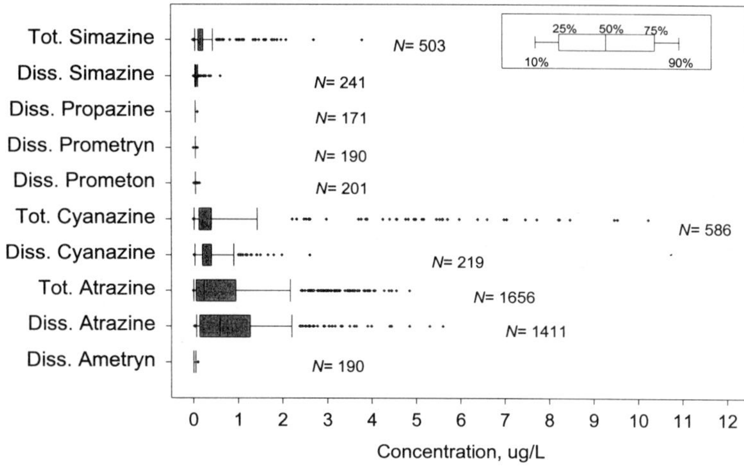

Figure 3. Concentration and distribution of triazine herbicides from 1975 – 1997.

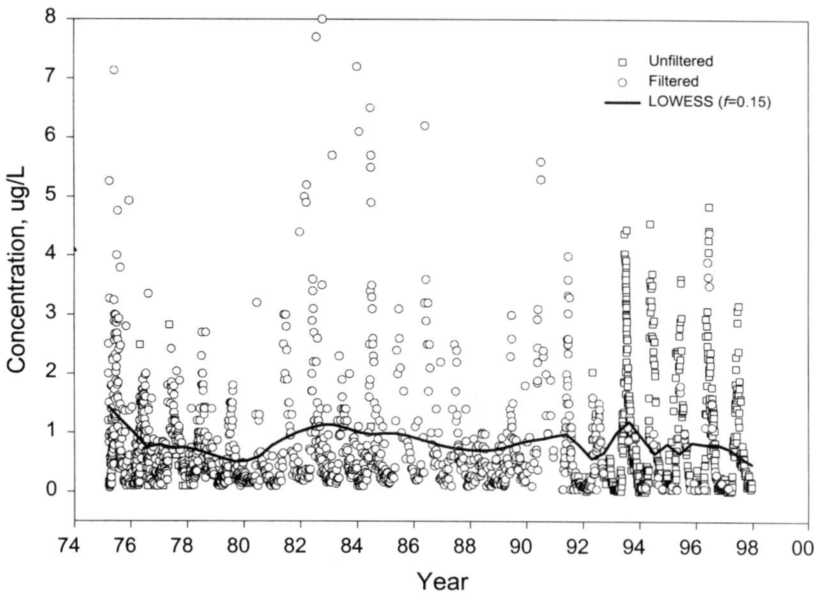

Figure 4. Filtered and unfiltered atrazine levels(weekly means) from 1975-1997

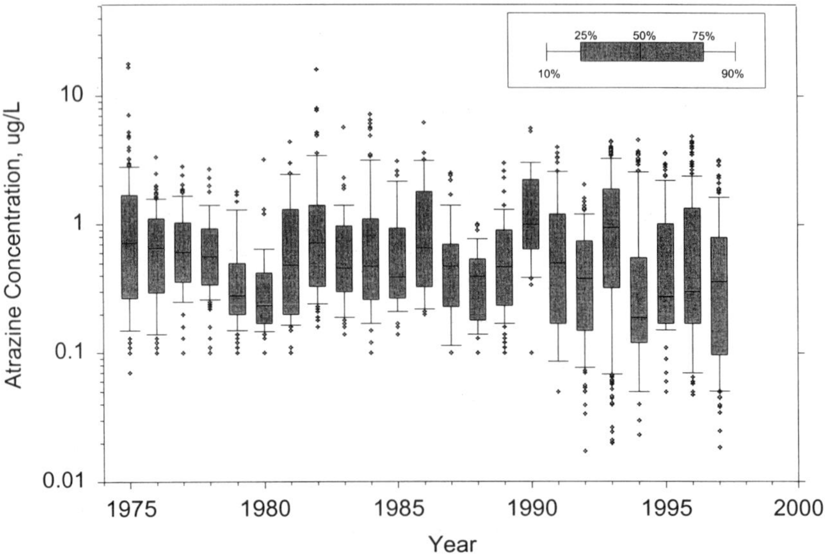

Figure 5. Annual atrazine concentrations from 1975-1997

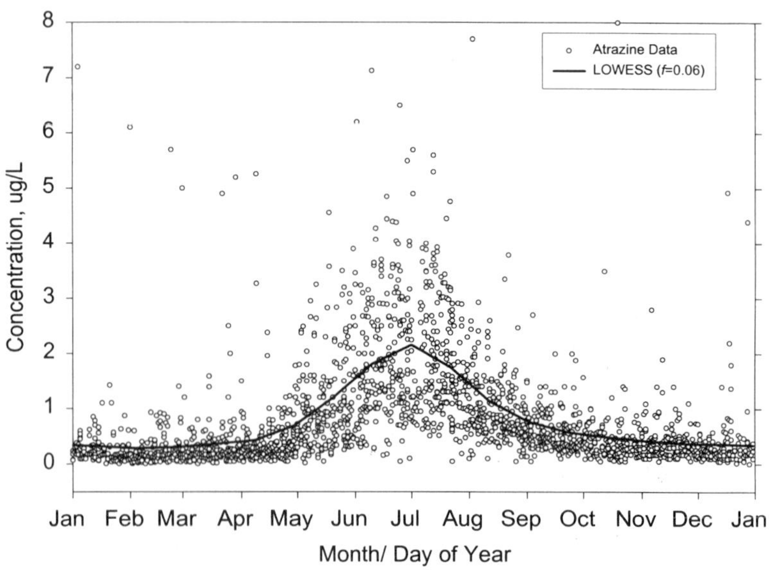

Figure 6. Seasonal distribution of atrazine from combined data (filtered and unfiltered) from 1975-1997

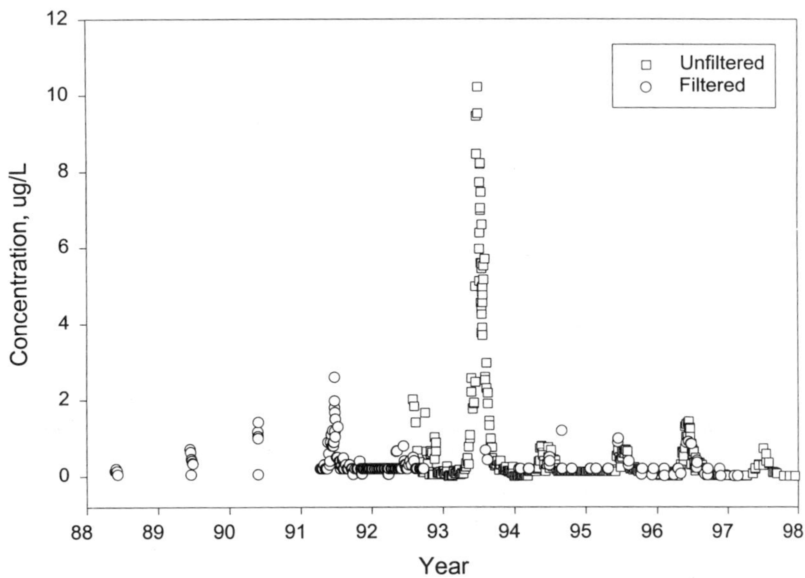

Figure 7. Weekly mean concentrations of filtered and unfiltered cyanazine from 1989 - 1997

Seasonal concentrations of cyanazine were similar to atrazine with increases during the June to August period and relatively trace levels during the remainder of the year (Figure 8). Although plotted together, comparability of filtered and unfiltered fractions was not established.

Simazine (Figure 9) from 1987-1998 show a pattern of increasing concentration beginning in 1992 with the most striking increases beginning in 1996. The causes for this increase in simazine are unknown since there are no unusual rainfall events or flooding in the upper Mississippi River after 1993.

A summary of five year intervals showing atrazine median concentrations along with the 75th percentile, 95th percentile, and maximum concentration are provided in Table 1. For atrazine, median values, 75th percentile values and 95th percentile values are similar with the exception of 95th percentile for atrazine for 1980-84 and 1990-94 periods.

Maximum concentrations of atrazine have been declining particularly since 1985. A summary of three-year interval cyanazine and simazine concentrations is also shown in Table 1. For cyanazine the 75th, 95th, and maximum concentration values decreased during the 1995 1997 interval. For simazine, the 95th percentile and maximum concentration value increased from 1995 – 1997.

The reference doses for atrazine, cyanazine, and simazine are 0.035, 0.002, and 0.005 mg/kg-bw/day respectively. The MCLs for atrazine and simazine are 3 ppb and 4 ppb respectively. Although the is no MCL for cyanizine, a lifetime health advisory could be calculated using the provisions of the SDWA and would result in an allowable level of 5 ppb to 1 ppb depending on the selection of an additional uncertainty factor for equivocal evidence of carcinogenicity (USEPA, Group C classification). The MOE values in Table 2 indicate that for atrazine, cyanazine, and simazine, the values fall well below MOE 1.0. This indicates that none of the chemicals exceed the safe lifetime exposure dose (RFD). Even if the maximum values detected were encountered for a lifetime in drinking water, they would not exceed the safe dose. Using a worst case additive toxicity model for all three chemicals under the assumption of similar modes of toxicity, the cumulative MOE or Hazard Index (HI) would not exceed 1.0 and thus be considered safe. With regard to the ratio of the median concentration to the MCL, the regulatory standard is not exceeded for atrazine and cyanazine. However, some 95th percentile concentrations and maximum concentrations do exceed the MCL. This poses a legal and regulatory issue that may result in an array of remedies under the SDWA.

The risks from consumption of triazine herbicides in the lower Mississippi River were determined. The Louisiana Department of Environmental Quality (*24*), reported that from 1990 – 1994 atrazine was detected in the fillets of fish 1.3% of the time (2 detections) with a maximum concentration of 0.058 mg atrazine/kg fillet. Using this worst case situation, the RFD for atrazine would not be exceeded. Considering the low

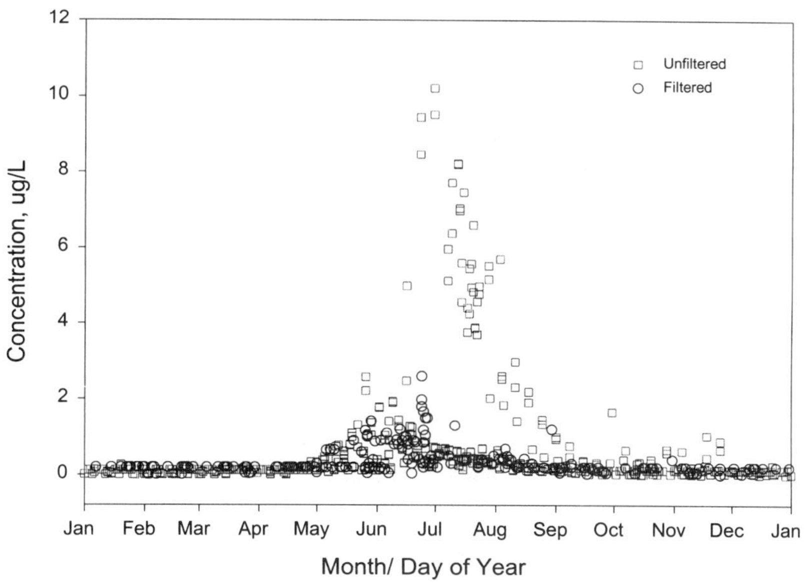

Figure 8. Seasonal distribution of cyanazine concentrations combined 1989 -1997

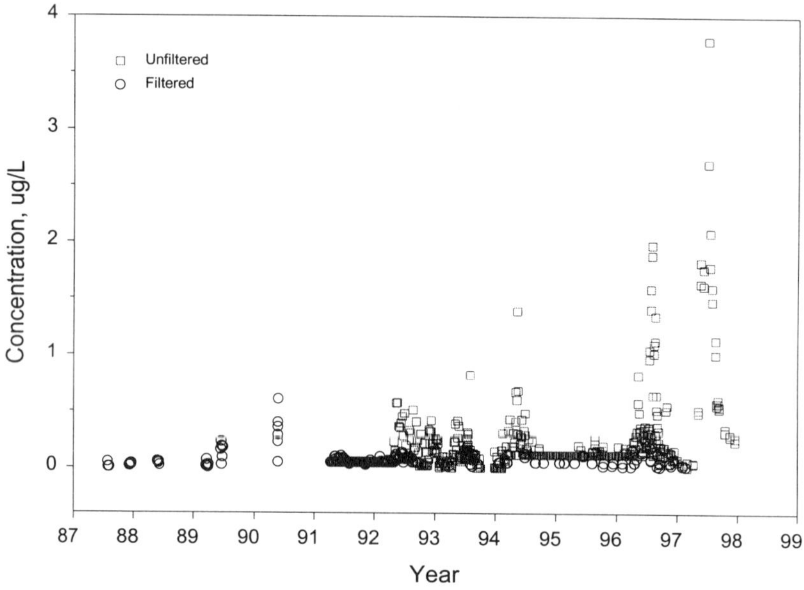

Figure 9. Weekly mean concentrations of filtered and unfiltered concentrations of simazine from 1988 –1997

Table 1. Summary Data on Selected Triazine Herbicides

Date Range	Concentration (ug/L) Median	75%	95%	Max
Atrazine, Unfiltered and Filtered				
75-79	0.565	1.08	2.47	17.8
80-84	0.44	1.1	3.5	16
85-89	0.46	0.8	2.4	6.2
90-94	0.46	1.21	3.33	5.6
95-97	0.31	1.1	2.66	4.9
Overall	0.47	1.1	2.92	17.8
Cyanazine, Unfiltered:				
92-94	0.151	0.69	5.97	10.2
95-97	0.124	0.3	0.86	1.44
Simazine, Unfiltered				
92-94	0.12	0.2	0.43	1.39
95-97	0.12	0.27	1.48	3.79

frequency of detection, the chance exposure to significant levels of atrazine through fish consumption is remote. Similar data on fish tissue levels for cyanazine and simazine were not available.

To determine risks to aquatic life in the lower Mississippi River, study results were compared to Canadian Government guidelines for atrazine (2 ppb), cyanazine (2 ppb), and simazine (10 ppb). All (Table 1) atrazine, cyanazine, and simazine median concentrations were below the Canadian guidelines. Some 95th percentile and maximum concentrations of atrazine and cyanazine exceeded the Canadian guidelines.

We have characterized available historical water quality data from 1975 – 1997 for triazine herbicides in the lower Mississippi River south of Memphis, TN . The most abundant of the triazine compounds was atrazine followed by cyanazine and simazine. The levels of atrazine and cyanazine followed seasonal patterns associated usage of these herbicides in the upper Mississippi River Basin. The use of these two herbicides in crops in the lower Mississippi River has little influence on atrazine and cyanazine levels due to the confinement of the lower Mississippi River by the levee system and the lack of flooding events which would breach the levee system. Upper Mississippi River Basin rain and/or flooding events in 1990 and 1993 resulted in minor increases in atrazine and cyanazine concentrations in the lower Mississippi

Table 2. Systemic Toxicity of Drinking Water Contaminated with Atrazine, Cyanazine, and Simazine.

Date	*MOE Median*	*MOE75th percentile*	*MOE 95th percentile*	*MOE Max*	*Median/ MCL*
		Atrazine (Unfiltered and Filtered)			
75-79	5×10^{-4}	9×10^{-4}	2×10^{-3}	2×10^{-2}	0.19
80-84	4×10^{-4}	9×10^{-4}	3×10^{-3}	1×10^{-2}	0.15
85-89	4×10^{-4}	7×10^{-4}	2×10^{-3}	5×10^{-3}	0.15
90-94	4×10^{-4}	1×10^{-3}	3×10^{-3}	5×10^{-3}	0.15
95-97	3×10^{-4}	9×10^{-4}	2×10^{-3}	4×10^{-3}	0.10
Overall	4×10^{-4}	9×10^{-4}	2×10^{-3}	2×10^{-2}	0.17
		Cyanazine (Unfiltered)			
92-94	2×10^{-3}	1×10^{-2}	8×10^{-2}	2×10^{-1}	NA
95-97	2×10^{-3}	4×10^{-3}	1×10^{-2}	2×10^{-2}	NA
		Simazine (Unfiltered)			
92-94	7×10^{-4}	1×10^{-3}	3×10^{-3}	8×10^{-3}	0.03
95-97	7×10^{-4}	2×10^{-3}	9×10^{-3}	2×10^{-2}	0.03

River. Although simazine concentrations in the river were low, we are unable to explain increases in the 1996 – 1997 period. We found no indication of increased usage of simazine during the 1996 – 1997 period. Periodic analysis of pollutants such as triazines herbicide data from large databases such as Tulane University River Database Project may provide information to make water quality management and herbicide usage decisions to minimize environmental impacts on the lower Mississippi River.

REFERENCES

1. USEPA. *Drinking Water Health Advisory: Pesticides – Atrazine;* Lewis Publishers Inc., Chelsea, Michigan, 1989, p 819.
2. *Farm Chemicals Handbook*, Meister, T.T Ed.; Meister Publishing Company, Willoughby, OH, 1987.
3. USEPA. *The Triazine Pesticides, Office of Pesticide Programs*, U.S. Environmental Protection Agency, Washington, D.C. 1997, p3.
4. EXTOXNET. *Extension Toxicology Network, Pesticide Information Profiles – Atrazine;* Oregon State University, 1999, p7.
5. DeLeon, I.R.; Christian, J.B.; Peuler, E.A.; Antoine, S.R.; Schaeffer, J.; Murphy, R.C. *Chemosphere* **1986**, 15, 795-805.
6. Larson, S.J.; Capel, P.D.; Goolsby; D.A.; Zaugg; S.D.; Sandstrom, M.W. *Chemosphere* **1995**, 31, 3305-3321.
7. Moody, J.A. and Goolsby, D.A. *Environ. Sci. Tech.* **1993**, 27, 2120-2126.

8. Goolsby, D.A.; Battaglin, W.A.; Thurman, M.E. River. *U.S. Geological Survey Circular* **1993**, 1120-C, p19.
9. Pereira, W.E. and Hostettler, F.D. *Environ. Sci. Tech.* **1993**, 27, 1542-1552.
10. Schottler, S.P.; Eisenreich, S.J.; Capel, P.D. *Environ. Sci. Tech.* **1994**, 28, 1079-1089.
11. Renner, R. *Environ. Sci. Tech.* **1996**, 30, 241.
12. McMillin, D.J. and Means, J.C. *J. Chromatography.* **1996**, 754, 169-185.
13. Preslan, J. E.; Belhouche, B.; Swalm, C. M.; Hughes, J. M.; Chen, H-L.; Henry, M.; Lin, D.; Bakeer, R.M.; Demtchouk, I.; Anderson, M. B.; Regens, J. L.; Means, J. C.; Bollinger, J. E.; Steinberg, L. J.; Luna, R.; Hernandez, R.; Hartley, W. R.; and George, W. J. *Environ. Prog.* **1997**, 16,145-163.
14. Moody, J. A.; Ed. *US Geological Survey Open File Report 94-523*, **1995**.
15. Pereira, W.E.; Moody, J.J. Hostettler, F.D.; Rostad, C.E.; Leiker, T.J. *US Geological Survey Open File Report 94-376, **1995**.*
16. Coupe, R. H.; Goolsby, D. A.; Iverson, J. L.; Markovchick, D. J. and Zaugg, S. D. *US Geological Survey Open File Report* 93-657, **1995**.
17. Tierney, D. *Ciba-Geigy Corporation Agricultural Group, Technical Report No. 6-92*, Greensboro, North Carolina, **1992**.
18. Sabins, D. S. *Clean Enough? A conference on Mississippi River water quality (Proceedings),* Lake Pontchartrain Basin Foundation, Metairie, **1997**, p32.
19. Wells, F. C. *US Department of the Interior, Water Resources Technical Report No. 21*, Baton Rouge, Louisiana, **1980**.
20. USACOE. *US Army Corps of Engineers report*, Vicksburg, Mississippi, **1976**, p3.
21. Everett, D.E. *Hydrologic and quality characteristics of the lower Mississippi River*, Louisiana Department of Public Works, Technical Report No. 5, Baton Rouge, Louisiana. **1971**, p 48.
22. Helsel, D. R. and Hirsch, R. M. *Statistical Methods in Water Resources. Studies in Environmental Science* 49 Elsevier, New York, **1993**, p529.
23. USEPA. *Guidance for Assessing Chemical Contamination Data for Use in Fish Advisories*, Vol. II, EPA 823-B-94-004, Washington D.C. **1994**.
24. LADEQ. *Mississippi River Toxics Inventory Project , Louisiana* Department of Environmental Quality, Baton Rouge, Louisiana, 1995, p29.

Analysis of Pesticide Residues in Wildlife

Chapter 18

Ion Trap GC-MS Analysis of Tissue Samples for Chlorinated Compounds

Pamela P. Hamlett, Gary L. Steinmetz, and David M. Klein

Texas Parks and Wildlife Environmental Contaminants Laboratory, 505 Staples Road, San Marcos, TX 78666 (phone 512–353–3486; email david.klein@tpwd.state.tx.us)

Currently, ion traps mass spectrometers are used in the Texas Parks and Wildlife (TPW) Lab to obtain full scan mass spectra of compounds at a detection limit comparable to selected ion monitoring (SIM) data from bench top quadrapole instruments. This combined with the ion trap's chemical ionization capability, tandem mass spectrometry, and simultaneous full scan with tandem mass spectrometry greatly enhances the quality of tissue analysis data. Ion trap identification/verification of both false-positive and false-negative results from electron capture detectors (ECD) and SIM are also discussed.

Introduction

The Texas state legislature has given Texas Parks and Wildlife (TPW) trustee responsibility for all public waterways, wildlife and both fresh and coastal water fishes of Texas. The TPW maintains a chemistry laboratory to:

1. provide information about fish and wildlife kills
2. support criminal investigations of polluters
3. obtain background information about contaminant levels in Texas
4. provide analysis in support of related projects for local, state and federal government entities

The majority of the samples submitted to the laboratory are fish and wildlife tissues although water, soil, and sediment samples are also received. Many of these samples are analyzed for chlorinated organic residues including pesticides, fungicides, herbicides and poly-chlorinated biphenyls (PCBs). Classical methods of sample preparation include Soxhlet extraction and Kuderna-Danish solvent evaporation. Both of these techniques are too time consuming and have been replaced by validated methods using static solvent extraction and automated nitrogen evaporation of solvents. The sample preparation scheme used by TPW is shown in Figure 1.

Experimental

Samples were frozen at –20° C. The samples were thawed and homogenized with a Robot Coupe RSI 6V. 10 g of sample was weighed out and mixed with 100 g of anhydrous sodium sulfate. This mixture was frozen overnight to remove all traces of moisture from the sample. Extraction was with three portions of dichloromethane. Evaporation was done with a Zymark Turbovap. Samples were defatted on a Waters HPLC system set up with a gel permeation column. Samples that were analyzed by electron capture were further purified by liquid chromatography using a mixed alumina and silica column.(3)

Gas chromatography with electron-capture detection was performed on DB 5 and DB 1701 30 m X 0.25mm X 0.25 um column in a Hewlett-Packard 5890 gas chromatograph. SIM were recorded on a Hewlett-Packard 5970 mass selective detector. Ion trap mass spectral detection was performed on a Thermoquest GCQ. For PCB analysis a 60 m X 0.25 mm X 0.25 um DB-5 column.

Safety Considerations

Chemical analysis poses hazardous situation and appropriate safety precautions should be taken. No one should work in a chemistry lab without specific education on methods, techniques and the hazards of the chemicals that must be handled. Therefore, any laboratory doing analysis of hazardous substances must have a working chemical hygiene plan with a designated and trained chemical hygiene officer. This is not a recommendation, as it is required by law under the 29 CFR 1910.1450 which states, "This section shall apply to all employers engaged in the laboratory use of hazardous chemicals. Employers shall have developed and implemented a written Chemical Hygiene Plan no later than January 21, 1991."

A specific note of caution concerns the hazards of working with pesticide extracts. Extracting tissue involves concentrating contaminants from a large quantity of sample into a small volume of solvent. The concentrated extract poses a special risk in handling and proper personal protection is required. Latex gloves are NOT appropriate for this work. Only nitrile gloves or those rated to resist organic solvents should be used. Finally, anyone working with concentrated dioxin extracts and standards should obtain specific training before working with concentrates of these extremely toxic compounds.

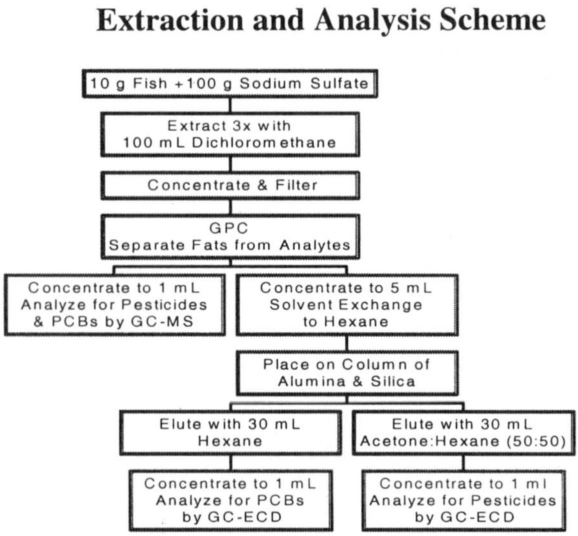

Figure 1. Sample Preparation Scheme

Electron Capture Detectors

Electron-capture detectors have been the standard for gas chromatographic analysis of chlorinated pesticides and PCBs (1, 2). While these detectors are very sensitive, they suffer from a lack of selectivity. Confirmation requires a second column and detector. Due to overlapping retention time of many compounds, a time consuming liquid chromatography step is routinely used to separate the non-polar PCBs from more polar chlorinated pesticides (Figure 1) (3). Current analytical requests seek part per billion (ppb) or lower detection limits. Concentrating samples to provide this "ultra-trace" level of detection produces gas chromatograms where interfering compounds from the tissue matrix can equal or exceed those of the chlorinated residues.

Selected Ion Monitoring

Due to the complex backgrounds of tissue samples, the gas chromatograph-mass spectrometer (GC-MS) has been used to provide data with less false positive identifications. Full scan mass spectra from bench top quadrapole instruments lack the sensitivity required so the technique of selective ion monitoring (SIM) GC-MS is used to obtain a lower detection limit (4). In SIM only two or three ions are detected for each compound. The presence and ratio of the ions are used to identify analytes. However, there are some limitations inherent to SIM analysis. First, it is can only be used for known analytes. Ions of interest must be assigned to retention time windows and only compounds which are in a time windows will be detected. Therefore, unknown compounds will give no signal with this system. While this technique is useful at eliminating background, it provides no information about unknown compounds which may be important for environmental evaluations.

Another area that presents difficulty is the ion ratio(s) used to confirm an analyte. This ratio has a large window of acceptance. In EPA methods, the ion ration can vary by 20 - 30% depending on the method followed (e.g.. EPA 625 or 8270 C). This practice can give rise to a false positive detection if the background includes ions close to the m/e of the secondary confirmation ion(s). If there is an interfering background ion it may cause false negative reports by obscuring the correct ratio of a confirmation ion. False negatives can also arise if the retention time of the analyte changes and the ions are no longer in the correct retention time window. SIM has given rise to both false positive and negative identifications in the TPW when samples were re-analyzed by more sensitive ion-trap instruments.

Analyte Quantitation

Quantitation by SIM is usually performed from calculations on the most abundant ion. Often the confirmation ions fall below a 5:1 signal to noise ratio yet this problem is rarely addressed in SIM data. With greater sensitivity than full scan mass spectra, SIM provides analyte concentrations with lower detection limits. Because of this, confirmation of suspected false positive or false negative reports is not possible without re-analyzing the sample on a more sensitive instrument. There are two instruments that have significantly higher sensitivity than quadrapole instruments operated in the SIM mode. One is the ion trap and the other is a magnetic sector mass spectrometer. Sector instruments are extremely sensitive and selective, however, the expense and difficulty of use have kept these mass spectrometers beyond the realm of most analytical laboratories.

Ion Trap Detectors

Ion trap detectors (ITD), are relatively inexpensive and readily available from several manufacturers. The ITD is robust enough for everyday use and though an ITD is more expensive than an electron capture detector, the quality of the data is far superior. While a selective detector provides only a retention time and some information about a compound the ion trap can provide a full mass spectrum of a compound.

Early ion traps had a limitation because samples with a complex matrix, a so-called "dirty sample," would not give "classical" mass spectra. This was resolved with Automatic Gain Control was incorporated into the second generation of ion traps.(5) External ionization now available from Thermoquest is another technique available to further enhance the quality of the mass spectra obtained from an ion trap. These innovations along with (-) chemical ionization make the ion trap a tool worth considering for the analysis of chlorinated compounds in wildlife tissue samples.

Application of Modern Ion Traps

The current ion traps have significant advantages over SIM and selective detectors for residue analyses. The current state-of-the-art ITD is capable of multiple mass spectra or MS^n. This includes tandem mass spectroscopy (MS-MS or MS^2) which is especially useful when analyzing dirty matrices and in cases where an unambiguous confirmation is required. Tandem mass spectroscopy isolates a single parent ion in the trap while all other ions are ejected. The isolated ion then undergoes a collisionally induced dissociation (CID) giving rise to ions known as daughter or secondary ions. This is known as "tandem in time." "Tandem in space" can also be done with a triple quadrapole instrument but the cost is significantly higher and the instruments are not capable of simultaneous full scan – tandem mass spectroscopy.

Guidelines for the use of MS^2 for quantitation require at least two daughter ions be detected, each with a signal to noise ration of not less than 5:1 (6). Tandem mass spectroscopy is demonstrated in Figure 2. This technique has been successfully applied to several pesticide residue problems. One group is using MS-MS extensively for the determination of residues in fruits and vegetables (7). Another interesting application has been in the investigation of polychlorinated dibenzodioxins (8). Several of these compounds are extremely toxic with the level of interest in the part per trillion range. Until the use of MS-MS with ion traps, the only EPA approved method of analysis for these compounds was with a magnetic sector instrument in what is known as High Resolution Gas Chromatography-High Resolution Mass Spectrometry (HRGC-HRMS) (9). With MS^2, detection limits in the range of 25 ppt have been obtained, depending on the isomer (8). While this is still more than an order of magnitude greater than that of HRGC-HRMS, the ion trap is more readily available and significantly reduces the cost of these analyses. This places analysis of dioxins within the realm of possibility for many labs which would never have the funds or expertise to have a magnetic sector mass spectrometer on site.

Application of GC-Ion Trap MS in the TPW Laboratory

At the TPW Lab the goal is not to "err on the side of caution" but to provide reliable analytical data. To meet that goal several samples have been analyzed by ECD, SIM, and ion trap detectors. There is not sufficient space to review all of the results but some selected examples will help to illustrate the advantages obtained from the use of the ion trap detector.

Dieldrin is a chlorinated pesticide prepared from epoxidation of aldrin. This epoxide compound has a low LC 50 in fish (0.037 mg/L) (10). ECD analysis indicated that this compound was present in a fish tissue sample at 40 ppb. Further analysis by SIM also confirmed that dieldrin was present. However, when the sample was re-analyzed by ion trap GC-MS full scan, no dieldrin was detected. To ensure the ion trap was not giving a false negative, the extract was fortified with dieldrin to give an equivalent of 5 ng/g tissue level. Dieldrin was then readily detected (Figure 3).

Fish kills are a frequent occurrence during the hot summer months in southern states. Most of these events can be traced to elevated temperature and low dissolved oxygen. In some cases information is obtained about chemical use that leads to an investigation of a specific compound. One such case involved the reported application of the fungicide chlorothalonil. Extracts from dead fish tissues were analyzed by GC-MS-MS and chlorothalonil was identified at a concentration of 7 ng/g. (Figure 4). To further confirm this identification, negative chemical ionization was done followed by MS-MS to unambiguously identify the fungicide.

Simultaneous Full Scan and Tandem Mass Spectrometry

One of the most interesting applications of modern ion traps is the recent technological innovation which allows the instrument to perform full scan mass

EI MS-MS Mass Spectrum

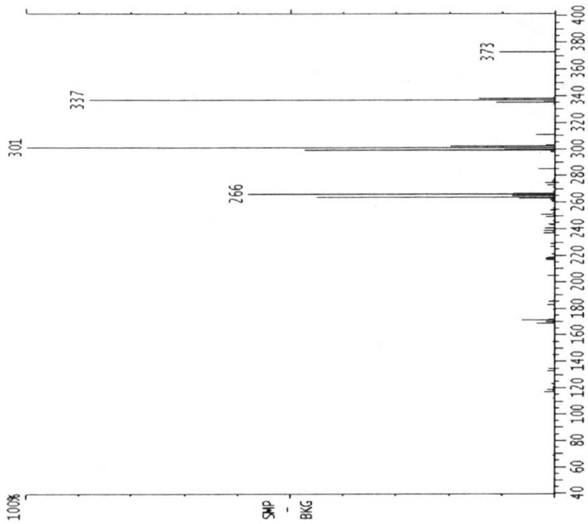

EI Single Stage Mass Spectrum

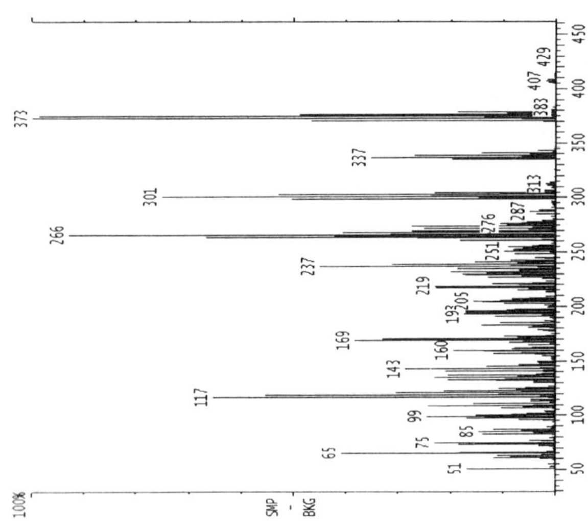

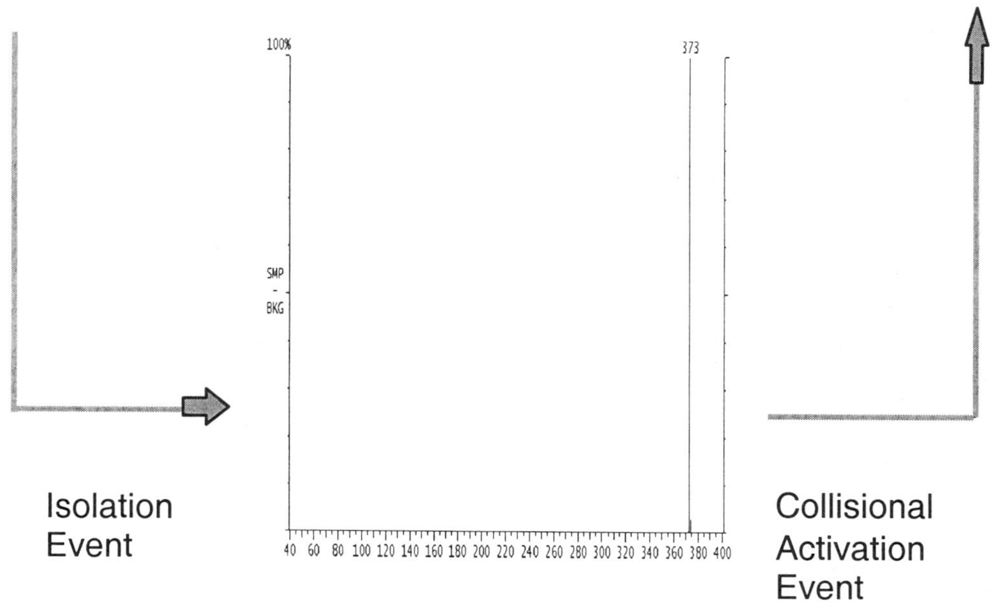

Figure 2. E.I mass spectrum where the m/e 373 ion is isolated after a CID gives rise to the MS-MS Spectrum

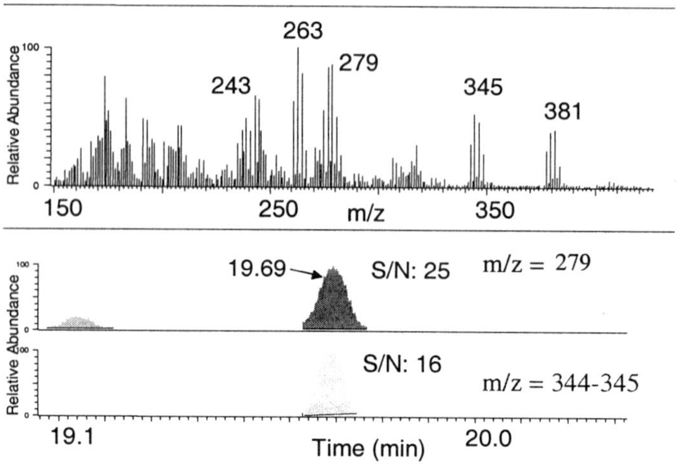

Figure 3. Dieldrin Spiked in Fish Tissue at 5 ng/g

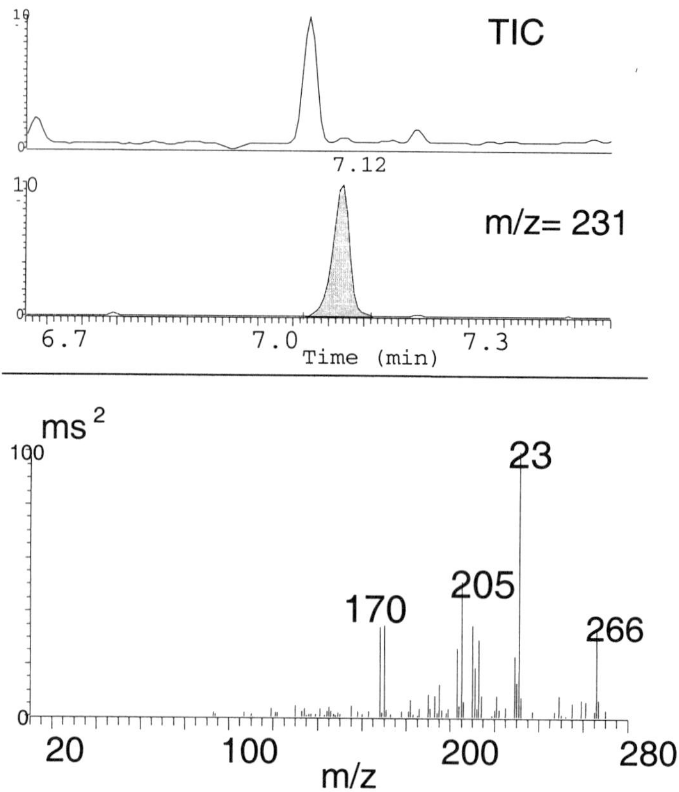

Figure 4. EI MS-MS of Sample with chlorothalonil detected

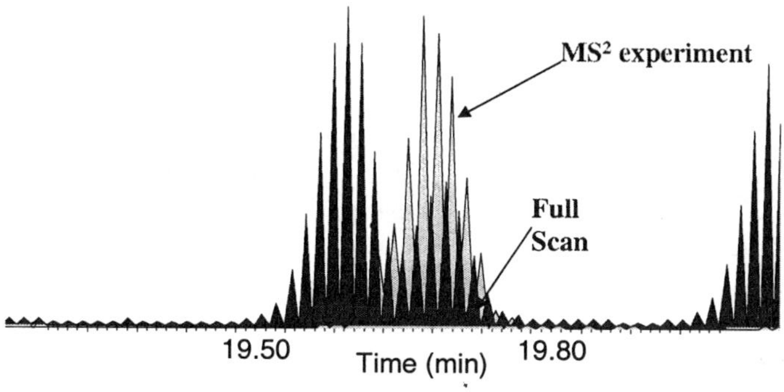

Figure 5. Chromatogram showing alternating full scan and MS^2 for simultaneous analysis across peak.

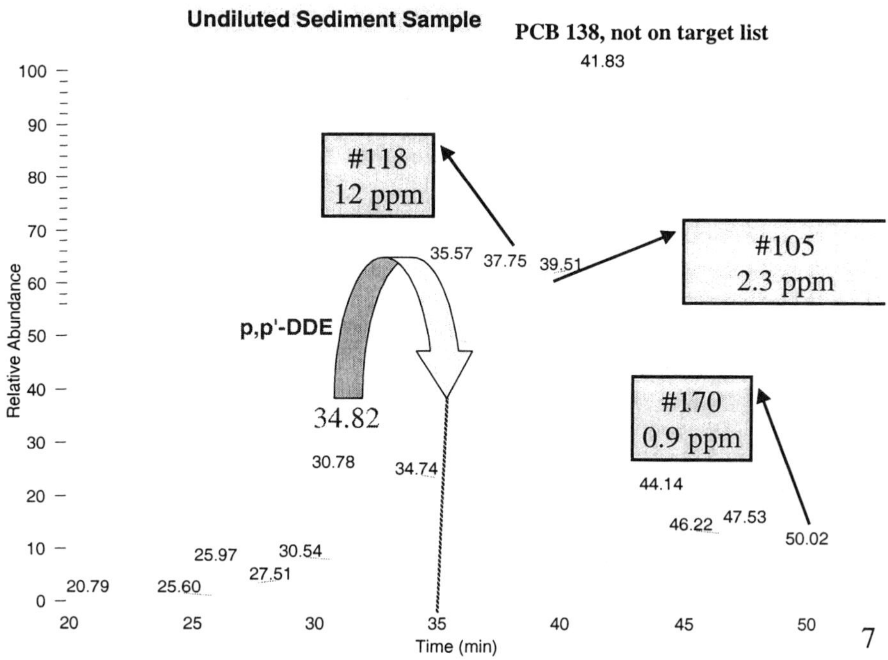

Figure 6. PCB and DDE in simultaneous mode

spectrometry simultaneously with MS-MS. The instrument does this by alternating a full scan spectrum with an MS-MS scan. There are four to five scans per second. The software separates the scans and provides two separate chromatograms for the sample. (Figure 5)

This technique was used to investigate a challenging problem: how to identify trace amounts of pesticides in the presence of significant amounts of PCBs. Historically, the only method for this analysis was to use liquid chromatography to separate the compounds followed by separate gas chromatographies. As stated previously, this is a time consuming step which greatly increases waste solvents and adds additional sample transfers with the potential of sample loss at each transfer. With the advent of simultaneous full scan and MS-MS it is possible to analyze extracts for PCBs and DDE with only a gel permeation clean up to remove fats as illustrated in Figure 6. In this chromatogram, p,p'- DDE is seen by MS-MS as less than baseline in the presence of ppm levels of various PCB isomers. In this case the ion trap is functioning as a chromatograph to separate and isolate the DDE parent ion. The ion trap then performs MS-MS for identification and confirmation of DDE in this sample at the retention time of 34.82.

Conclusion

Ion trap detectors have proven to be sensitive, selective, and reliable for investigation of residues in tissue extracts. At ultra-trace levels of interest there are too many interferences from the matrix to provide acceptable residue confirmation by electron capture or selected ion monitoring detection methods. An ion trap is more expensive than an electron capture detector but the data obtained is significantly more reliable. With the advent of simultaneous full scan and tandem mass spectrometry, ultra trace levels of residues can be detected without giving up the ability to look for large amounts of unknown or unexpected residues as is the case in selected ion monitoring and without giving up any sample through-put.

Acknowledgement

We would like to thank the U.S. Fish and Wildlife Service and the Sport Fish Restoration Fund for financial support of this work.

References

1. *Official Methods of Analysis of the Association of Official Analytical Chemists*; Helrich, K. Ed;, Fifteenth Edition; Association of Official Analytical Chemists, Gaithersburg, MA, 1990, Vol. 1, pp 283-284.

2. *Methods for Organic Chemical Analysis of Municipal and Industrial Wastewater;* Longbottom, J.E. and Lichtenberg, J.J.: Eds.; Organochlorine Pesticides and PCBs – Method 608, U.S. Environmental Protection Agency: Cincinnati, OH, 1982; pp 608.1 – 608.11.
3. Methods of Analysis by the U.S. Geological Survey National Water Quality Laboratory—Determination of Chlorinated Pesticides in Aquatic Tissue by Capillary-Column Gas Chromatography with Electron-Capture Detection; Leiker, T.J.; Madsen, J.E.; Deacon, J.R.; Foreman, W.T., Eds., U.S. Geological Survey, Denver, CO, 1995; pp. 11-12.
4. *Methods for Organic Chemical Analysis of Municipal and Industrial Wastewater;* Longbottom, J.E. and Lichtenberg, J.J.: Eds.; Organochlorine Pesticides and PCBs – Method 625, US Environmental Protection Agency: Cincinnati, OH, 1982; pp 625.1 – 625.17.
5. Huston, C.K.; *J. Chromatogr.* **1992**, 606, 203-209.
6. Goodenowe, D.; Duffy, M.; Establishing LC-MS Confirmation Criteria; 47th American Society for Mass Spectrometry Conference, Dallas, TX,
7. Sheridan, R.S.; Meola, J.R. *J. AOAC Int.* **1999**, Vol. 82, 00.
8. Hayward, D.G.; Hooper, K.; Andrzejewski, D. *Anal. Chem.* **1999**, 7, 212-220.
9. *Guidance for Assessing Chemical Contaminant Data for Use in Fish Advisories*, Bigler, J. and Greene, A. Eds; Volume 1: Fish Sampling and Analysis; Office of Science and Technology, Office of Water, U.S. Environmental Protection Agency, Washington, DC, 1993, p 8.6.
10. *Farm Chemicals Handbook*, R.T., Ed., Meister Publishing Company: Willobughby, OH, 1997; p. C 128

Chapter 19

Recent Advances in Analytical Techniques to Investigate Pesticide Poisoning of Wildlife

Ainsley Jones, S. McGaw, L. Ross, C. McCoy, G. Turnbull, A. J. A. C. Charlton, and Simon Hird

Central Science Laboratory, Ministry of Agriculture, Fisheries and Food, Sand Hutton, York Y041 1LZ, United Kingdom

In the UK, suspected cases of pesticide poisoning of terrestrial wildlife (including beneficial insects) and companion animals are investigated by the Wildlife Incident Investigation Scheme. To ensure that analyses for pesticides are performed to the desired standard in the most cost effective manner, there is an ongoing program to develop new and improved methods. Some of the recent work in this program has investigated improved methods of detection and extraction of pesticides. Liquid chromatography-mass spectrometry (LC-MS) of anticoagulant rodenticides and carbamate pesticides is discussed. The use of gas chromatography-mass spectrometry-mass spectrometry (GC-MS-MS) is discussed in cases for which extremely selective detection is required. Investigations into supercritical fluid extraction (SFE) of organophosphate and organochlorine insecticides are described.

Introduction

Post-registration monitoring of wildlife poisoning from pesticides in the UK is undertaken by the Wildlife Incident Investigation Scheme (WIIS). The Scheme is a reactive one which investigates suspected cases of poisoning reported by members of the public (1). In order to perform such forensic toxicology investigations, suitable analytical methods are required.

Our intention in investigating analytical methods and techniques is to develop approaches to analysis that provide results of suitable quality in the most cost-effective manner. The quality requirements for analytical methods in this work can be considered under 4 parameters:

- selectivity
- sensitivity
- robustness
- quantitation

Selectivity is extremely important in this work. The results may be used by regulators as a basis on which to amend approvals or the results may be used in criminal prosecutions for pesticide misuse. We aim for virtually unequivocal identification of pesticides and make extensive use of mass spectrometric techniques for analysis or confirmation.

Extremely sensitive analysis methods are not always required in this work. Vertebrate animals poisoned by fast acting pesticides such as anticholinesterase insecticides usually have large residues in the stomach and reaching the required detection limits is not problematic. Honeybees are however much more susceptible to insecticide poisoning than vertebrates and body residues at time of death are much lower. This, combined with the difficult nature of the samples, which contain a high level of wax, means that achieving required detection limits is much more challenging. For slower acting pesticides such as bipyridylium herbicides and anticoagulant rodenticides, there are unlikely to be residues in the stomach at death and analysis of internal tissues is necessary. Concentrations in such tissues will be much lower than stomach contents levels found with rapid acting compounds and reaching the required detection limits may be difficult.

Robustness of methods is another important factor for this work. Methods need to be reliable in routine use but should also be able to cope with wide variation in matrix composition. Stomach contents samples are obviously extremely variable but there can be considerable variation in the composition of internal organs, especially with regard to fat content.

The ability to accurately measure concentrations is, in general, of less importance, than some of the above factors. A toxicological assessment requires that the dose of toxic agent be compared with known toxicity data. However, the analytical measurement may be a poor indicator of dose. In the case of rapidly acting compounds, the amount measured in the stomach is only the unabsorbed portion of the dose and the amount adsorbed remains unknown. For slower acting compounds it is practicality impossible to relate tissue concentration to dose. In addition there may be a considerable time-lag between the time of death and the animal being discovered, allowing time for significant degradation of the pesticide residue present at death. Even if the dose can be accurately estimated, toxicity information is often not available for the pesticide and species combination involved and the assessment has to be done by extrapolation from toxicity studies on other species. When all these areas of uncertainty are kept in mind, there is little benefit in producing accurate estimates of residue concentrations. In practice the assessment of whether an animal has been poisoned is often based on other evidence such as brain cholinesterase levels or post-mortem findings such as haemorrhaging from anticoagulants or oedema of the lungs from paraquat. The role of chemical analysis is largely to identify the exact causative agent of such effects.

The methods in use in WIIS in 1996 have been described by Brown *et al* (2). In this paper we describe some recent investigations to attempt to develop improved analytical methods for such forensic toxicology investigations. Much of the effort has involved the search for more selective detection methods. In addition to improving the certainty of compound identification, more selective detection methods can also mean that time-consuming and expensive clean-up stages can be minimised or eliminated. GC-MS has been used for many years in WIIS to confirm pesticide residues (2). Its selectivity makes it extremely useful in providing compound identification at a very high level of certainty. However many compounds are insufficiently volatile to be analysed by GC. LC-MS employing atmospheric pressure ionisation techniques such as electrospray ionisation (ESI) and atmospheric pressure chemical ionisation (APCI) is a highly selective detection technique that is rapidly gaining acceptance as a technique in pesticide analysis and there have been a number of reports of its application to complex matrices such as food (3-9). We have evaluated LC-MS for confirmation of pesticides not amenable to GC analysis. For our work, one of the most important group of pesticides in this category is anticoagulant rodenticides. Multi-residue methods for the analysis of these compounds have been described based on HPLC with fluorescence or UV detection (10-16) but no mass spectrometric methods have been described. A further area in which LC-MS has been used for confirmatory analysis of wildlife samples is oxime carbamate pesticides. Of these, aldicarb is of particular interest due to its high mammalian toxicity. Most carbamate pesticides are of the N-phenyl methyl type and can be analysed by GC-MS but aldicarb and its metabolites are much less amenable to GC-MS. LC-MS methods with atmospheric pressure ionisation have been described for carbamates in complex matrices such as food (4,18) and sediment (19).

For compounds amenable to GC analysis, even the selectivity of GC-MS is not sufficient in all cases. The types of samples we examine vary greatly and confirmation can prove extremely difficult in some samples. GC-MS-MS is a technique which offers improved selectivity over GC-MS (17) and we have investigated its use in such circumstances.

Another time consuming step in analytical procedures is the extraction of analytes into a suitable liquid and we have investigated alternatives to traditional solvent extraction methods. For complex matrices solvent extraction usually results in extracts containing large amounts of co-extractives and it is to remove these that sample clean-up is employed. More selective extraction would also reduce the requirement for sample clean-up and we have also investigated the extraction selectivity of alternative extraction techniques. Supercritical fluid extraction is an alternative to classical, solvent-based extraction techniques. It can be faster and easier to automate than traditional techniques and does not require large volumes of extraction solvents. It has been applied to the extraction of pesticides from complex matrices such as fruit and vegetables (20-24), flour

(25), eggs(26), meat (27, 28), animal tissues (30) and honeybees (31). In addition to the above advantages, SFE can offer more selective extraction giving rise to cleaner extracts (20, 30). The usefulness of SFE was evaluated for extraction of organophosphate (OP) pesticides from animal stomach contents and organochlorine (OC) pesticides from animal livers.

Materials and Methods

Liquid Chromatography-Mass Spectrometry (LC-MS)

LC-MS was performed on a Micromass Quattro I tandem quadropole (Micromass, Manchester, UK) equipped with a model 231 autosampler (Gilson, Villiers le Bel, France) and a model 600 HPLC pump (Waters, Milford, MA)

Anticoagulant Rodenticides

Data was from the Quattro LC-MS operated in negative ion electrospray ionisation (ESI) mode. Full scan and selected ion monitoring data were collected using a dwell time for each channel of 0.1 s, an interchange delay of 0.02 s and a mass span of 0.2 Da. Eight channels were monitored at *m/z* 541.2, 527.1, 525.1, 523.1, 521.1, 443.2, 307.1 and 291.1, corresponding to the molecular ions of flocoumafen, bromadiolone (Br^{79} isotope), bromadiolone (Br^{81} isotope), brodifacoum (Br^{79} isotope), brodifacoum (Br^{81} isotope), difenacoum, warfarin and coumatetralyl, respectively. The HPLC column was a 150 mm x 2.1 mm Genesis C8 4 µm (Jones Chromatography, Hengoed, UK) with a ternary gradient composed of component A (water), component B (methanol) and component C (0.05 M ammonium acetate solution). This was programmed at a flow rate of 0.2 mL/min as follows. At t= 0 the proportions of A, B and C were 70:20:10 respectively. This was linearly programmed so that at 8 min proportions were 10:80:10 and this was held until 18 min. The components were then linearly programmed back to the initial conditions by 19 min and held for a further 9 min. Six microliter injections were made. Extracts were prepared according to the method of Jones (16).

Aldicarb and Metabolites

Data was from the Quattro LC-MS operated in positive ion ESI mode. Two channels, corresponding to the $[M+H]^+$ ion and $[M+NH_4]^+$ ion were monitored for each of aldicarb, aldicarb sulfoxide and aldicarb sulfone. The channels for aldicarb were monitored from a run time of 8 min. to 18 min. For the sulfoxide and sulfone the channels were monitored from 0 to 8 min. The dwell time for each channel was 0.1 s, the interchange delay was 0.02 s and the mass span was 0.2 Da. The HPLC column was a 150 mm x 2.1 mm Hypersil Carbamate column with a binary gradient composed of component A (0.1% v/v acetic acid in 0.05 M ammonium acetate solution) and component B (0.1% v/v acetic acid in acetonitrile). This was programmed at 0.2 mL/min as follows. At t= 0, %A=85 then linearly programmed to 90% B at 11 min and held until 16 min. The components were then linearly programmed back to the initial conditions by 17 min and held for a further 2 min. Seven microliter injections were made. Extracts were prepared as described by Brown (2).

Gas Chromatography-Mass Spectrometry-Mass Spectrometry(GC-MS-MS)

Analyses were performed with a Finnigan MAT GCQ ion-trap mass spectrometer fitted with an A200S autosampler (Finnigan MAT, Hemel Hempstead, UK) with electronic pressure control (EPC). The column was coupled to a 4 mm uniliner direct injection liner (Thames Restek, Windsor, UK). One-microliter injections were made into the injector operated at 80 °C; initial oven temperature was 35 °C held for 1 min, then linearly increased at 25 °C/min to 280 °C, held for 10 min. The column was 30 m x 0.25 mm i.d. coated with BPX5 at 0.25 μm film thickness (SGE Europe, Milton Keynes, UK). The EPC was used to provide a constant linear velocity of helium carrier gas at 40 cm/sec. In order to ensure fast initial injection flow conditions the carrier gas was programmed for an initial pressure of 30 psi, held for 1 min. Typical mass spectrometer operating conditions were: full-scan acquisition mode from *m/z* 50-200 at 2 scans/sec, ion source temperature 180°C, electron impact (EI) ionization at 70 eV with the multiplier tube voltage set at 1450 V.

Supercritical Fluid Extraction (SFE)

SFE was performed using a Model 7680T (Hewlett-Packard, Bracknell, UK) with automated restrictor and solid sorbent collection system. Extraction parameters for OPs were 187 bar pressure and 60°C (CO_2 density, 0.7 g/mL); 2-min equilibration time followed by 40 ml of CO_2 at 1.6 mL/min; 50 °C restrictor temperature. Collection was on an octadecylsilica (ODS) trap at 25 °C and the analytes eluted with 1.5 mL acetonitrile at 0.5 mL/min into 2-ml sealed vials. The trap was then rinsed to waste with 4 mL ethyl acetate followed by 2 ml acetonitrile. OCs were extracted at 122 bar pressure and 60 °C (CO_2 density, 0.45 g/mL); 2-min equilibration time followed by 40 mL of CO_2 at 1.6 mL/min; 50 °C restrictor temperature. Collection was on a diol trap at 25 °C and analytes eluted with 1.5 mL acetonitrile at 0.5 mL/min into 2-ml sealed vials. The trap was rinsed to waste with 4 mL ethyl acetate followed by 2 mL acetonitrile. After extraction, extracts were evaporated to dryness under a stream of nitrogen and redissolved in 5 mL of hexane:acetone (50:50).

A 1-g sub-sample of stomach contents or liver was thoroughly homogenised with 2 g of Hydromatrix diatomaceous earth by grinding in a glass pestle and mortar. After allowing 15 min. for all water to be absorbed this mixture was packed into a 7-mL extraction vessel and any remaining space filled with Hydromatrix.

SFE extracts were analysed on a Hewlett Packard model 5890 GC equipped with an on-column injector connected to a Finnigan model ITD800 ion-trap mass spectrometer (Finnigan MAT, Hemel Hempstead, UK). One-microliter on-column injections were made into the injector. Initial oven temperature was 60 °C held for 1 min, then linearly increased at 25 °C/min to 260 °C, held for 3 min; initial injection temperature was 63 °C and the injector was programmed to be 3 °C above the oven temperature. The column was 25 m x 0.25 mm i.d. coated with BPX5 at 0.25 μm film thickness, retention gap was 1 m x 0.53 mm i.d. of deactivated uncoated silica column. He head pressure was 15 psi. Mass spectrometer operating conditions were: full-scan acquisition mode from m/z 50-400, electron impact ionisation at 70 eV and a typical multiplier tube voltage was 1700 V.

Results and Discussion

Liquid Chromatography-Mass Spectrometry (LC-MS)

Anticoagulant rodenticides fall into two categories, 4-hydroxy coumarin based and indanedione based. Initial work concentrated on identifying suitable mass spectrometry operating conditions by flow injection analysis of standard solutions in methanol:water. Both electrospray ionisation (ESI) and atmospheric pressure chemical ionisation (APCI) were investigated. In positive ion ESI and APCI modes the response was low for all rodenticides with the spectra of coumarin rodenticides dominated by adducts such the protonated molecular ion $[M+H]^+$, sodium adduct $[M+Na]^+$ and water methanol adduct $[M+(water+methanol)2H]^+$. Spectra for indanedione rodenticides were more complex and not characterised by the presence of easily identifiable adduct ions. In negative ion mode spectra were characterised by deprotonated molecular ions $[M-H]^-$ as would be expected for weakly acidic compounds which can easily lose a proton. The results indicated that ESI in negative mode was the ionisation mode of choice for all the rodenticides of interest.

When coupled to HPLC however, it did not prove possible to obtain the desired sensitivity using the conditions we use for screening analysis of coumarin anticoagulants. These conditions involve addition of acetic acid to a water/methanol mobile phase on a reverse phase column. This low pH is intended to suppress ionisation and keep the compounds in the protonated form to improve chromatography but appeared to have the effect of also suppressing $[M-H]^-$ ion formation in the source. By employing a base-deactivated column based on ultrapure silica it proved possible to achieve satisfactory chromatography of coumarin rodenticides at a suitably high pH to allow the desired sensitivity to be realised. For indanedione rodenticides it did not prove possible to achieve adequate chromatography on reverse phase columns at suitably high pH. Ion-pair chromatography using both tetra butylammonium and triethylamine based systems was briefly investigated for these compounds but sensitivity was poor and chromatography unstable.

Figure 1 shows LC-MS chromatograms from a liver sample containing 0.22 mg/kg brodifacoum and 0.10 mg/kg difenacoum. Detection limits routinely achieved in liver tissue range from 0.006 μg/g for difenacoum to 0.02 μg/g for bromadiolone.

Figure 2 shows LC-MS chromatograms from a sample containing 23 mg/kg of aldicarb and 7.5 mg/kg aldicarb sulphoxide. Limits of detection we routinely achieve, based on the respective $[M+NH_4]^+$ ions, are approximately 5 mg/kg for aldicarb and 1mg/kg for aldicarb sulphone and sulphoxide. Note that for aldicarb the response for the $[M+H]^+$ ion is much lower than for the $[M+NH_4]^+$ ion and in most cases only the latter ion can be detected. The selectivity of the detection system suggests that by greater concentration of the sample extract, lower detection limits could be achieved. We have not investigated this as residues detected are normally well above these limits, however 10-fold lower limits should be easily achievable and 50-fold lower might be possible.

Gas Chromatography-Mass Spectrometry- Mass Spectrometry (GC-MS-MS)

Figure 3 gives an example of such a situation from the suspected poisoning of a red kite. A screening analysis suggested that the sample contained fenthion but confirmation was not possible by GC-MS as the mass spectrum of the fenthion peak in the sample extract suffered from severe interferences. Improving the selectivity could, in theory, be accomplished by developing an improved clean-up procedure but such an approach is impractical for one-off situations such as these. Figure 3 also shows the MS-MS spectra of the same standard solution and sample extract obtained by selecting the parent ion at *m/z* 278 and generating the daughter spectra by collisionally induced dissociation with collision energy applied at an amplitude of 1.3 V. Under these conditions the sample extract shows almost no interferences in the mass spectrum.

Supercritical Fluid Extraction

The SFE condition used were those previously developed for extraction of pesticides from honeybees (30). Extraction efficiency was assessed my measuring recoveries of OPs and OCs from spiked samples (stomach contents for OPs and liver for OCs) by comparing peak areas with calibration solutions. Up to three ions were used for quantitation of each compound and these are given in Table I. The stomach contents samples used for spiking were from a range of animals: cat, dog, hare, kittiwake and badger. The four liver samples used for spiking came from a cat, dog, sparrowhawk and duck. Results of recovery experiments are given in Tables II and III.

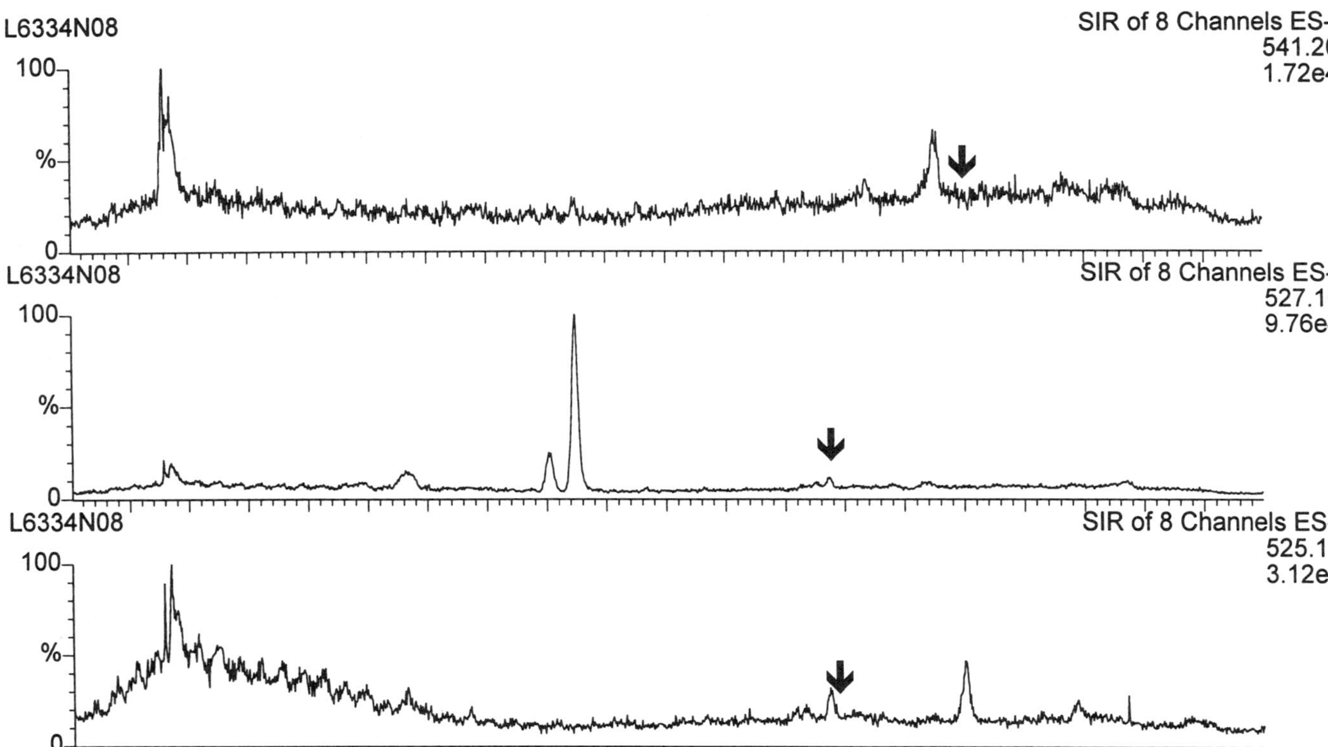

Figure 1. LC-MS chromatograms from extract of buzzard liver containing 0.22 mg/kg brodifacoum and 0.10 mg/kg difenacoum. Channels displayed are (top to bottom) flocoumafen, bromadiolone (Br^{81}), bromadiolone (Br^{79}), brodifacoum (Br^{81}), brodifacoum (Br^{79}), difenacoum, warfarin and coumatetralyl. Retention times of rodenticides not present are marked with arrows.

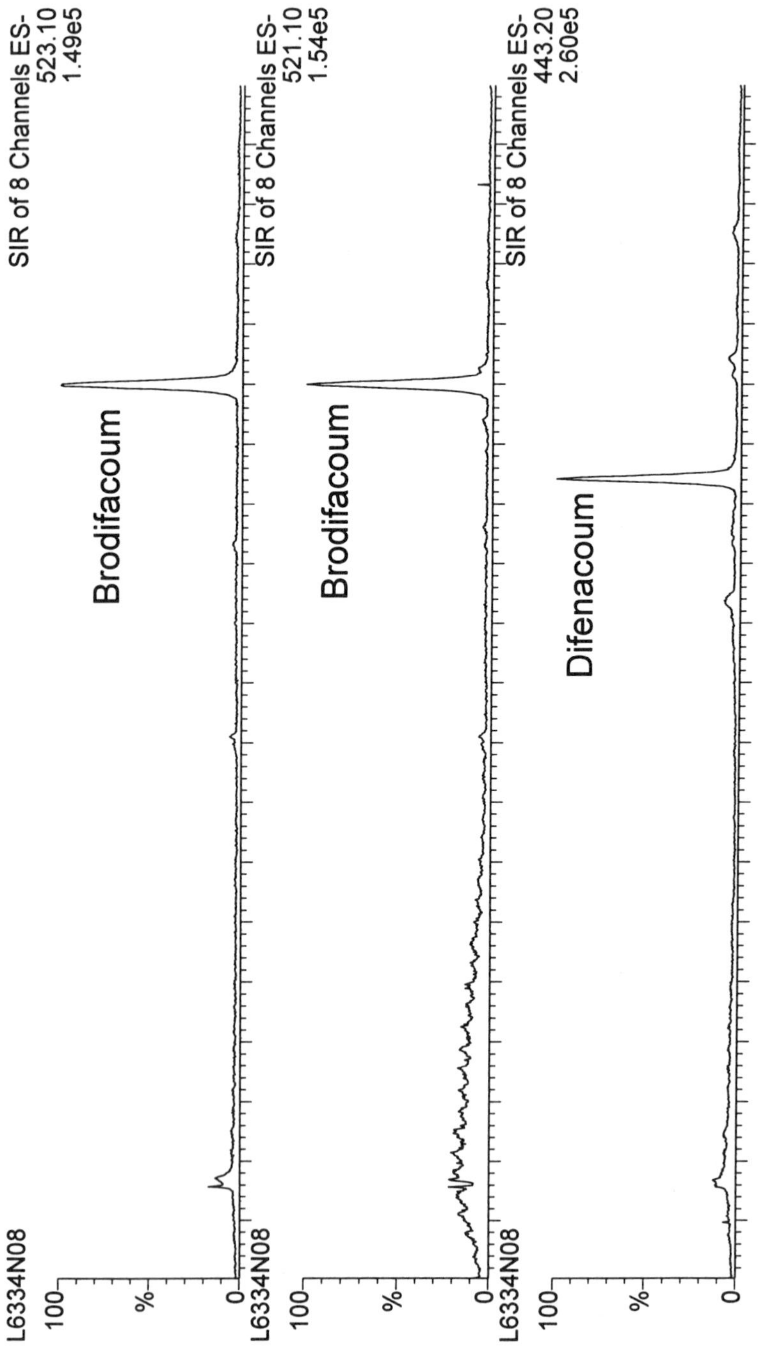
L6334N08
SIR of 8 Channels ES-
523.10
1.49e5
100
%
0
Brodifacoum
L6334N08
SIR of 8 Channels ES-
521.10
1.54e5
100
%
0
Brodifacoum
L6334N08
SIR of 8 Channels ES-
443.20
2.60e5
100
%
0
Difenacoum

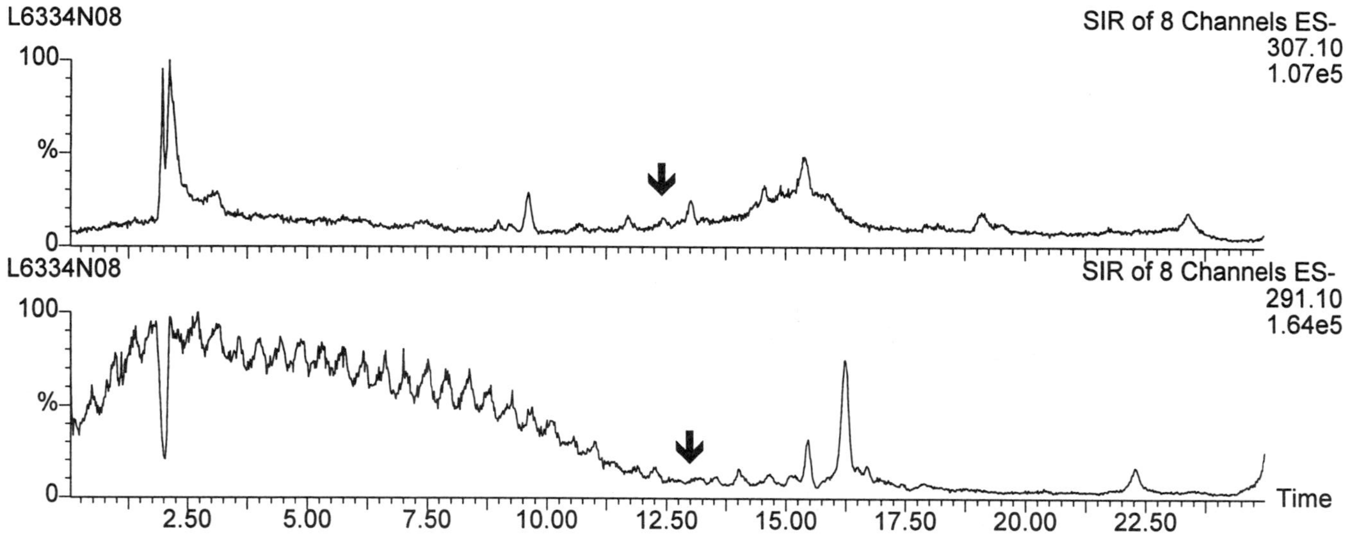

Figure 2. LC-MS chromatograms from extract of cat stomach contents containing 23 mg/kg aldicarb and 7.5 mg/kg aldicarb sulfoxide. Channels displayed are the $[M+H]^+$ ions and $[M+NH_4]^+$ ions for aldicarb, aldicarb sulfoxide and aldicarb sulfone. Retention time of aldicarb sulfone, which is not present is marked with arrows.

Continued on next page.

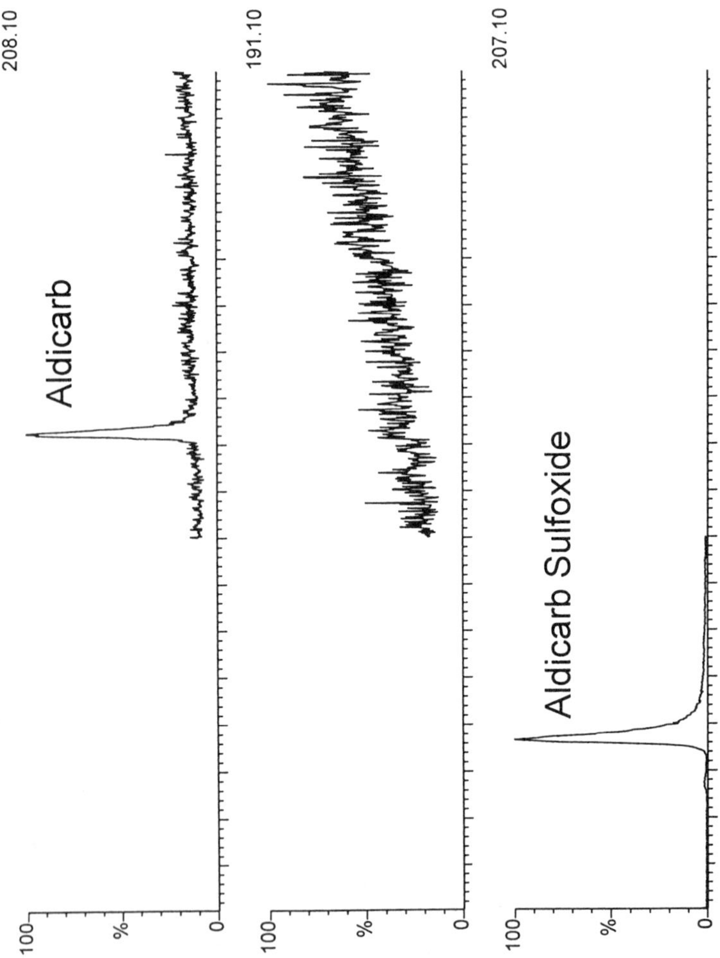
208.10
Aldicarb
191.10
207.10
Aldicarb Sulfoxide
100
%
0
100
%
0
100
%
0

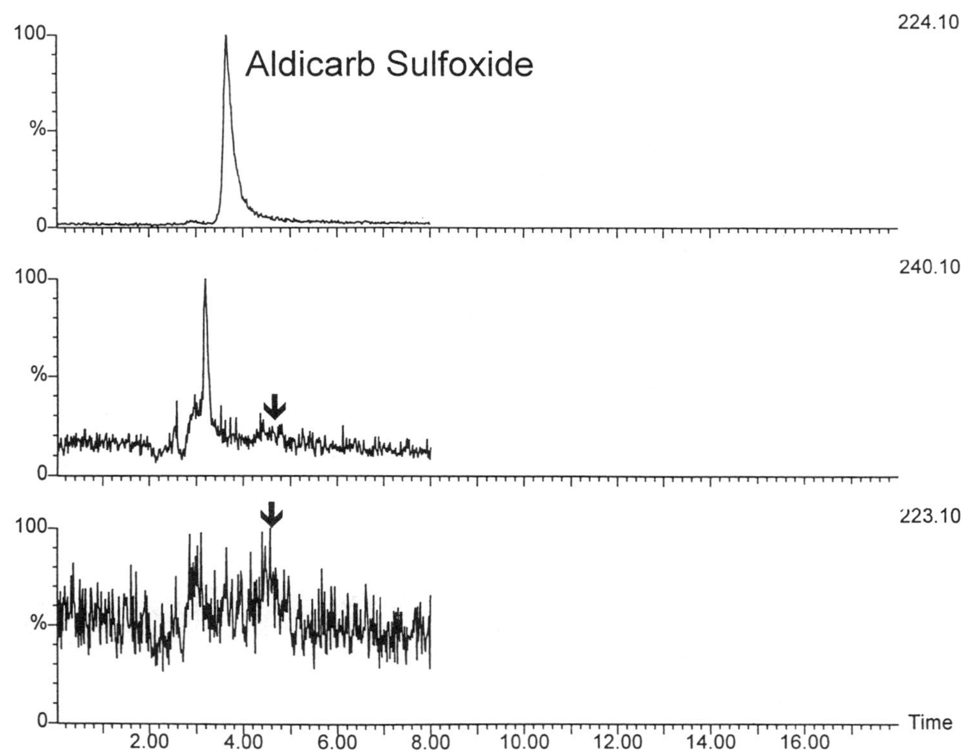

Figure 2. *Continued.*

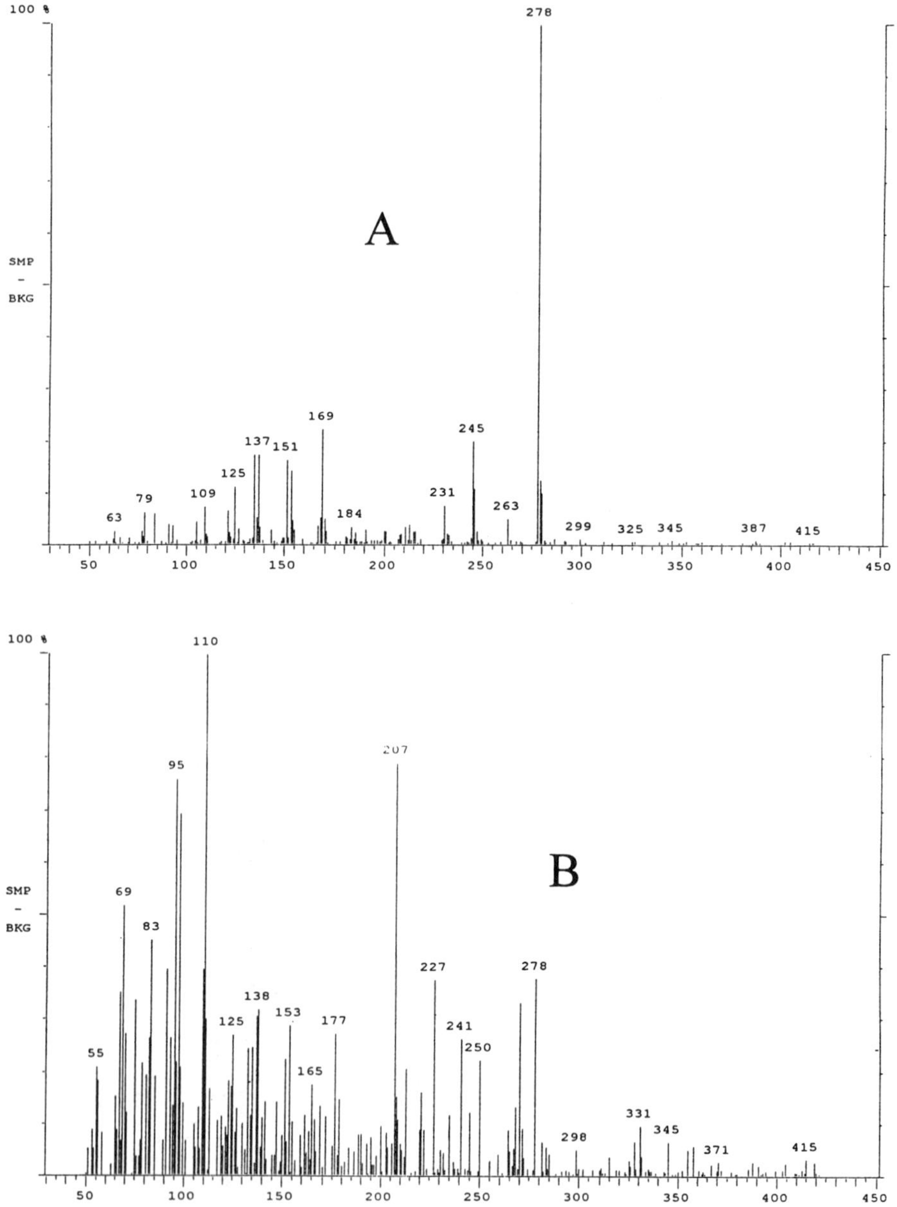

Figure 3. Background subtracted mass spectra; (A) full scan spectrum from 0.2 ug/mL standard solution of fenthion; (B) full scan spectrum from red kite stomach contents extracts containing 0.14 μg/mL of fenthion; (C) MS-MS daughter ion spectrum with *m/z*=278 as parent ion from 0.2 ug/mL standard solution of fenthion. (D) MS-MS daughter ion spectrum with *m/z*=278 as parent ion from red kite stomach contents extracts containing 0.14 μg/mL of fenthion.

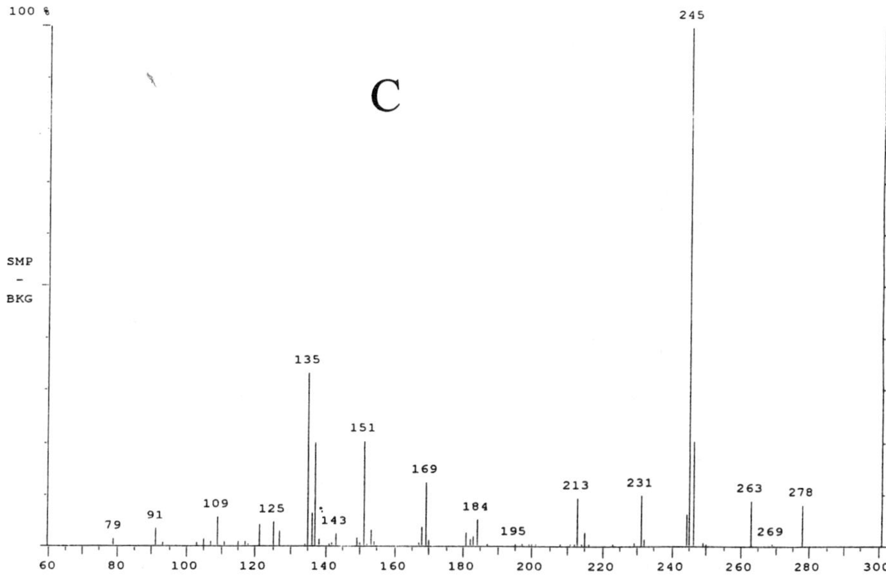

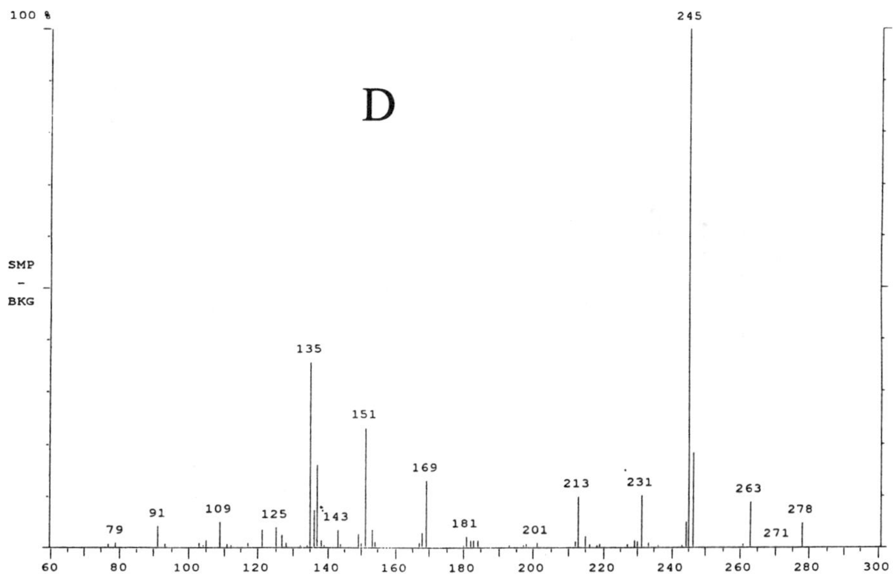

Figure 3. *Continued.*

Table I. Ions Used for Quantitation of Pesticides

	m/z 1	*m/z 2*	*m/z 3*		*m/z 1*	*m/z 2*	*m/z 3*
Azamethiphos	109			Mephosfolan	140		
Azinphos-methyl	160	132	104	Mevinphos	127	164	193
Bromophos-methyl	331	125		Naled	109	185	
Carbophenothion	157	199	342	Omethoate	156	110	
Chlorfenvinphos	267	323		Parathion	291	109	
Chlorpyrifos	352	314	258	Phorate	75	231	260
Chlorpyrifos-methyl	125	288		Phosalone	121	182	367
Crufomate	256	182		Phosmet	160	133	77
Demeton-S-methyl	142	169	109	Pirimiphos-methyl	276	290	
Diazinon	276	179	304	Propetamphos	138	194	222
Dichlorvos	109	185		Pyrazophos	221	265	232
Dimethoate	87	125		Quinalphos	156	298	146
Disulfoton	89	274		Thiometon	125	88	
Ethoprophos	243	158		Triazophos	162	172	257
Etrimphos	292	277	181	gamma-HCH	217	221	185
Famphur	218	125		Aldrin	263	293	66
Fenitrothion	260	277	109	Heptachlor epoxide	353	263	237
Fenthion	278			Dieldrin	263	277	
Fonofos	109	137	247	Endrin	345	251	
Heptenofos	215			pp-DDT	235	165	199
Iodofenphos	377	125	250	Endosulfan	195	241	207
Malathion	127	173					

Table II. Recoveries of Organochlorine Insecticides from Spiked Liver

	Amount Added (μg/g)	*Mean Recovery*	*Range*	*Amount Added (μg/g)*	*Mean Recovery*	*Range*
Aldrin	3.7	82	46-105	37	78	76-81
Dieldrin	6.5	73	37-90	65	82	77-85
ppDDT	2.9	86	53-109	29	83	80-90
Endrin	8.4	55	39-71	84	76	58-108
Endosulfan	2.7	a		27	95	83-109
gamma HCH	2.3	88	84-90	23	85	81-90
Heptachlor epoxide	8.2	a		82	85	59-99

[a] below limit of detection

Table III. Recoveries of Organophoshorous Insecticides from Spiked Stomach Contents, Means and Standard Deviations of 6 Results

	Amount Added (µg/g)	*Mean Recovery*	*Range*	*Amount Added (µg/g)*	*Mean Recovery*	*Range*	*Amount Added (µg/g)*	*Mean Recovery*	*Range*
Azamethiphos	4.8	72[a]	51-93	48	120[a]	100-136	240	114[a]	100-126
Azinphos-methyl	3.9	135	84-213	39	161	100-315	195	165[b]	125-257
Bromophos-methyl	4.3	98	65-125	43	127[b]	85-270	215	112[b]	93-142
Carbophenothion	5.3	106	63-155	53	155[b]	97-385	265	117[b]	92-167
Chlorfenvinphos	4.6	97	53-128	46	102	84-120	230	110	92-148
Chlorpyrifos	6.9	106	68-135	69	94	70-129	345	97[c]	80-108
Chlorpyrifos-	1.2	89	81-112	12	97	73-110	60	106[c]	84-136
Cruformate	4.3	117	30-183	43	100	74-117	215	110[c]	88-125
Demeton-S-methyl	5.2	81	54-111	52	86	65-99	260	123	92-182
Diazinon	4.6	87	67-118	46	91	63-119	230	103[c]	93-118
Dichlorvos	5.7	72	46-101	57	77	49-92	285	62	22-79
Dimethoate	5.2	187	79-613	52	104	64-156	260	112	101-153
Disulfoton	5.0	98	63-130	50	101	63-169	250	101[b]	71-126
Ethoprophos	4.8	86	67-118	48	136[b]	81-318	240	234[b]	97-321
Etrimphos	5.2	87	64-101	52	95	69-116	260	119[c]	87-178
Famphur	4.8	129	86-199	48	112	38-211	240	110[b]	80-128
Fenitrothion	6.7	122	71-217	67	194[b]	94-495	335	149[b]	130-162
Fenthion	7.6	80	52-101	76	87	62-98	380	101	74-153
Fonofos	5.6	104	65-153	56	109	66-175	280	114	75-143
Heptenofos	5.0	93	70-132	50	115	58-168	250	104[b]	91-128
Iodofenphos	3.3	99	60-124	33	115	72-196	165	106[b]	101-113
Malathion	8.8	131	56-257	88	102	70-134	440	125	98-192
Mephosfolan	5.6	92	15-173	56	97	68-124	280	95[b]	59-109
Mevinphos	6.3	82	63-98	63	82	68-99	315	97	82-134
Naled	9.3	27	0-53	93	39[b]	0-101	465	64[c]	38-77
Omethoate	5.6	57	9-101	56	49	19-94	280	75[c]	54-87

Parathion	5.3	93	64-135	53	112	65-185	265	111 [b]	71-172
Phorate	5.2	86	59-108	52	146 [b]	83-303	260	101 [b]	82-141
Phosalone	4.8	124	55-206	48	112	91-157	240	124	101-183
Phosmet	4.3	117	86-149	43	118	97-173	215	116 [c]	91-137
Pirimiphos-methyl	5.8	89	58-104	58	91	67-109	290	109	80-160
Propetamphos	3.6	98	65-111	36	136 [b]	90-302	180	143 [b]	117-172
Pyrazophos	6.4	174	76-220	64	110	95-158	320	109 [c]	96-119
Quinalphos	5.5	104	65-146	55	154 [b]	83-387	275	95 [b]	66-116
Thiometon	5.9	79	47-102	59	105	61-179	295	90 [b]	71-124
Triazophos	3.4	117	40-192	34	107	89-144	170	113 [c]	105-121

[a] mean of 3 results; [b] mean of 5 results; [c] mean of 4 results

In the majority of cases mean recoveries for OPs are greater than 70%. Recoveries of omethoate and naled were consistently low. The results for omethoate confirm earlier work (30) that showed that SFE without modifier was unsuitable for quantitative determination of omethoate although the method is still suitable for screening purposes. Naled is a compound which decomposes to dichlorvos during gas chromatography. The low and variable recoveries are probably due to the variable degree of decomposition between different samples. Recoveries of dichlorvos were also low in a number of cases. This compound is highly volatile and losses probably occur during the evaporation of the acetonitrile extract. In a number of cases mean recoveries of OPs are above 120%. In most cases, high mean recoveries were caused by high recoveries from one or two individual determinations. This is unlikely to be caused by chromatographic interference since this would result in a change in the relative ratios of the measured ions, compared to the ratios in standards. Chromatographic interferences would also be more significant at lower spiking concentrations. However these individual, high, recovery results were observed at all 3 concentrations studied and in the majority of cases ion ratios were not significantly different to those for standard solutions. It appears that, in the majority of cases, the high recoveries observed for some compounds are probably due to an enhanced response in sample extracts caused by sample components blocking active sites on the column. These active sites can cause degradation of some OPs and this degradation is reduced in the presence of sample components. Similar enhanced responses for OPs have also been reported (20). On-column injection was employed in this work to minimise such effects but they are still significant for some OPs. There were however a small number of cases in which high recoveries did appear to be caused by chromatographic interferences, mostly for compounds for which one or more low mass ions were used for quantitation. The amount of interference varied between samples and this is not unexpected as stomach contents can vary greatly in composition.

Elevated recoveries were not observed for organochlorines which are less liable to degradation during GC analysis and hence do not show an enhanced response in the presence of sample components. Chromatographic interferences are also less likely as it was generally possible to use higher mass ions to quantitate.

The selectivity of SFE extraction relative to solvent extraction was assessed by evaporating to dryness aliquots of sample extracts and weighing the resulting residue of non-volatile (mainly fatty) co-extracted material. Results from 5 different samples of stomach contents showed that ratios of the weight of co-extractives in SFE extracts relative to solvent extracts varied between 0.14 and 0.41. The reason why SFE results in cleaner extracts is not clear although similar findings have been reported by ourselves for honeybees (30) and by others for food samples (20). It is probably due in part to the on-line clean-up which is an inherent part of the collection procedure on the ODS trap. However it is also possible that the extraction procedure itself is more selective.

Conclusions

The 2 LC-MS methods described have already proved extremely useful for confirmatory analysis and have been introduced into routine use. Previously described procedures for these chemicals (2, 10-14, 16) have been used in this laboratory and were sometimes subject to chromatographic interferences which varied considerably between samples and led to variation in the achievable limits of detection. Although, in most cases, no interferences were seen, a proportion of samples analysed gave considerable problems in this regard. The selectivity of LC-MS results in a much reduced problem with interferences and the LC-MS methods described appear to be very robust.

GC-MS-MS has also proved very valuable. It is particularly useful in "one-off" cases in which a residue in a particular sample proved difficult to confirm due to the nature of the sample matrix or the low level of chemical present. If GC-MS-MS was not available the only approach in such circumstances would often to be to develop a suitable clean-up for that particular sample, something that, even in those cases in which it is possible, is extremely time-consuming and expensive.

SFE has proved suitable for extraction of OP and OC insecticides from animal tissue samples. In addition to providing more selective extraction, SFE offers other advantages such as ease of automation and minimisation of solvent use.

References

1. Fletcher, M.R.; Hunter, K.; Barnett, E.A. *Pesticide poisoning of animals 1997: Investigations of suspected incidents in the United Kingdom:* MAFF Publications: London, 1998.
2. Brown, P.; Charlton, A.; Cuthbert, M.; Barnett, L.; Ross, L.; Green, M.; Gillies, L.; Shaw, K.; Fletcher, M. *J. Chromatogr.* **1996**, *754*, 463-478.
3. Barnes, K.A.; Startin, J.R.; Thorpe, S.A.; Reynolds, S.L.; Fussell, R.J. *J. Chromatogr.* **1995**, *712*, 85-93.
4. Newsome, W.H.; Lau, B.P.Y.; Ducharme, D.; Lewis, D *J. Assoc. Off. Anal. Chem.* **1995**, *78*, 1312-1316.
5. Vollmer, D.A.; Vollmer, D.L.; Wilkes, J.G. *LC-GC*, **1996**, *14*, 216-224.
6 Vollmer, D.A.; Vollmer, D.L. *LC-GC*, **1996**, *14*, 236-242.
7. Stout, S.J.; DaCunha, A.R.; Picard, G.L.; Safarpour, M.M. *JAOAC Int.* **1998**, *81*, 685-690.
8. Lacassie, E.; Dreyfuss, M.F.; Daguet, J.L.; Vignaud, M.; Marquet, P.; Lachatre, G. *J. Chromatogr.* **1999**, *830*, 135-143.
9. Startin, J.R.; Hird, S.J.; Sykes, M.D.; Taylor, J.C.; Hill, A.R.C. *Analyst* **1999**, *124*, 1011-1015.
10. Hunter, K. *J. Chromatogr.* **1983**, *270*, 267-276.

11. Hunter, K. *J. Chromatogr.* **1983**, *270*, 277-283.
12. Hunter, K. *J. Chromatogr.* **1984**, *299*, 405-414.
13. Hunter, K. *J. Chromatogr.* **1985**, *321*, 255-272.
14. Felice, L.J.; Chalermchaikit, T; Murphy, M.J. *J. Anal. Toxicol.* **1991**, *15*, 126-129.
15. Chalermchaikit, T; Felice, L.J.; Murphy, M.J. *J. Anal. Toxicol.* **1993**, *17*, 56-61.
16. Jones, A. *Bull. Environ. Contam. Toxicol.* **1996**, *56*, 8-15.
17. Schachterle, S.; Brittain, R.D.; Mills, J.D. *J. Chromatogr.* **1994**, *683*, 185-193.
18. Nagai, S.; Ito, S.; Tomioka, M.; Ishizuka, T. *Chromatography*, **1998**, *19*, 290-291.
19. Honing, M.; Riu, J.; Barcelo, D.; vanBaar, B.L.M.; Brinkman, U.A.T. *J. Chromatogr.* **1996**, *733*, 283-294.
20. Lehotay, S.J.; Eller, K.I. *JAOAC Int.* **1995**, *78*, 821-830.
21. Pearce, K.L.; Trennery, V.C.; Were, S. *J. Agric. Food Chem.* **1997**, *45*, 153-157.
22. Stefani, R.; Buzzi, M.; Grazzi, R. *J. Chromatogr.* **1997**, *782*, 123-132.
23. Tena, M.T.; Rios, A.; Valcarel, M.; SanchezAlarcon, M. *Chromatographia*, **1997**, *46*, 524-528
24. Anastassiades, M.; Schwack, W. *J. Chromatogr.* **1998**, *825*, 45-54.
25. Kim, D.H.; Heo, G.S., Lee, D.W. *J. Chromatogr.* **1998**, *845*, 63-70.
26. Fiddler, W.; Pensabene, J.W.; Gates, R.A.; Donoghue, D.J. *J. Agric. Food Chem.* **1999**, *47*, 206-211.
27. Juhler, R.K. *Analyst*, **1998**, *123*, 1551-1556.
28. Nam, K.S.; King, J.W. *J. Agric. Food Chem.* **1994**, *42*, 1469-1474.
29. Voorhees, K.J.; Gharaibeh, A.A.; Murugaverl, B. *J. Agric. Food Chem.* **1998**, *46*, 2253-2359.
30. Jones, A.; McCoy, C. *J. Agric. Food Chem.* **1997**, *45*, 2143-2147.

Chapter 20

Using Chorioallantoic Membranes for Non-Lethal Assessment of Exposure and Effect in Oviparous Wildlife

George P. Cobb, Donald M. Norman[1], Pattie D. Houlis, and Tim A. Bargar[2]

The Institute of Environmental and Human Health, Texas Tech University and The Texas Tech University Health Sciences Center, Lubbock, TX 79416

Assessing chemical exposure during ecotoxicology studies presents unique analytical problems. Sacrificing enough organisms to obtain sufficient numbers of tissue samples for reliable exposure assessment may impact community dynamics. Also, lethal sampling is not appropriate when evaluating threatened or endangered wildlife. Chorioallantoic membranes (CAMs) may be used as a surrogate tissue for evaluation of chlorinated hydrocarbon exposure in oviparous species. Residues in CAMs are highly correlated with residues in eggs,(for total PCBs ,r^2=0.592 to 0.782; p= 0.02 to 0.0001). Residues in CAMs also correlate with monooxygenase activity in hatchling and maternal liver. As chlorination within a given class of compounds increases, more of the compound is retained in the egg yolk and is ultimately resorbed into the hatchling. Interestingly, lower chlorinated PCBs are preferentially transferred from maternal reserves to eggs. The CAM technique has been validated with chicken eggs in laboratory studies. Contaminant exposure in heron, alligator and sea turtle populations have been successfully monitored using this approach in field studies.

[1]Current address: Pacific NW Pesticide Consulting Group, 2112 NW 199th Street, Shoreline, WA 98117.
[2]Current address: National Ocean Service, National Oceanographic and Atmospheric Administration, Charleston, SC 29412.

Introduction

Assessment of wildlife exposure to chlorinated organic contaminants has historically used residue analysis [1-8]. There is also a significant move toward biomarkers of exposure and effect [9-14]. While providing important information, most residue and biomarker approaches are lethal to the organisms being studied. Lethal sampling can have deleterious effects on wildlife populations and removes individual organisms from further study. Such population alterations bias any reproductive, growth, predation or habitat use endpoints that may be desired. Lethal sampling techniques are not possible with threatened or endangered species. These obstacles can be prevented by collecting tissue samples non-lethally and by drawing strong correlations between chemical concentrations and biomarker response [12-14].

Nonlethal Monitoring of Eggs

The first nonlethal tissue sampling from eggs utilized eggshell membranes from peregrine falcon (*Falco peregrinus*) eggs. Eggshell membranes are located between egg contents and the egg shell. These membranes were used to estimate chlorinated hydrocarbon content in museum specimens [15] and freshly hatched eggs [16]. In a modification of this approach, Norman utilized chorioallantoic membranes (CAMs) of great blue heron *(Ardea herodias fannini)* eggs from the Puget Sound Ecosystem [17]. CAMs are highly vascular membranes that are formed just before organogenesis. CAMs perform gas exchange, nutrient transport from the yolk, and excretory containment functions [18]. The CAM remains in the egg after hatching or is loosely attached to the hatchling's abdomen. The nonessential nature of the post hatching CAM allows noninvasive tissue collection for subsequent chemical analysis [19].

Study Species

Great blue herons (*Ardea herodias*) are accepted as good indicators of contaminant presence in estuarine ecosystems [20,21]. The heron's diet and resident nature contribute heavily to its utility as an indicator species. Based on these characteristics, three heron colonies in The Puget Sound ecosystem monitored for Chlorinated hydrocarbon uptake [17,19,22].

Effects of chlorinated pesticide exposure on American alligators (*Alligator mississippiensis*), has been evaluated [23,24], but alligator exposure to PCBs is poorly defined [25-27]. To better characterize this exposure, Bear Island (BI) and Yawkey Wildlife Center (YWC), were monitored as reference and contaminated areas, respectively. BI is in the ACE Basin, a National Estuarine Research Reserve, at the confluence of the Ashepoo, Combahee, and Edisto Rivers. YWC is located on the shore of Winyah Bay, one hundred and fifty miles northeast of BI. These two study areas are situated in the tidal areas of two distinctly separate watersheds. From

1989 through 1992, South Carolina Department of Health and Environmental Control advised against fish and shellfish consumption from Winyah Bay due to chlorinated hydrocarbon contamination [28,29]. Alligators consume primarily aquatic prey which allows accumulation of chlorinated hydrocarbons. The different historical PCB concentrations at the two study sites allowed comparison of PCB uptake by alligators.

Data concerning contaminants in sea turtles are rare. Most research concerning PCB effects on reptiles has utilized more readily available turtle species [30,31]. Mean PCB concentrations have been determined in snapping turtles (*Chelydra serpentina*) from the Hudson River and from Lake Ontario, Canada [31,32]. Little data exist concerning PCBs in loggerhead sea turtles (*Caretta caretta*) [33,34]. We have found no studies correlating PCB concentrations with adverse effects in sea turtles. Utilization of CAMs as tissues for chemical analysis would allow residue and biological effect data to be collected for sea turtles and other threatened species.

Laboratory Studies

Even though field research in our laboratory has demonstrated that PCB and DDE concentrations in CAMs correlate to concentrations in whole eggs [19] and that CAMs collected from hatched eggs can be used to assess contaminant uptake by oviparous wildlife near hazardous waste sites [22,27,35], the utility of the CAM technique has been questioned. The basis for this question is a study in which concordance analysis showed partitioning of chlorinated organic compounds between CAM and egg tissue did not follow passive diffusion models [36]. This finding implicates active transports in Chlorinated hydrocarbon movement among egg tissues. Therefore, laboratory dosing studies were conducted to better define the effect of chlorination and other structural characteristics on contaminant distribution between CAM and egg tissue. The following text describes the toxicological foundation and field applicability of CAM monitoring.

METHODS

Sample Collection and Handling

Three wildlife species were evaluated during field studies: great blue herons, American alligators, and loggerhead sea turtles (18,19,24,35). All eggs were collected with the supervision of State or Federal wildlife officials. Heron eggs were collected by climbing trees and retrieving eggs directly from nests. Heron CAMs were retrieved directly from the ground below colonies. Twenty-one alligator eggs were collected from YWC, and twenty eggs were collected from BI. Three eggs were collected from 13 of 14 nests. Eggs were collected at least 48 hours following hatching of siblings from a given nest. Twenty one unviable loggerhead sea turtle eggs were collected from the Hatchling Headstart Program at the Cape Romain Island National Wildlife Refuge.

Eggs were kept on ice in the field and placed in glass containers for storage at <-10C until processing could be performed. Eggs were opened with scissors and teflon coated forceps. While still partially frozen, CAMs were removed from the egg contents and analyzed separately. Egg contents were placed into acetone rinsed glass containers for weighing, homogenization and storage at <-10C. CAMs were weighed, homogenized, and extracted immediately upon removal from the egg [17,19,22,27,35].

Test Solutions and Dosing in Laboratory Studies

Excretion Study Test Solutions

Eighteen actively laying chickens (*Gallus domesticus*) were divided into 6 groups of three. Six dosing solutions, including a control, were prepared. Four dosing solutions contained one of the following: 2,3,3',4,4' pentachlorobiphenyl (PCB105), 2,3,3',4,4',5 hexachlorobiphenyl (PCB156), 2,3,3',4,4',5,5' heptachlorobiphenyl (PCB189), or endosulfan. The fifth solution contained a mixture of all chemicals. The sixth dosing solution was a corn oil control. The alpha and beta isomers of endosulfan were present in the same formulation and, while quantified separately, were considered one test substance for dosing purposes. Chemical concentrations in the first four dosing solutions were: a) 1350 ng/µL for PCB105, b) 1280 ng/µL for PCB156, c) 1160 ng/µL for PCB189, and d) 1350 ng/µL and 410 ng/µL for a-endosulfan and ß-endosulfan, respectively. Chemical concentrations in the fifth dosing solution were 1330 ng/µL , 1210 ng/µL , 1160 ng/µL , 1340 ng/µL , and 410 ng/µL for PCB105, PCB156, PCB189, a-endosulfan, and ß-endosulfan, respectively.

Dose Response Study Test Solutions

Sixteen actively laying adult hens were divided equally into three dose groups and one control. PCB105, PCB156, PCB189, a-endosulfan, and ß-endosulfan concentrations were analytically determined in all dosing solutions. Low dose hens received 6.3, 5.8, 5.7, 6.4, and 2.2 µg respectively. The medium dose group contained 11.7, 10.8, 10.1, 12.1, and 3.8µg, respectively. The high dose group received 17, 15.7, 14.8, 18.2, and 5.7 µg of the respective toxicants. Sub-cutaneous injections containing 100µl of a given solution were administered at study initiation and once every four days thereafter. Hens were artificially inseminated three times. Following initial, artificial insemination, fertile eggs were collected and incubated at 99°F until hatch (~ 21 days). Eggs were checked daily until day 19 of incubation and more frequently thereafter. CAMs from hatched chicks were collected within 12 hours post-hatch and placed into solvent rinsed scintillation vials for storage at -20°C until analysis. Livers were collected from euthanized hatchlings and hens and stored at -80C for monooxygenase activity determination.

Extraction and Analyses

Extraction

Solvents were pesticide grade and solid reagents were heated or extracted to minimize contaminants as described in previous publications [17,19,22,27,35].

Standards were 99% pure. Tissues were weighed and mixed with sodium sulfate before homogenization and Soxhlet extraction [17,19,22]. Extracts were evaporated and a portion was retained for lipid determination. The analytes were isolated from interfering compounds using liquid chromatography, and desired fractions were concentrated to 2 mL. Chlorinated hydrocarbon recovery from herring gull eggs (surrogate for heron eggs) was 93±5% [17,19]. PCB recoveries for alligator eggs were 88±10% [27,35] while chicken egg recoveries were 92±12.

Chemical Analyses

Two microliters of each CAM extract were analyzed while one microliter of each yolk extract was analyzed. Samples were analyzed with a Hewlett Packard 5890 GC equipped with an ECD and a 30m DB-1701 column or a 60m DB-5 column, both 0.25mm ID with 0.25um film thickness. Temperature programs spanned 100°C to 285°C. The data obtained from the GC/ECD were adjusted for the calculated recovery of the recovery standard in each sample. Individual congeners or insecticides were quantified [17,19,22,27,35].

A Hewlett-Packard 5890- 5988A GC-MS, operated in electron impact/selected ion monitoring mode, was used to confirm peak identity in 10% of reptile samples. Total PCB concentrations were determined by summing measured individual PCB congeners. DDT, DDE and DDD concentrations were often summed to obtain a total exposure profile of DDT and its degradates this quantity was designated DDX.

Cytochrome P450 Assays

A portion of chicken liver (1.6-2.4g for hens, 0.4-1.0g for chicks) was homogenized in chilled buffer and centrifuged at 10,000 x g to remove cellular debris. The supernatant was centrifuged at 15,000 x g to remove cellular debris, transferred to an ultra-centrifuge tube, and centrifuged at 105,000xg to isolate microsomes. Centrifugation was performed at 4°C. The microsomal pellet was then suspended in buffer and frozen at -80°C until assay. A 1:20 dilution of the microsomal solution was assayed to determine ethoxyresorufin dealkylase (EROD) activity[37].

Results

Herons in Puget Sound

Initial Investigations

Concentrations of Aroclor 1242, Aroclor 1254, and Aroclor 1260 were each found to be different ($p < 0.005$) between the Samish Island (SI) and Dumas Bay (DB) colonies. Heron eggs from DB contained 21 $\pm$ 8.1 (SD) ppm of total PCB by weight while Samish eggs contained 3.5 $\pm$ 0.77 (SD) ppm. Whole eggs from DB and SI had total DDE concentrations of 1.7 $\pm$ 0.97 (SD) ppm and 0.38 $\pm$ 0.22 (SD) ppm, respectively. Significant differences ($p < 0.01$) exist between DDE concentrations in eggs from these sites. Contaminant concentrations in CAMs from DB and SI were statistically different ($p<0.05$). Data describing chlorinated hydrocarbon partitioning within an egg were used to transform chlorinated hydrocarbon concentration data from salvaged CAMs into estimates of chlorinated hydrocarbon concentrations in whole eggs (Table I). CAMs provided good predictions of contaminant concentrations in whole eggs from the SI colony.

Table I. Estimation of Chlorinated Contaminant Concentrations in Heron Eggs Using Concentrations[a] in Chorioallantoic Membranes from Salvaged Heron Eggs.

Contaminant	*Egg/CAM Partitioning Coefficient*	*Concentration in Salvaged CAMs from Samish Island*	*Estimated concentration in eggs from Samish Island[b]*	*Measured concentration in eggs from Samish Island*
Cl_5-Benzene	0.744±0.311	1.44±3.19	1.07±2.40	1.15±0.52
Cl_6-Benzene	0.910±0.309	7.98±11.6	7.26±11.4	18.0±15.1
p,p'-DDE	1.31±0.812	336±419	440±613	380±223
PCB 118	1.03±0.431	57.6±69.3	57.9±75.5	69.8±23.9
PCB 138	1.20±0.692	68.2±126	81.6±157	98.1±40.9
PCB 170	1.16±0.653	20.5±31.4	23.7±38.8	31.8±10.9
PCB 180	1.08±0.595	59.9±76.9	64.3±89.8	79.2±27.0

[a] expressed as ng/g tissue
[b] calculated as (salvaged CAM)*(Contents/CAM)

Assessing Chlorinated contaminants at an NPL Site

When evaluating the possible chlorinated hydrocarbon uptake at a heron rookery on Naval Air Station, Whidbey Island (NASWI), PCB and OC residues were found in all CAMs. CAMs contained 0.247 ±0.048 ppm (mean ± S.E.) and 0.171± 0.042 ppm of total PCB and DDX, respectively [22].

Study Comparisons

Based on the contaminant data above, SI was considered a reference site and DB was considered a positive control. Chlorinated hydrocarbons in CAMs from SI and DB are compared to those from NASWI (Figure 1). Mean DDE and PCBs are highest in CAMs from DB. Contaminant concentrations in CAMs from DB were greater than from NASWI ($p<0.05$). Mean DDE concentrations in CAMs from

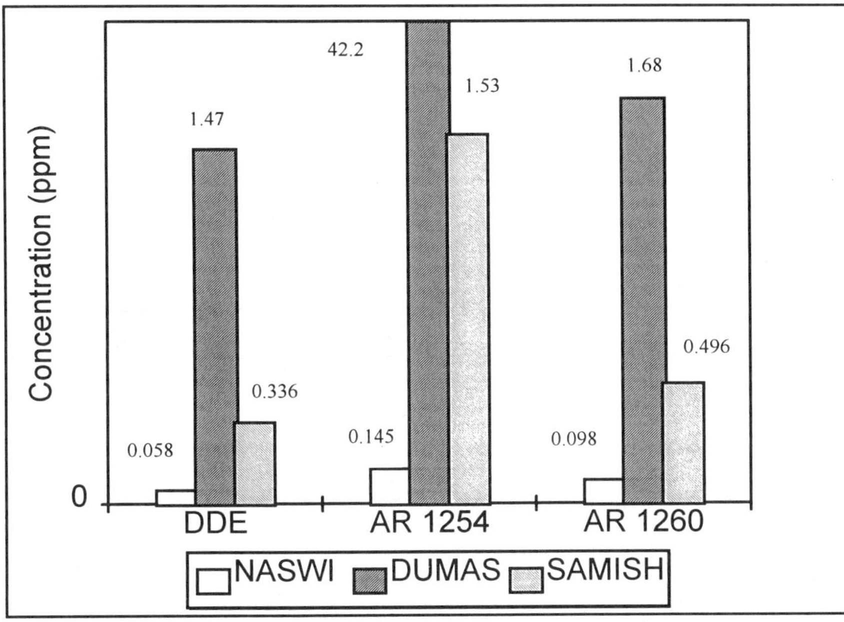

Figure 1. Concentrations of DDE, Aroclor 1254, and Aroclor 1260 in chorio-allantoic membranes of great blue herons (Ardea herodias fannini) *from NAS Whidbey Island, Dumas Bay and Samish Island.*

NASWI were statistically similar to those at SI ($p>0.1$). These data indicate minimal contaminant impacts at the Whidbey Island colony, and this assessment is supported by the increase of active nests from 60 to 100 during the time period 1990 to 1993.

Alligators

Comparison of PCB Concentrations

PCB concentrations in CAM and egg differed between sites (Table II). Alligator eggs from Yawkey Wildlife Center (YWC) and Bear Island (BI) contained total PCB concentrations of 3763±1175 and 386±81 ng/g, respectively. PCB concentrations in CAMs were 1671±483 and 711±250 ng/g for YWC and BI alligators, respectively. Concentrations of tetraCB through decaCB were higher in YWC eggs than in BI eggs. PentaCBs through decaCB in CAMs were also higher in YWC eggs than in BI eggs. Total PCB concentrations in CAMs and eggs were significantly correlated for BI and YWC alligators (Table III). TetraCBs through octaCBs and decaCB showed significant egg:CAM correlations in eggs from BI (Table III). HexaCB concentrations in CAMs and eggs were correlated for eggs from YWC [27].

Table II. Total PCB and PCB Homologue Concentrations[a] in American Alligators (*Alligator mississippiensis*) from Coastal South Carolina, USA

	CAM			*Egg*			*CAM:egg Ratio*	
Contaminant	*Bear Island (N = 15)*	*Yawkey W.C. (N = 14)*	p^b	*Bear Island (N = 19)*	*Yawkey W.C. (N = 19)*	*p*	*Bear Island*	*Yawkey W.C.*
Dichlorobiphenyl	1.2±0.6	8.7±5.1	0.3	0.04±0.03	0.65±0.61	0.4	3.0	13
Trichlorobiphenyl	41±26	26±10	1.0	12±2.7	23±6.8	0.06	3.4	1.1
Tetrachlorobiphenyl	159±67	292±99	0.4	32±5.3	213±89	0.001	5.0	1.4
Pentachlorobiphenyl	177±61	445±91	0.02	108±22	994±288	0.0001	1.6	0.45
Hexachlorobiphenyl	179±79	476±77	0.04	113±29	1407±467	0.0001	1.6	0.34
Heptachlorobiphenyl	120±40	311±77	0.02	86±19	832±272	0.0001	1.4	0.37
Octachlorobiphenyl	28±9	100±29	0.01	29±6.8	262±82	0.0001	0.97	0.38
Nonachlorobiphenyl	3.7±1.5	12±2.9	0.02	6.9±4.8	31±9.8	0.0002	0.54	0.39
Decachlorobiphenyl	26±21	53±19	0.03	5.9±2.1	41±7.7	0.0001	4.4	1.3
Total PCBs	711±250	1671±483	0.02	386±81	3763±1175	0.0001	1.84	0.44

[a] expressed as ng/g lipid

[b] p value as determined by ANOVA of square root transformed concentrations

Table III. Correlations between PCB Concentrations[a] in Egg versus CAM of American alligators (*Alligator mississippiensis*) from Bear Island and Yawkey Wildlife Center, South Carolina, USA.

PCB group	*Bear Island*				*Yawkey Wildlife Center*			
	r^2	*p*	*slope*	*y-intercept*	r^2	*p*	*slope*	*y-intercept*
Dichlorobiphenyl	0.045	0.5	-0.048	0.14	0.384	0.6	-0.096	0.54
Trichlorobiphenyl	0.042	0.5	-0.065	3.4	0.011	0.8	-0.088	5.1
Tetrachlorobiphenyl	0.461	0.02	0.19	3.3	0.076	0.6	0.32	9.3
Pentachlorobiphenyl	0.721	0.0009	0.49	2.8	0.246	0.2	1.2	7
Hexachlorobiphenyl	0.923	0.0001	0.58	2.7	0.682	0.01	2	-1.9
Heptachlorobiphenyl	0.681	0.002	0.44	3.5	0.123	0.4	0.87	14
Octachlorobiphenyl	0.680	0.002	0.34	2.5	0.32	0.1	1.2	5.5
Nonachlorobiphenyl	0.050	0.5	0.04	0.94	0.346	0.1	1.2	2.1
Decachlorobiphenyl	0.510	0.01	0.023	1.8	0.356	0.09	0.36	3.7
Total	0.668	0.002	0.39	6.4	0.592	0.02	0.95	14

[a] Square root of ng/g lipid

PCB Distribution

Total PCB concentrations in eggs and CAMs indicate that alligators accumulate greater PCB concentrations when inhabiting areas of known chlorinated organic contamination (e.g. YWC) than when inhabiting more pristine areas (BI). YWC eggs contain higher concentrations of tetraCB through nonaCB homologues when compared to concentrations from BI. CAMs from YWC contained higher penta through nona chloro congeners than did CAMs from BI. Both data sets imply that female alligators in areas near YWC are exposed to higher PCB concentrations. The fact that eggs from the two sites contain similar concentrations of lower chlorinated PCBs is not surprising as those congeners are the most water soluble, are poorly retained in sediment, and are the more easily metabolized PCB congeners [11]. These properties allow more rapid removal of lower chlorinated congeners from individual organisms and from ecosystems, resulting in lower tissue and environmental PCB concentrations.

Partitioning from Egg to CAM

HexaCBs were the most prevalent PCB homologues in both CAM and egg tissues from both sites. However, concentrations of penta and hexaCBs were quite similar. The predominance of pentaCBs and hexaCBs in egg tissues is characteristic of commonly used Aroclors [38-40] and weathered PCBs [41,42]. DiCBs through tetraCBs represented larger percentages of total PCBs in CAMs than in eggs. PentaCBs through nonaCBs contributed more to PCB concentrations in eggs than in CAMs.

PCBs were present in the egg before the CAM developed, hence egg:CAM partitioning data suggests preferential migration of diCBs through tetraCBs to the CAM or more pronounced metabolism of these lower chlorinated congeners in the egg. PCBs with low chlorine content should reach CAMs more readily, as they are more water soluble and are metabolized by liver enzymes, which are not concentrated in the CAM.

PCB concentrations in CAM and egg tissues were more strongly correlated at BI than at YWC (III). Partitioning of pentaCBs through octaCBs, and total PCBs between CAM and egg at BI were as well defined (r^2=0.668 to 0.923) which provides a better estimate of total chlorinated hydrocarbon concentrations in eggs than does maternal blood in the case of raptors (r^2=0.611 to 0.669) [11]. Also, slopes of regressions for YWC alligators were greater than slopes for BI alligators. That is, any given PCB concentration change in CAMs are associated with larger PCB concentration changes in eggs at YWC. However, hexaCBs and total PCBs were the two significant regressions for YWC eggs.

Turtles

Loggerhead sea turtle eggs from Cape Island contained PCB concentrations of 1.2 ± 0.3 μg/g (mean ±SE) in egg and 10.1± 5.4 μg/g in CAM (Table IV). Partitioning of PCBs between CAM:egg differed according to homologue group. The CAM:egg partition ratio for total PCBs was 8.5 which is higher than found in heron or alligator eggs (Tables I, II and IV). PentaCBs and hexaCBs were the most concentrated homologues in both tissues (Table IV). PCB homologue concentrations declined dramatically as chlorination was increased or decreased from the pentaCB or hexaCB concentration maxima. Concentration declines about the maxima were not as dramatic for less chlorinated homologues in CAMs. Therefore, diCBs through tetraCBs comprised higher percentages of total PCBs in CAMs than in eggs, while pentaCBs through nonaCBs contributed a larger percentage in eggs than in CAMs.

Table IV. PCB Concentrations[a] in Chorioallantoic Membranes and Eggs of Loggerhead Sea Turtles from Cape Romain National Wildlife Refuge.

Contaminant	*CAM (n=16)*	*Egg (N=16)*	*CAM:egg Ratio*
Dichlorobiphenyl	236±146	9.3±4.2	25
Trichlorobiphenyl	889±345	8.7±1.6	100
Tetrachlorobiphenyl	2410±1298	120±31	20.1
Pentachlorobiphenyl	3000±1640	446±138	6.7
Hexachlorobiphenyl	2710±1540	427±106	6.3
Heptachlorobiphenyl	763±406	173±46	4.4
Octachlorobiphenyl	197±107	30.3±5.9	6.5
Nonachlorobiphenyl	37±21	3.7±0.7	10
Decachlorobiphenyl	55±20	5.4±0.8	10
Total PCBs[b]	10100±5466	1190±311	8.5

[a] (ng/g lipid) [b] excluding decachlorobiphenyl

Total PCB and homologue concentrations in turtle eggs were similar to concentrations found in alligator eggs from BI. Correlations between total PCB concentrations in egg versus CAM were high (r^2=0.783) (Table V). Significant slopes for PCB congener partitioning between egg and CAM ranged from 0.11 to 0.26, suggesting preferential accumulation of PCBs in lipids of CAM tissue. As with

Table V. Correlations Between PCB Concentrations[a] in CAM and EGG of Loggerhead Sea Turtles (*Caretta caretta*) from Cape Island, South Carolina, USA.

PCB group	r^2	*p*	*slope*	*intercept*
DiCB	0.069	0.3	-0.018	2.14
TriCB	0.223	0.06	-0.033	3.37
TetraCB	0.631	0.0002	0.11	5.83
PentaCB	0.783	0.0001	0.26	7.68
HexaCB	0.75	0.0001	0.22	9.96
HeptaCB	0.461	0.004	0.23	6.95
OctaCB	0.383	0.01	0.14	3.57
NonaCB	0.105	0.2	0.049	1.59
DecaCB	0.032	0.5	0.0075	4.86
Total PCBs[b]	0.782	0.0001	0.2	16.3

[a] square root of ng/g lipid [b] excluding decachlorobiphenyl

the alligator study, PCB partitioning between CAM and egg tissues suggests preferential migration of diCB through tetraCBs into the CAM or more pronounced metabolism of these lower chlorinated homologues by the embryo.

Laboratory Studies

PCBs were not detected in eggs laid by control hens. PCB transfer into eggs correlated inversely with the extent of chlorination (Figure 2). Significant transfer differences were found among each homologue group. Dosing with four chlorinated hydrocarbons reduced the transfer of the heptachlorobiphenyl but not the lower chlorinated congeners. The mixture also reduced the accumulation of α-endosulfan. These data demonstrate the effect of mixture composition and chemical structure on the toxicokinetics of chlorinated hydrocarbons.

Chlorinated hydrocarbon partitioning between the embryo and CAM was consistent for the different toxicants (Figure 3). This indicates the consistency of CAM:egg partitioning ratios in pipping eggs, which addresses one of the major questions regarding the utility of CAM analyses in field studies. These data suggest that some of the variance in the field may be explained by the timing of collection and by the different chemical mixtures that are present in study areas.

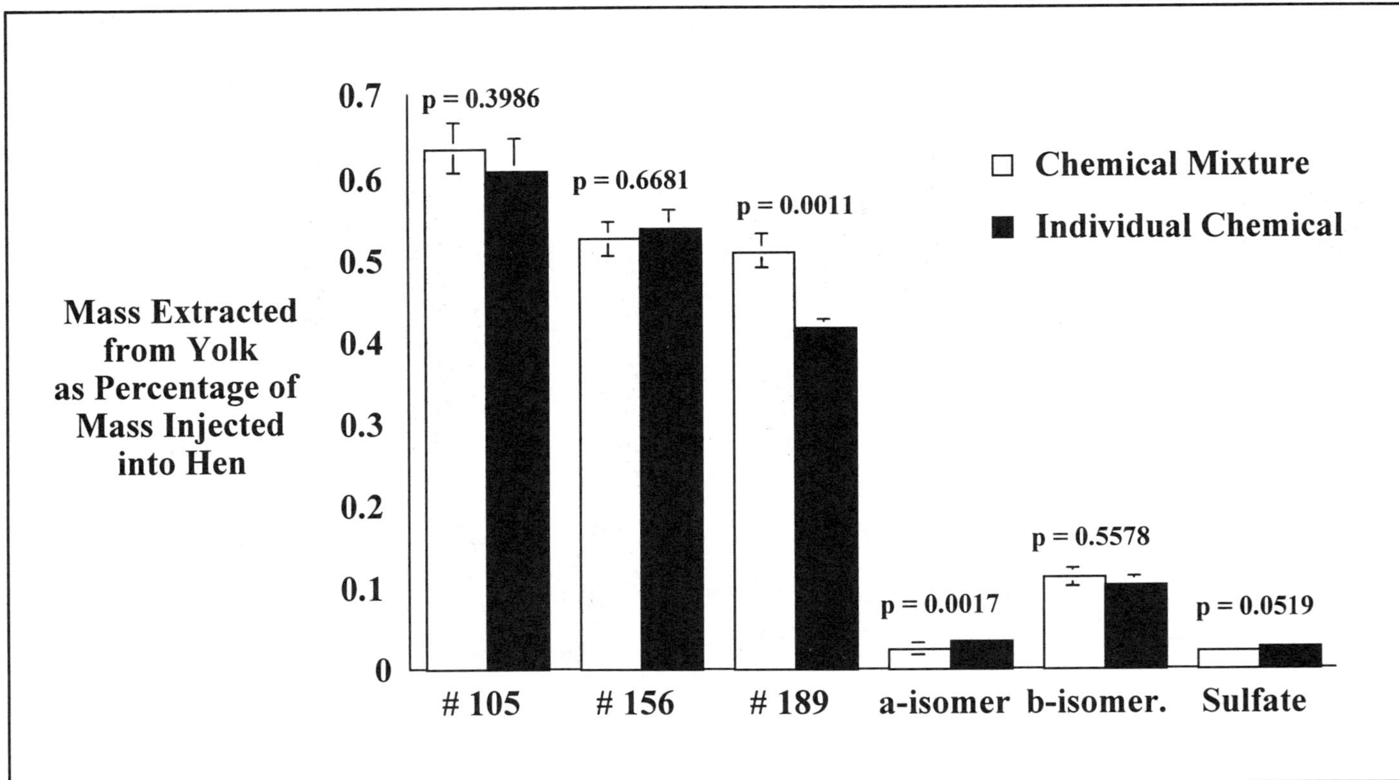

Figure 2. *Deposition of individual and mixed chlorinated hydrocarbons from hens* (Gallus domesticus) *into whole eggs.*

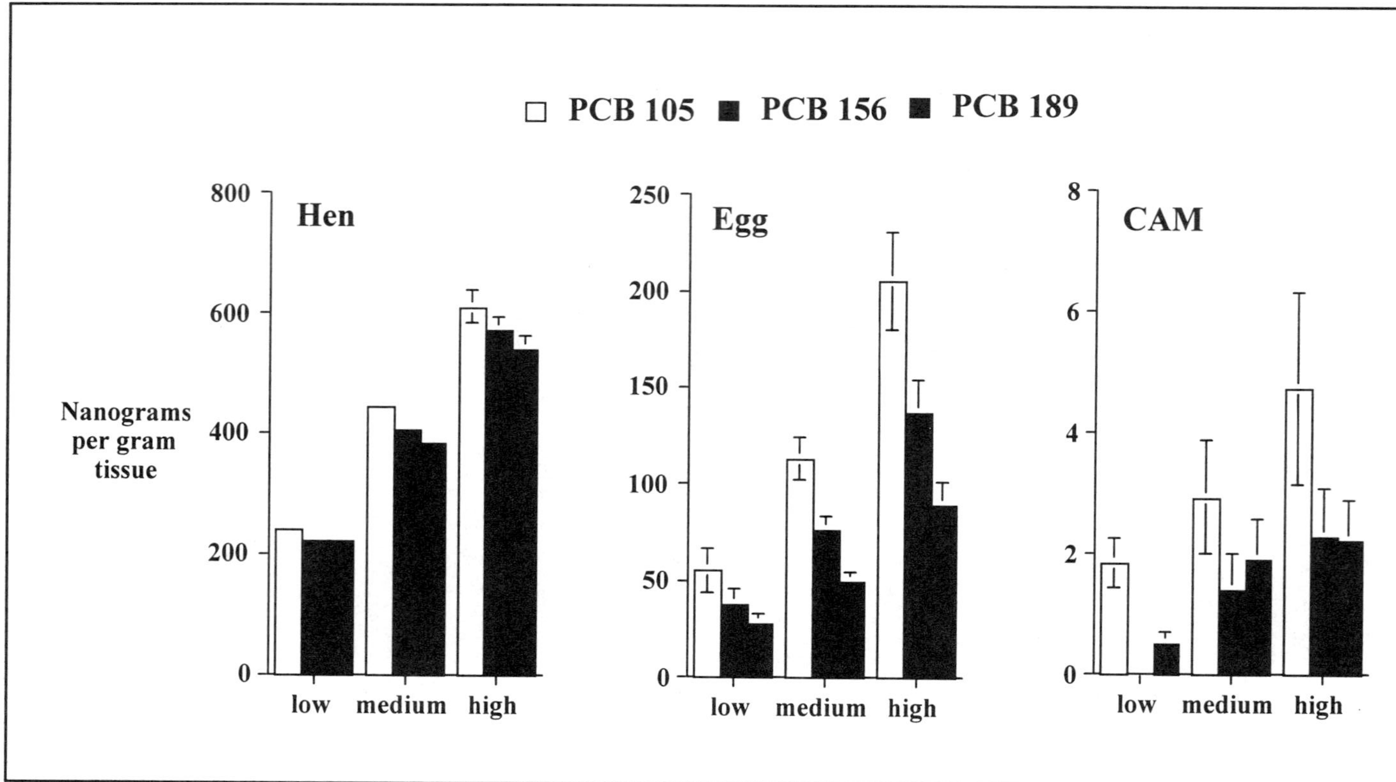

*Figure 3. Effect of increasing PCB dose on individual PCB distribution among hen (*Gallus domesticus*), egg yolk, and CAM.*

One important finding from the laboratory dosing study was the correlation between PCB concentrations in CAMs and PCB concentrations in hens. This relationship extends the extrapolation of CAM content to encompass adult females in a study population. Significant regressions were obtained for PCB content in CAMs and EROD activity in embryo livers (Figures 4,5) as well as in hen livers (Figure 6). These correlations allow CAM monitoring to be used to estimate biomarker response in avian species. The important aspect of this research is that normally lethal biomarkers can be quantitatively evaluated through this non-lethal technique.

Chlorinated hydrocarbon concentrations in CAMs also show strong relation to concentrations in maternal tissues and to EROD activities. This represents a critical tie between residue data from CAMs analyses and population level exposures and effects. Such data will allow non-lethal, non-invasive evaluation of chlorinated hydrocarbon exposure and effect in oviparous species.

ACKNOWLEDGMENTS

Tom Murphy of the South Carolina Department of Natural Resources provided all alligator eggs for this project. This contribution is greatly appreciated. Sarah Dawsey and Barbara Chapman of the Cape Romain National Wildlife Refuge graciously provided transportation and logistical support during sea turtle collections. Virginia Waldrop, Francis Harper and M. Warren Miller provided valuable assistance in analytical procedures. The US Fish and Wildlife Service and the Washington Department of Wildlife provided funding for initial heron research. Funding for the

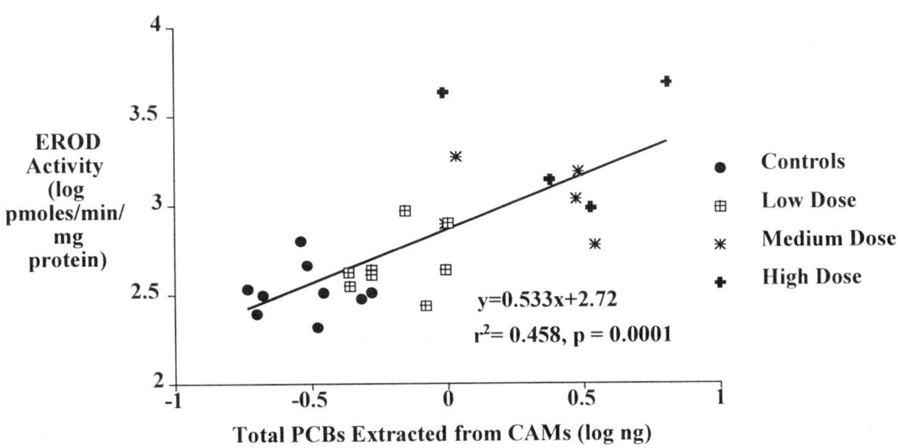

***Figure 4**. Regression of ethoxyresorufin-O-deethylase activity in chick* (Gallus domesticus) *liver versus PCB concentration in chorioallantoic membranes (CAMs). The CAM and chick comprising a data point are from a single egg.*

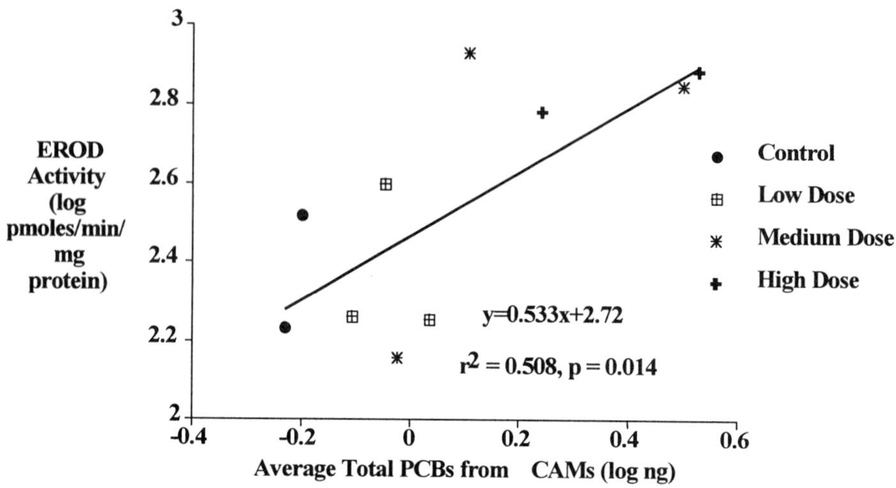

Figure 5. *Regression of ethoxyresorufin-O-deethylase activity in hatchling chicken* (Gallus domesticus) *liver versus PCB concentration in chorioallantoic membranes (CAMs). Mean EROD and PCB data from all eggs within a given clutch comprise one data point.*

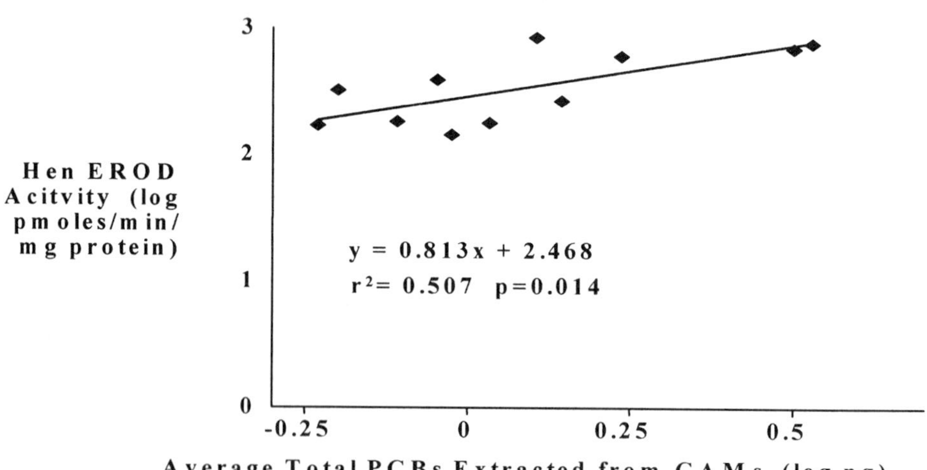

Figure 6. *Regression of ethoxyresorufin-O-deethylase activity in hen* (Gallus domesticus) *liver versus PCB concentration in chorioallantoic membranes (CAM). EROD from an individual hen and PCB concentration in all CAMs from eggs within that hen's clutch comprise one data point.*

hazardous waste site assessments was provided by The National Institute of Environmental Health Sciences (Grant ES04696). The US Navy provided the funding for the heron assessment at Whidbey Island. The South Carolina Hazardous Waste Management Research Fund provided funds that were used in analytical method development for this project. Through a grant from The South Carolina Research and Education Foundation, The South Carolina Governor's School for Science and Mathematics provided financial and logistical support for Michael O'Quinn, a high school intern, to participate in this project.

References

1. Blus, LJ. *Environ. Pollut.*, 1982. 28: 15-33.
2. Elliot, JE, RJ Norstrom, JA Kieth. *Environ. Pollut.* 1988, **52**: 81-102.
3. Kubiak, TJ, HJ Harris, LM Smith, TR Schwartz, DL Stalling, JA Trick, L Sileo, D Docherty, TC Erdman. *Arch. Environ. Contam. Toxicol.*, 1989, **18**: 706-727.
4. Spiech, SM, J Calambokidas, DW Shea, J Peard, M Witter, DM Fry. Colonial Waterbirds. 1992 **15**: 103-112.
5. Yamashita, N, S Tanabe, JP Ludwig, H Kurita, ME Ludwig, R Tatsukawa. *Environ. Pollut.* 1993, **79**: 163-173.
6. Bayer, WN, GH Heintz, AW Redmond-Norwood Environmental Contaminants in Wildlife: Interpreting Tissue Concentrations. CRC Press: Boca Raton, FL. 1996; 495 pp.
7. Henny, CJ, DL Meeker. *Environ. Poll.* 1981, **25**: 291-304.
8. McCarthy, JF LR Shugart. Biomarkers of Environmental Contamination. Lewis Publishers: Boca Raton, FL. 1990; 457.
9. Huggett, RJ, RP Kimerle, PM Mehrle, HL Bergman. Biomarkers: Biochemical, Physiological, and Histological Markers of Anthropologenic Stress. Lewis Press: Boca Raton, FL. 1992; 347 pp.
10. Dickerson, R.L, M.J. Hooper, G.P. Cobb, N.W. Gard, R.J. Kendall. *Environ. Health. Persp.* 1994, **102**:65-69.
11. Safe, S. *Crit. Rev. Toxicol.* 1994, **24**: 87-149.
12. Fossi, M.C.C.Leonzio. Nondestructive Biomarkers in Vertebrates. Lewis Publishers. Boca Raton, FL. 1994; 345 pp.
13. Cobb, GP, MJ Hooper. *in* R.J. Kendall, T. Lacher, eds. *The Population Ecology and Wildlife Toxicology of Agricultural Pesticide Use: A Modelling Initiative for Avian Species*. SETAC Special Publication Lewis Publisher, Chelsea, MI. 1993.
14. Friend, M, MA Haegle, DL Meeker, R Hudson CH Baer. *Animals as Monitors of Environmental Pollution*. National Academy of Sciences, Washington DC. 1979; pp. 319-326.
15. Peakall, DB. *Science.* 1974., 183: 673.
16. Peakall, DB, TS Lew, AM Springer, WW Walker, RW Risebrough, JG Monk,

WM Jarman, BJ Walton, LM Reynolds, RW Fyfe, LF Kiff. Arch. Environ. Contam. Toxicol. 1983, 12:523-528.

17. Norman, DM Master's Thesis, Western Washington University, Bellingham, WA. 1992; 248 pp.
18. Hamilton, HL Lillie's Development of the Chick. Holt and Company: New York, NY. 1952, 624 pp.
19. Cobb, GP, DM Norman, RJ Kendall. *Environ. Pollut.* 1994, **83**:299-309.
20. Calambokidis, J, S Speich, J Peard, GH Steiger, JC Cubbage, DM Frye, LJ Lowenstine. NOAA Technical Memorandum. National Oceanic and Atmospheric Administration, Rockville, MD. 1985; 159 pp.
21. Moul, IE. Masters Thesis. University of British Columbia. 1990; 59 pp.
22. Cobb, GP, DM Norman, LW Brewer. *Chemosphere.* 1995, **30:**151-164.
23. Guillette, LJ, TS Gross, GR Mason. *Environ. Hlth. Perspect.* 1994, **102**:680-688.
24. Heinz, GH, HF Percival, ML Jennings. *Environ. Monit. Assess.* 1991, **16**:277-285.
25. Delany, MF, JU Bell, SF Sundlof. *Journal of Wildlife Diseases.* 1988, **24**: 62-66.
26. Hall, RJ, TE Kaiser, WB Robertson PC Patty. *Bull Environ. Contam. Toxicol.* 1979, **23**: 87-90.
27. Cobb, GP, PD Wood, Michael O'Quinn. *Environ. Toxicol. Chem.* **1997**,16:1456-1462.
28. SCDHEC. Sampit River Advisory Update. South Carolina Department of Health and Environmental Control. Columbia, SC. 1992; 8pp.
29. Bidleman, TF, AC Svastits. *NOAA, Coastal Energy Impact Program Grant No. CEIP82-08/20361*, January, 1983, p. 7-8.
30. Bergeron, J.M., D. Crews, J.A. McLachlan. *Environ. Health Perspec.* 1994, 102(9): 780-781.
31. Bishop, C.A., R.J. Brooks, J.H. Carey, P. Ming, R.J. Norstrom, D.R.S. Lean. *J. Toxicol. Environ. Health.* 1991, 33: 521-547.
32. Bishop, C.A., G.P. Brown, R.J. Brooks, D.R.S. Lean, J.H. Carey. *Arch. Environ. Contam. Toxicol.* 1994, 27: 82-87.
33. Clark, D.R., A.J. Krynitsky. *Pesticides Monitoring Journal.* 1980, 14(1): 7-10.
34. McKim, J.M. Jr., K.L. Johnston. *Bull. Environ. Contam. Toxicol.* 1983, 31: 53-60.
35. Cobb, G.P., P.D. Wood. *Chemosphere.* 1997, 34:539-549.
36. Pastor , D., X.Ruiz, L. Joyner, J. Albagiges. *Environ. Toxicol. Chem.* 1996, **15**:167-171.
37. Trust, KA, A Fairbrother, MJ Hooper. Environ. Tox. and Chem. 1994, 13(5):821-830.
38. Mullin, M.D., C. M. Pochini. *Environ. Sci. Technol.* 1984, **18**:468-476.
39. Albro, P.W., C.E. Parker. *J. Chromatog.* 1979, **169**:161-166.

40. Albro, P.W., J.T. Corbett, J.L. Schroeder. *J. Chromatog.* 1981, **205**:103-111.
41. Dunnivant, F.M. Master's Thesis. Clemson University: Clemson, South Carolina. 1985, 168 pp.
42. Skerfving, S., B.-G. Svensson, L. Asplund, L. Hagmar. *Clin. Chem.* 1994, **40**: 1409-1415.

Chapter 21

Chemical and Biochemical Evaluation of Swainson's Hawk Mortalities in Argentina

Edward J. Scollon[1], Michael I. Goldstein[2], Melissa E. Parker[2], Michael J. Hooper[1], Thomas E. Lacher[2], and George P. Cobb[1]

[1]The Institute of Environmental and Human Health, Texas Tech University, Lubbock, TX 79409–1163
[2]Department of Wildlife and Fisheries Sciences, Texas A & M University, College Station, TX 77843–2258

Several isolated incidents of Swainson's hawk (*Buteo swainsoni*) mortality were reported in the Pampas regions of Argentina during the austral summer of 1996. Eighteen sites containing 5,093 dead hawks were identified. Four sites, I, II, III, and IV, containing 387, 103, 3024, and 595 hawk mortalities, respectively, were investigated. Landowner and pesticide applicator interviews indicated use of the organophosphate pesticides (OP) monocrotophos at sites I and IV and dimethoate at site II. The agent used at site III was unknown due to conflicting reports. However, brain and plasma cholinesterase activity were sufficiently depressed in the hawks found at site III to indicate exposure to an OP. Brain and plasma acetylcholinesterase (AChE) taken from several dead and incapacitated birds were depressed 90% and 50%. Dimethoate, methamidiphos, methyl parathion, and chlorpyrifos were suspected, but only monocrotophos was found in several gastrointestinal (GI) tracts. No other parent OPs were detected from any of the sites. Dimethylphosphoroate, the primary metabolite of a dimethylphosphate insecticide, was detected in GI tracts from sites I and IV. The absence of dialkylthiophosphates precluded the use of any phosphorothioate or phosphorodithioate insecticide. Based upon these findings, monocrotophos was responsible for the mortalities at sites I, III, and IV, and dimethoate for site II.

The Swainson's hawk, (*Buteo swainsoni*), is a neotropical migrant, breeding in the western portions of the United States, Canada, and Mexico and overwintering in the pampas regions of Argentina. The grassland and desert habitats of its breeding grounds provide a variety of prey items including insects, small mammals, reptiles and birds (1). However, the pampas of Argentina, which are primarily agricultural, provide an overwhelming abundance of insect prey. Swainson's hawks seasonally shift their diets to take advantage of the plentiful insect populations (2).

Swainson's hawks are gregarious, gathering in large flocks prior to their annual southward migration. Flocks containing several hundred thousand birds have been reported passing over their southern migration routes (1). They remain in these large groups throughout the wintering period. Common agricultural practices such as cultivating, harvesting, or pesticide application, tend to disturb resident insects. Swainson's hawks take advantage by descending in large numbers upon the available prey (1). If insecticides are being applied, or have recently been so, the hawks are subject to direct exposure from application mists, dermal absorption through feet and legs, or oral exposure via their insect prey. The propensity of Swainson's hawks to aggregate in large flocks makes them susceptible to large-scale mortality on the wintering grounds when exposed to insecticides in this manner. Woodbridge and colleagues in 1995, and Goldstein and colleagues in 1996 and 1997 documented that such incidents had occurred in the pampas of Argentina (2-6).

Organophosphate insecticides have largely replaced the once common organochlorine pesticides in the pampas region. Chlorinated organic pesticides and their metabolites tend to be persistent in the environment. For instance, the half-life of p,p'-DDT and its metabolites ranges from 2 to 15 years depending on the environment (7). Their persistence, high affinity for lipophilic tissue, and resistance to metabolism, causes these compounds to biomagnify in top predators. Negative population impacts have been demonstrated in sensitive species (8). Organophosphate pesticides (OPs) have the advantage that they do not persist in the environment and therefore do not bioaccumulate. However, they have the disadvantage that they traditionally exhibit much higher acute toxicities to nontarget species. Five OPs were used in the vicinity of the four hawk mortality sites, monocrotophos (O,O-dimethyl(E)-1-methyl-2-(methylcarbamoyl)vinylphosphate), methamidiphos (O,S-dimethyl phosphoroamidothioate), dimethoate (O,O-dimethyl-S-methylcarbamoylmethyl phosphorodithioate), methyl parathion (O,O-dimethyl-O-(4-nitrophenyl) phosphorothioate), and chlorpyrifos (O,O-diethyl-O-(3,5,6-trichloro-2-pyridinyl) phosphorothioate) (Figure 1).

The breakdown of OPs into alkylphosphate metabolites is characterized by a specific pattern. Dialkyldithiophosphoroate insecticides can be metabolized into a dialkyldithiophosphate, dialkylthiophosphate, or dialkylphosphate (Figure 2a). In a similar manner, dialkylthiophosphoroate insecticides can be metabolized into a dialkylthiophosphate or dialkylphosphate (Figure 2b). Dialkylphosphoroate

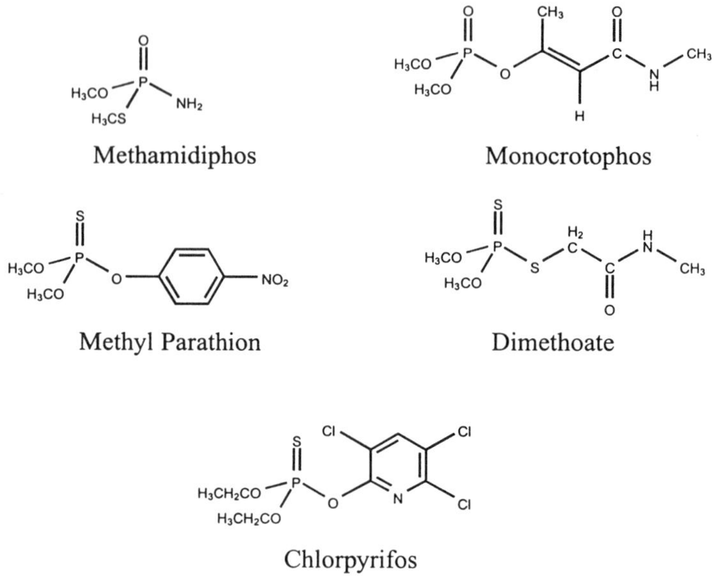

Figure 1. Organophosphorous insecticides suspected in Swainson's hawk mortality sites I, II, III, and IV based on landowner and farmer surveys.

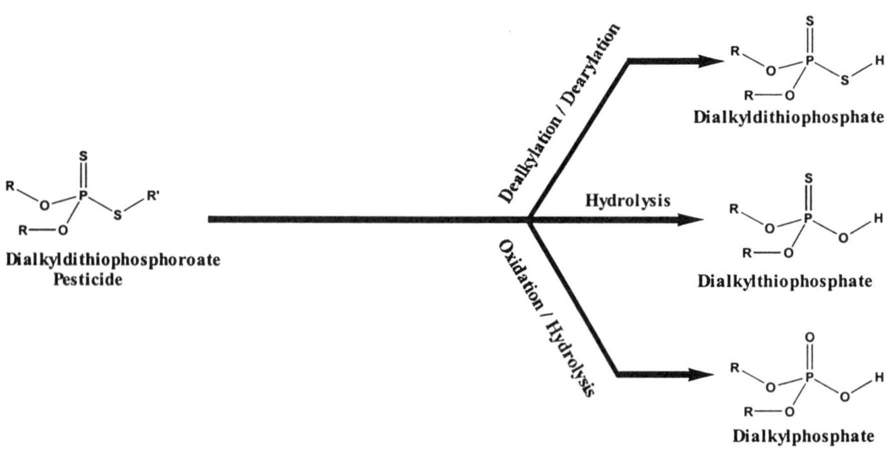

Figure 2a. Metabolic pathway of a dialkyldithiophosphate ester and its respective alkylphosphate metabolites.

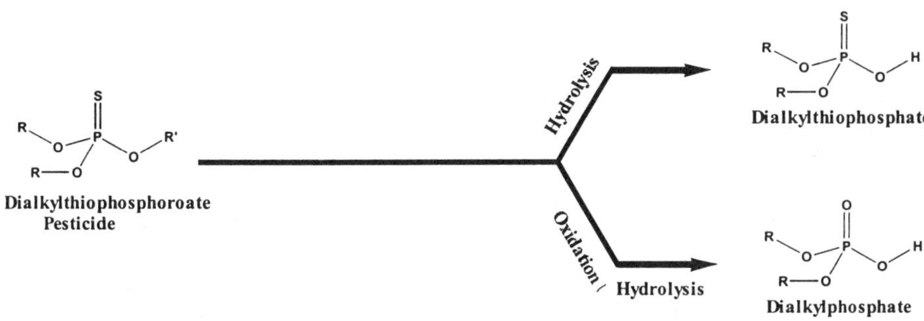

Figure 2b. Metabolic pathway of a dialkylthiophosphate ester and its respective alkylphosphate metabolites.

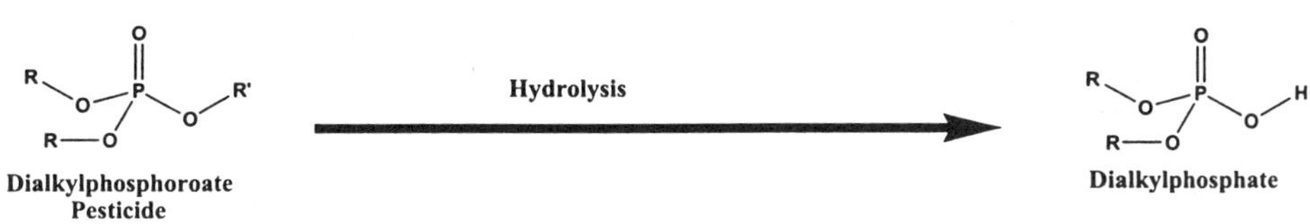

Figure 2c. Metabolic pathway of a dialkylphosphate ester and its respective alkylphosphate metabolites.

insecticides can only be broken down into dialkylphosphates (Figure 2c). Putting the current list of pesticides into perspective, dimethoate is a dimethyldithiophosphoroate, methamidiphos and methyl parathion are dimethylthiophosphoroates, chlorpyrifos is a diethylthiophosphoroate, and finally, monocrotophos is a dimethylphosphate, lacking thiol groups altogether. If the hawks are exposed to dimethoate, the metabolic products would include dimethyldithiophosphate (DMDTP), dimethylthiophosphate (DMTP), and dimethylphosphate (DMP). Note that the OP degradation does not occur so rapidly that all of the parent compound would be converted to DMP over the course of this study. If the hawks had ingested dimethoate, DMDTP or DMTP would be present. The presence of DMTP or DMP in the absence of DMDTP would indicate the use of methamidiphos, methyl parathion or monocrotophos. The presence of DMP and the absence of DMDTP or DMTP in the GI samples would indicate the use of monocrotophos. The presence of any diethylphosphate metabolites would implicate the use of chlorpyrifos.

In a forensic study of this type, it is imperative to link the causative agent with a physiological response. Organophosphate pesticides are anticholinerase agents. They bind to the serine active sites of acetylcholinesterase, rendering the enzyme nonfunctional. The role of acetylcholinesterase is to promote the hydrolysis of the neurotransmitter, acetylcholine, into its acetyl and choline constituents. In the absence of acetylcholinesterase, acetylcholine builds up within the neural synapse causing stimulatory paralysis. Death ultimately results from depression of the respiratory centers of the brain (9).

The nature of the anticholinesterase agent bound to the serine active site determines the extent to which the agent can be removed and the acetylcholinesterase reactivated. Carbamates, another type of anticholinesterase pesticide, bind to the serine active sites but can be removed from the active site by simple dilution and incubation (10). Cholinesterases inhibited by diethyl OPs may be reactivated with 2-PAM (2-pyridinealdoxime methochloride) (11). Esters containing diisopropyl and dimethyl alkyl groups age quickly, making them unresponsive to reactivation.

Materials and Methods

Eighteen Swainson's hawk mortality incidents, accounting for approximately 5,000 hawks, were found during the 1995-96 austral (southern hemisphere) summer within the pampas region of Argentina (5). A forensic evaluation of sites I through IV was initiated by investigators to determine the cause of the kills. Local residents indicated that the deaths were usually associated with the application of organophosphorous insecticides in hawk foraging areas. The investigation was based on 1) landowner and field worker surveys regarding the pesticide application prior to the hawks dying; 2) analytical determination of parent pesticides and their metabolites in hawk tissues; and 3) biochemical analysis of brain and plasma cholinesterase activity to link a physiological effect with possible pesticide exposure (5).

Surveys

Landowners and field workers within close proximity of the mortality sites were interviewed. They informed investigators of the timing and nature of pesticide applications. Despite the fact that several reports were deficient or contradictory, the list of probable agents was narrowed down to the five organophosphorous pesticides previously discussed (5).

Analytical

Parent OP and metabolite analyses were performed on hawk gastrointestinal contents, specifically, from the proventriculus. Feather and footwash rinses were collected from affected hawks for OP analysis. Residue determination was initially performed by gas chromatograph (GC) using a flame photometric detector (FPD). Analyte confirmation was performed by dual column chromatography or mass spectrometry (MS).

Sample splits were sent to Novartis Crop Protection, Basel, Switzerland, for monocrotophos analysis more detailed than our laboratory was able to provide (5). The specifics are outlined in the discussion.

Sample Extraction and Cleanup

GI tract. Two gram samples were extracted with 10 ml of 9:1 acetone:hexane using a blender (VirTis Co., Gardinier, NY, USA). The samples were homogenized for 1 minute, after which the eluant was poured through a 5 g granular anhydrous sodium sulfate column into a flat bottom flask. The extraction was repeated a second time. Following the second elution, the sodium sulfate column was rinsed with 25 ml of 9:1 acetone:hexane. The extract was reduced to 2 ml by vacuum rotary evaporation.

The concentrate and rinses (3 x 2 ml 9:1 acetone:hexane) were transferred to the head of a 5 g silica gel column previously conditioned with 9:1 acetone:hexane. The samples were eluted with 35 ml of 9:1 acetone:hexane into a flat bottom flask and again reduced to 2 ml by vacuum rotary evaporation. The concentrate and rinses were brought to a known volume of 5 ml and eluted through a gel permeation chromatography column. Fractions containing the OPs were collected in flat bottom flasks, reduced to 1 ml under vacuum rotary evaporation, and combined with rinses for a final volume of 2 ml. The extracts were loaded into 2 ml autosampler vials for GC analysis.

Alkylphosphate metabolites. The original technique for alkyl phosphate analysis is outlined in Weisskopf and Seiber (12). However, several modifications were incorporated to account for differences in sample matrix.

Three ml of deionized water were combined with 0.5 g of the proventriculus contents in polystyrene centrifuge tubes. The homogenate was adjusted to a pH of 4.0 with 1 M sodium hydroxide or 7% acetic acid and vortexed for 20 seconds every 5 minutes for 45 minutes. The slurry was saturated with ammonium sulfate (1.4 g;

Mallinckrodt Analytical Reagent), vortexed for 30 seconds, and then centrifuged for 7 minutes at 2500 rpm.

The supernatant containing the alkylphosphate metabolites was placed onto a 1 g cyclohexyl column preconditioned with 6 ml of 9:1 acetone:hexane, 1:4 methanol:acetone, deionized water, and ammonium sulfate saturated deionized water each. Elution through the column was accomplished using a vacuum manifold. The column was rinsed with 2 ml of 1:9 acetone:hexane and aspirated for 5 minutes to remove residual water.

The alkyl phosphate metabolites were eluted off the cyclohexyl columns with 3.5 ml 1:4 methanol:acetone. The eluant was brought to a known volume of 2 ml and mixed with 0.6 g powdered anhydrous sodium sulfate. The solutions were removed from the sodium sulfate and transferred to 2 ml GC autosampler vials. Immediately prior to running on the GC, 15 μl of the derivatizing agent, tetrabutyl ammonium hydroxide (TBAH) (1 M in methanol), was added to each vial.

Feathers. Feather samples consisting of 3 to 4 coverts from each of the backs of both wings, the back and the breast, were placed in Erlenmeyer flasks along with 35 ml of 2:1 acetone:hexane. The sample was placed on an orbital shaker and extracted for 30 minutes. The extracted solvent and two 10-ml rinses of 2:1 acetone:hexane were eluted through a 2-g column of granular anhydrous sodium sulfate into a flat bottom flask. The column was rinsed with an additional 10 ml of 2:1 acetone:hexane. The eluant and rinses were reduced to 2 ml using a vacuum rotary evaporator. The extract was filtered through 0.45 μm PTFE Acrodisc filters into 2 ml autosampler vials for GC analysis.

Footwashes. Ethanol footwashes were reduced to 2 ml, filtered as above, and injected directly onto the GC as described below.

Instrumental Analysis

Quantitation was performed using a Hewlett Packard Model 5890 II gas chromatograph equipped with an autosampler and flame photometric detector (GC/FPD) in the phosphorous mode. When matrix interference was minimal, dual column confirmation was provided by fused silica capillary columns coated with DB-1701 or DB-210 stationary phases (J & W Scientific, Rancho Cordova, CA, USA). The GCs were calibrated with a four point external curve for monocrotophos, methamidiphos, dimethoate, and methyl parathion.

Pesticide recovery was determined by spiking clean crickets with a mixture containing dimethoate, methyl parathion, methamidiphos, and monocrotophos. Crickets were used as a surrogate for the grasshoppers commonly found in the hawk GI tract. The crickets were spiked at two concentrations, 0.1 ppm and 2 ppm. Recoveries ranged from 86% to 123% for the 0.1 ppm spikes and 76% to 121% for the 2 ppm crickets (Table I).

The method level of detection (MLOD) for hawk digesta from site III was 0.037 μg/g. The digesta from sites I and IV were thicker and contained greater amounts of matrix interference. The MLOD at these sites was correspondingly higher at 0.2 μg/g.

Table I. Percent Recovery of Organophosphate Pesticides from Spiked Crickets.

	Percent Spike Recoveries	
	0.1 PPM	*2.0 PPM*
Dimethoate	86.7 ± 8.5[a]	79.5 ± 6.9[a]
Methyl Parathion	108 ± 10	76.9 ± 6.1
Methamidiphos	123 ± 6.6	121 ± 16
Monocrotophos	121 ± 6.9	87.6 ± 9.6

[a] mean ± standard error

When matrix interference was high, GC-mass spectrometry (GC/MS) was used to confirm GC/FPD detections. GC/MS was performed on a Hewlett Packard Model 5988A quadropole mass spectrometer operated in the electron impact mode. Full spectrum scans were run for most samples. In samples from areas suspected of monocrotophos contamination, three characteristic ions (m/z = 127, 192, and 223) were monitored to improve detection limits.

Biochemical Analysis

Sample Preparation

Blood samples were taken from moribund birds euthanized at the field sites as well as birds active enough to be released after capture. The blood was drawn with heparinized needles from the humoral or femoral veins. Blood samples were centrifuged and the plasma removed. The plasma was stored frozen until laboratory analysis, which occurred within two weeks of collection. The samples were thawed and diluted ten-fold for use in cholinesterase (ChE) and reactivation analysis.

Brain tissue was removed from the skulls and diluted 1:4 (weight:volume) with 0.05 Tris buffer. The samples were homogenized for 30 seconds in a tissue homogenizer (VirTis Co., Gardinier, NY, USA). The resulting homogenate was diluted 20-fold, for total dilution of 100-fold, prior to ChE analysis.

Cholinesterase Activity Determination

Cholinesterase (ChE) activity was determined by the Ellman assay (13). This assay measures the hydrolysis of acetylthiochloline (AThCh), a thio ester analogue of acetylcholine. Hydrolyzed thiocholine attacks DTNB (5,5'-dithiobis(2-nitrobenzoic acid)), present in the assay medium. Hydrolyzed DTNB releases a chromophore, TNB (5-thio-nitrobenzoic acid), which can be measured spectrophotometrically. The assay was modified according Hunt and Hooper (10) for use on a UVmax (Molecular Devices Corp., Palo Alto, CA, USA) 96-well plate reader set in kinetic mode at a wavelength of 405 nm with a run time of 2 minutes read at 8 second intervals.

Brain cholinesterase (ChE), predominately acetylcholinesterase (AChE), was measured directly. Plasma AChE activity was measured through the use of Iso-OMPA (tetraisopropyl-pyrophosphoramide, 1 x 10-4M FC), a butylcholinesterase (BChE) inhibitor. Iso-OMPA was added immediately prior to the 5 minute preincubation. The remaining activity following Iso-OMPA treatment was AChE. BChE activity was computed as the difference between total ChE and AChE activities.

Reactivation Analysis

Dilution and 2-PAM reactivation assessments were performed to test for the presence of carbamate and OP inhibited cholinesterase, respectively. The dilution reaction involved incubating split sample aliquots at 4°C and 37°C for up to 3 hours. Samples with 37°C incubated activity elevated more than 20% over the 4°C activity were considered to contain carbamate inhibited cholinesterase (10). The 2-PAM reactivation technique included the addition of 2-PAM to one of the incubated aliquots. Samples that showed an increase of 5% or greater in the presence of 2-PAM are considered to contain OP inhibited cholinesterase, most likely of diethyl nature (11).

Results

Site I

Landowner and worker surveys indicated that monocrotophos had recently been applied at or adjacent the Swainson's hawk feeding grounds prior to their deaths (5).

Eleven GI tracts were collected from this site; none contained detectable levels of parent OPs. However, GI samples sent to Novartis Crop Protection Laboratories, Basel, Switzerland, for confirmation, contained detectable concentrations of monocrotophos (0.1±0.03 μg/g) in 6 of the 11 samples. Two of three GI samples analyzed for dialkylphosphate metabolites contained detectable levels of DMP.

Birds from sites I, II, and IV were not found until 9 to 10 days postmortem. Cholinesterase activity at the time of death can be accurately diagnosed up to approximately 4 days after death (14). After this period the data become questionable and were not included in this forensic evaluation.

Site II

The surveys indicated that dimethoate had been applied to the fields where the hawks were found. There were no discrepancies among the reports (5). The carcasses were dehydrated and scavenged preventing the collection of adequate samples for either analytical or biochemical analysis.

Site III

The surveys were confused and often contradictory with respect to pesticide application. The pesticide used at site was initially assumed to be unknown (4, 5). Nine GI tracts were collected from dead birds. Four of these contained detectable levels of monocrotophos (0.089 ± 0.017 μg/g). Two samples collected from euthanized survivors did not contain detectable levels of parent OPs. Only one sample was selected for alkyl phosphate analysis from this site, and it did not contain detectable levels of any metabolites. Novartis confirmation determinations reported detectable levels of monocrotophos (0.13± 0.03 μg/g) in 6 of the 11 samples.

Brain cholinesterase activity for 9 dead birds and 2 that were incapacitated and euthanized were depressed 95.4±1% and 70%, respectively. None of the samples showed any sign of reactivation with dilution and only 2 birds, one dead and one euthanized, were minimally reactivated when incubated with 2-PAM.

Plasma samples were taken from 8 birds, 2 euthanized at the site, and 6 which were later released. ChE was inhibited 60.1±12.5% in the released birds and 49% in those euthanized. None of the samples reactivated with simple dilution. However, plasma from one of the released hawks showed reactivation when incubated with 2-PAM.

Site IV

The landowners and field workers indicated that a monocrotophos application preceded the arrival of the hawks at this site (4, 5).

No detectable levels of the parent OPs were found in the ten GI samples collected. However Novartis laboratories detected monocrotophos (0.07±0.01 μg/g) in 7 of 9 samples. DMP was detected in 2 of 4 samples analyzed for dialkyl phosphate metabolites (Table II).

Feather and Footwashes

Feathers were collected from 26 birds across the four study sites. A single bird from site III had a detectable level of monocrotophos (~1 μg/g).

One footwash, performed on a single bird from site I, tested negative for OP analysis.

Discussion

Site I

Three hundred eighty seven hawks were found at site I. Chemical analysis of 11 proventriculus samples did not indicate the presence of any parent OP (5). Monocrotophos and methamidiphos were applied as oxons (P=O). In this activated form, the pesticides are able to interact with the serine esterases. The acidic environment of the proventriculi and the additional serine esterases, released by digesting grasshoppers, enhanced the degradation of these pesticides. Therefore, the probability of these reactive OPs remaining in high concentrations several days postmortem is extremely low. Methyl parathion, dimethoate and chlorpyrifos were applied in the inactive P=S form. These pesticides have to be bioactivated into oxons before they actively bind esterases. Additionally, the inactive states are hydrolyzed at much lower rates than the activated forms. Therefore, if the hawks had consumed grasshoppers treated with methyl parathion, dimethoate, or chlorpyrifos, the pesticides are more stable and would have persisted well beyond the time between death and GI sample analysis.

Table II. Swainson's hawk mortality incidents during January and February of 1996, in the pampas region of Argentina.
(Table modified from Goldstein et al. (5))

Site	# Dead	Survey[a]	ChE % Inhibition				GI Contents MCP μg/g		AP[g]	Indicated[h] Pesticide
			Brain		Plasma		TIEHH	Novartis[f]		
			Dead	Euthanized	Euthanized	Released				
I	387	MCP	NA	NA	NA	NA	0/11[e]	0.1±0.03[b] 0.05-0.19[c] N=6/11[e]	2/3	MCP
II	103	Dimethoate	NA	NA	NA	NA	NA	NA	NA	Dimethoate (?)
III	3,024	MCP	95.4±1[b] 89.4-98.3[c] N=9[d]	72.7 71.8-73.6 N=2	48.6 38.1-59.2 N=2	60.1±12.5 24.3-97.4 N=6	0.09±0.02 0.05-0.13 N=4/11	0.13±0.03 0.06-0.25 N=6/11	0/1	MCP
IV	595	MCP	NA	NA	NA	NA	0/10	0.07±0.01 0.05-0.12 N=8/10	2/4	MCP

NA- Not Available; MCP – monocrotophos; ChE – cholinesterase; TIEHH – The Institute of Environmental and Human Health;

[a]pesticide implicated by landowner/worker surveys; [b]mean±SE; [c]range; [d]# of samples analyzed; [e]# of samples containing monocrotophos/total # of samples analyzed; [f]Data from R. Tribolet, Novartis Crop Protection; [g]# of samples containing dectectable levels of DMP/total # of samples analyzed; [h]pesticide implicated in hawk mortalities

Despite the fact that our laboratory was unable to detect monocrotophos in the GI tracts, Novartis Crop Protection Laboratories consistently found monocrotophos residues in their samples. Novartis employed a method developed to analyze solely for monocrotophos. Their extraction and cleanup methods catered towards the specific solubility of the analyte. Our method was geared towards the extraction and cleanup of several OPs that spanned a broad range of solubilities. As such, we extracted many more non-pesticide compounds from the samples, increasing the baseline noise and our level of detection above the low concentrations of monocrotophos found in many of the samples.

The presence of DMP in the absence of thiol containing metabolites in the GI samples at this site can only occur with the use of monocrotophos as discussed in the introduction. Therefore, based upon our analytical and biochemical data, landowner surveys, and Novartis findings, it is most probable that the Swainson's hawks were killed by lethal concentrations of monocrotophos at site I.

Site II

One hundred three hawks were found at site II (5). The carcasses were desiccated and scavenged precluding the collection of samples. However, interviews with field workers indicated that dimethoate had been applied to the area just prior to the hawks foraging. The surveys were not contradictory in this regard. Therefore, in the absence of corroborating analytical and biochemical data, dimethoate was designated the likely agent in hawk mortalities at site II.

Site III

By far the most widespread mortality occurred at site III (5). Over 3,000 hawks were found dead or moribund. The presence of detectable monocrotophos in the GI tract indicated recent exposure. The absence of methamidiphos, methyl parathion, or dimethoate in the GI tracts supports the premise that monocrotophos was the only OP responsible for the mortalities at this site.

The extreme depression of brain ChE activity in the birds sampled dead (>90% inhibition) and live (>70% inhibition) from this site indicates that an anticholinesterase agent was responsible for the hawk mortalities. Plasma ChE activity was also significantly depressed compared to normal activities. However, the inhibition was much less than that observed in the brain. This may be attributed to the relatively rapid turnover of ChE in the plasma compared to the brain. Perhaps the plasma ChE was recovering following exposure to the anticholinesterase agent.

The lack of reactivation of ChE by dilution and incubation indicates that the ChE was not inhibited by a carbamate. The infrequent reactivation in the presence of 2-PAM indicates inhibition by an OP, in particular a dimethyl type OP, which ages quickly. Our findings and those of Novartis indicate that the Swainson's hawk at this site experienced a lethal exposure of monocrotophos.

Site IV

Nearly 600 hundred birds were found dead after foraging in an alfalfa field recently treated with monocrotophos. There was no discrepancy among the surveys with regard to the pesticide applied (5).

DMP was found in two of four samples checked for metabolites. The presence of DMP in the absence of DMTP and DMDTP implies the use of monocrotophos. Novartis Crop Protection also found monocrotophos in 7 of 9 GI samples analyzed. Chemical analysis of the GI tract in our labs did not detect any pesticide residues. However, due to the highly reactive nature of monocrotophos and the length of time between exposure and analysis, this result is expected. Our findings, in conjunction with those of Novartis, indicate that the hawks were exposed to monocrotophos.

Conclusions

Based upon our analysis of four Swainson's hawk mortality incidents in the pampas regions of Argentina during the austral summer of 1996, it is evident that current analytical and biochemical techniques can be used to forensically evaluate the impact of agricultural chemicals on wildlife (5). We were able to ascertain the presence of monocrotophos or a primary metabolite in fields where each of the kills occurred. The absence of four additional insecticides was demonstrated through analysis for parent OPs, primary hydrolytic metabolites, and ChE reactivation when appropriate samples were available. This study demonstrates the importance of incorporating multiple disciplines in environmental studies.

Acknowledgments

The authors would like to thank A. Lanusse and S. Salva for their endless hospitality; veterinarians J. Garat and C. Peregalli for their time and help in the field collections; M. Zaccagnini, S. Canavelli, and J. Panigatti of INTA for their support in Argentina; J. Rogers and C. Cline for their assistance in the lab; M. Fuller for satellite coordinates to locate flocks of hawks; B. Woodridge and M. Bechard for their foresight, motivation, and extensive knowledge of Swainson's hawk ecology: and R. Tribolet who performed the monocrotophos analyses at Novartis Crop Protection, Basel, Switzerland.

This study was made possible by support from the National Wildlife Federation, Ecorisk Incorporated, USDA Forest Service, Rachel Kolokoff, Archbold Tropical Research Center, National Fish and Wildlife Foundation, Novartis Crop Protection, and the USFWS Office of International Affairs.

Literature Cited

1. England, A.S.; Bechard, M.J.; Houston, C.S. In *The Birds of North America*, Pool, A., and Gill, G., Eds. The Academy of Natural Sciences: Philadelphia, PA, and The American Ornithologists Union: Washington, D.C., 1997.
2. Woodbridge, B.; Finley, K.; Seager, S.T. *J. Raptor Res.* 1995, 29, 202-204.
3. Goldstein, M.I.; Woodbridge, B.; Zaccagnini, M.E.; Canavelli, S.B.; Lanusse, A. *J. Raptor Res.* 1996, 30, 106-107.
4. Goldstein, M.I. *Toxicological Assessment of a Neotropical Migrant: Case Study of the Swainson's Hawk in Argentina*, M.S. Thesis, Clemson University, Clemson, SC, 1997.
5. Goldstein, M.I.; Lacher, T.E. Jr.; Woodbridge, B.; Bechard, M.J.; Canavelli, S.B.; Zaccagnini, M.E.; Cobb, G.P., III; Scollon, E.J.; Tribolet, R.; Hooper, M.J. *Ecotoxicology*, 1999, 8, 201-214.
6. Goldstein, M.I.; Lacher, T.E., Jr.; Zaccagnini, M.E.; Parker, M.L.; Hooper, M.J. *Ecotoxicology*, 1999, 8, 215-224.
7. Augustijn-Becjkers, P.W.M.; Hornsby, A.G.; Wauchope, R.D. *SCS/ARS/CES Pesticide Properties for Environmental Decision Making II. Additional Properties Reviews of Environmental Contamination and Toxicology.* 1994. Vol. 137.
8. Clark, K.E., *Arch. Environ. Contam. Toxicol.*, 1998, 61, 247-54.
9. Ecobichon, D.J., In *Casarett and Doull's Toxicology: The Basic Science of Poisons*, Klaassen, C.D., Ed.; McGraw-Hill: New York. 1996. p. 1111.
10. Hunt, K.A.; Hooper, M.J. *Anal. Biochem.*, 1993, 212, 335-343.
11. Wilson, B.W.; Hooper, M.J.; Hanson, M.E.; Nieberg, P.S. In *Organophophates: Chemistry , Fate, and Effects*, Chambers, J.E.; Levi, P.E., Eds.; Academic Press: Orlando, FL, 1992.
12. Weisskopf, C.P., and Seiber, J. N., In *Biological Monitoring for Pesticide Exposure: Measurement, Estimation, and Risk Reduction*; Wang, R.G.M.; Franklin, C.A; Honeycutt, R.C.; Reinert, J.C., Eds.; New Approaches to Analysis of Organophosphate Metabolites in the Urine of Field Workers; American Chemical Society: New Orleans, LA, 1989; pp 206-214.
13. Ellman, G.L.; Courtney, K.D.; Andres, V., Jr.; Featherstone, R.M. *Biochem. Phamacol.*, 1961, 7, 117-122.
14. Burn, J.D.; Leighton, F.A.; *J. Wildl. Dis.*, 1996, 32, 216-24.

Chapter 22

HPLC Determination of Carbofuran Residues and Its Distribution from Alginate Controlled-Release and Granular Formulations in a Model Rice–Fish Ecosystem

H. R. Soltan[1], A. S. Mare, A. Khamis, A. Mousa, and H. M. El-Maghraby[2]

Pesticide Chemistry Department, Faculty of Agriculture, Alexandria University, P.O. Box 21545, Aflatoun El-Chatby, Alexandria, Egypt

A high performance liquid chromatography (HPLC) procedure was used for the determination of carbofuran residues and its distribution in static model rice/fish ecosystems at the ng level. Alginate controlled release (CF7 and CF8) and granular (10G) formulations of carbofuran were applied to paddy water and then transferred to the components of the ecosystem. The distribution of carbofuran residues in these components was remarkably different. Generally, the granular formulation (10G) quickly released the active ingredient into water and the concentration peaked (0.593 ppm) 4 days after treatment. Thereafter the concentration gradually decreased with the time. In contrast, the controlled release formulations (CF7 and CF8) initially increased slowly within the first week then increased more quickly until the end of the experiment. Carbofuran residues in the paddy soil, released from the different formulations showed a similar trend to that which was observed in the paddy water. Carbofuran residues in water and/or soil in ecosystem treated with the granular formulation could be taken up by the root system and translocated to all part of rice plants more quickly than those treated with the alginate formulations, mainly, within the first week after treatments. Thereafter, the concentration of carbofuran declined rapidly or increased gradually with time in the ecosystem treated with the granular or alginate formulations, respectively. Exposure study for 28 days revealed that alginate based controlled-release formulations (CF7 and CF8) of carbofuran were less toxic to common carp than the commercial formulation. Fish treated with CF7 and CF8 accumulated lower residues than did the fish treated with the commercial formulation.

[1]Corresponding author: E-mail: chmistdept@alexcomm.net.
[2]Present address: Pesticides Laboratory, Central Agricultural, Dokky, Cairo, Egypt.

INTRODUCTION

Carbofuran is one of the most important broad spectrum insecticides, acaricides and nematicides which is effective through contact, stomach and systemic action. In many countries ecosystems for the production of rice and fish in paddy fields are common practice and many farmers depend on fish as their major protein source (*1*).

Unfortunately, most of the conventional formulations quickly release carbofuran into water, subjecting carbofuran to loss by environmental factors. The half-life of the compound in a model ecosystem was about 20 days (*2*) while this half-life in pH 8.5 aqueous media at 40 °C was 2.96 days (*3*). Carbamates in plants are less persistent but metabolites remain in crops as conjugates and thus might effect humans directly (*4*).

Carbofuran is a highly toxic compound to fish (*5*). Controlled release formulation technology applied in the present study may be a partial solution for making this pesticide more useful, less toxic, more effective in controlling pests and environmentally safe. Sodium alginate was selected as a natural polymer and kaolin as filling material to avoid any additional negative effects on the environment through toxic or non- biodegradable materials. Evaluation of alginate controlled - release formulation of carbofuran for adverse effects to fish and its distribution in rice/fish ecosystem is the goal of the present study

MATERIALS AND METHODS

Controlled Release Formulations of Carbofuran

Alginate controlled release formulations (CRF) used in the present studies were formulated and their release rate profiles determined (*6*). Among the prepared CRF, two formulations (CF7 and CF8) were selected for further study due to their appropriate release rates. The percentage of carbofuran and other ingredients used in these formulations are shown in Table I.

Table I. Composition of Controlled Release Formulations

Formulation Code No.	*% Composition*		
	Sodium alginate	*Kaolin*	*Carbofuran*
CF_7	49.84	45.31	4.85
CF_8	73.00	22.15	4.85

Distribution and Bioaccumulation of Carbofuran in a Model Rice/Fish Ecosystem

Model rice/fish ecosystem

The static model aquatic ecosystem used in this study consisted of two compartments. A nylon screen partition was installed between the two compartments (I and II) as shown in fig 1.

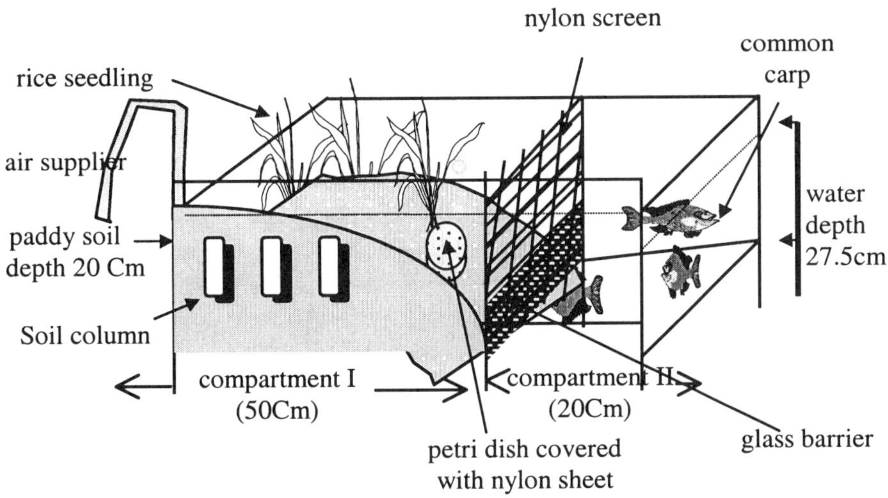

Figure 1. Micro-ecosystem designed to simulate a paddy/rice/fish ecosystem.

Ecosystem components and treatment

42 kg of paddy soil (air-dried) collected from pesticide free area located in Kafr El-Zyaat city, Gharbia Governorate was used. The soil characteristics are tabulated in Table II. Three quarters of the soil was added to each tank in compartment I (50 x 40 x 50cm) and flooded with an adequate amount of water. Seven soil sampling tubes (20 x 3 cm) covered from one side with 40 mesh nylon net were buried into the soil in each aquarium before treatment. The remaining one-quarter of soil was thoroughly mixed, slurried in water and then evenly spread into the compartment I. Controlled release formulations (CF7 and CF8) and granular formulation (10 G) were applied at a rate of 36.6 mg a.i./aquarium (0.8 kg a.i./faddan). Each formulation was placed in a petri dish covered with nylon netting in order to separate the carbofuran formulation and to facilitate quantification of the active ingredient remaining incorporated in the formulation at the end of experiment. The systems were operated with continuous aeration to provide homogeneity of carbofuran concentrations in the two compartments of aquarium. One day after treatment, fifteen 'hills' of plants (four 21 day- old rice seedlings), with a spacing of 10 x 10 cm between each hill, were transplanted in each tank (*1*). Five days after treatment, ten common carp (***Cyprinus carpio***) with an average length 6.2 cm and weight of 4.8gm were introduced into compartment II (20 x 40 x 50cm) to keep the fish from directly contacting with the carbofuran

formulations. The systems were operated under a water temperature around 24 °C and 12 hr light:12 hr dark cycle. The fish were fed daily. At different intervals after the treatments (2 hr, 1, 4, 7, 14, 21, and 28 day), water, paddy soil, rice and fish were sampled and analyzed for carbofuran residue levels.

Table II. Soil characteristics

Soil properties	*Sources of soil Kafr El-Zayaat soil*
1- Physical properties :	
a - Soil texture	Sandy clay loam
b - Water holding capacity (%)	53.30
2 - Chemical properties :	
a - pH value in water	7.90
b - $CaCO_3$ (%)	1.49
c - Total phosphorus (ppm)	9.94
d - Organic matter content (%)	1.96
e - Total nitrogen (%)	0.12

Samplings and Analysis

High-performance liquid chromatography of carbofuran

A Beckman 432 Liquid Chromatograph equipped with 110 B solvent delivery system, model 210 injector and model 166 detector was used in this study. An ultrasphere DDS C_{18} analytical column (30 cm x 4.6 mm i.d.) was used for reversed phase chromatography. Deionized water and HPLC grade solvent were degassed in an ultrasonic bath just before use. The mobile phase was methanol-water (1:1 V/V) with the flow rate of 0.8 ml/min. Fixed wavelength UV detection system at 220 nm was used. The retention time of carbofuran was 3.90 min. The peak area was measured using a Spectraphysics (SP 4100) computing integrator.

Paddy water

Carbofuran residues were extracted from water according to the method described by Jinhe *et al.* (1). At each sampling, water (100 ml) was randomly sampled using a 100 ml plastic syringe. The samples were adjusted to 0.25 M hydrochloric acid, and extracted in a 250 ml separatory funnel with methylene chloride (3 x 50 ml). The extracts were combined and dried by filtering through 20 gm of anhydrous sodium sulfate. Methylene chloride extracts were evaporated by rotary evaporation to dryness. Residues were re-dissolved in 5 ml methanol, quantitatively transferred to 5ml volumetric flask and then stored refrigerated at 4°C. Extracts were reduced under a stream of N_2 to exactly 1ml and analyzed by HPLC.

Paddy soil

The carbofuran residues were extracted from soil according to the method of Nicosia *et al.* (*7*). One of the soil sampling tubes was removed at each sampling period and allowed to dry at room temperature. Each soil sample (40 gm) was shaken for one hour with 125 ml 0.25 M HCl to extract carbofuran residues. Celite was added and samples were briefly shaken to homogenize, then filtered. The filter, celite and soil were extracted again with 100ml HCl, shaken 30 min, filtered, and rinsed three times with HCL. The extracts were brought up to a final volume of 400 ml and transferred to a 1L separatory funnel. Then samples were shaken three times with methylene chloride. The methylene chloride extracts were pooled, dried on anhydrous sodium sulfate, and evaporated by rotary evaporation to dryness. Carbofuran residues were re-dissolved in 5ml methanol, and then stored at 4°C. Extracts were reduced under a stream of N_2 to 1ml and analyzed by HPLC.

Rice plant

Carbofuran residues were extracted from plant according to the method reported by Ferreire and Seiber (*8*). One hill of rice plants was carefully removed, rinsed with running water until clean and then cut into small pieces with a single-edge razor blade. A 15 gm sample of leaf shoots and root pieces were blended with 50 ml of acetone for 5 min., and filtered through Whatman No. 1 paper. This procedure was repeated twice with two 25ml additional portions of acetone.

The combined filtrates were concentrated to 5ml by rotary evaporation and then mixed with 50 ml of coagulating solution (1.25gm of ammonium chloride and 1ml of concentrated phosphoric acid in one liter of water) and 0.5 gm of Hyflo Super cell. The mixture was agitated, allowed to stand for one hour and then filtered through a sintered glass funnel. The filtrate including a washing made with 50 ml of additional coagulating solution, was extracted three times with 30, 15 and 15ml portions of methylene chloride. The extracts were dried on 20 gm anhydrous sodium sulfate and concentrated to 2 ml. A 1.5 x 30 cm glass column was slurry packed with 15gm of silica gel in methylene chloride, followed by 1gm of anhydrous sodium sulfate, and then 0.2 gm of activated charcoal. The concentrated extract was added to the column by using a small volume of methylene chloride to complete the transfer and then eluted with 50 ml 2:8 of ethyl ether dichloromethane. The eluates were concentrated just to dryness by rotary evaporator and reconstituted in 5ml methanol for analysis by HPLC.

Fish

The method of Yu *et al.* (2) was applied with slight modification to extract carbofuran residues for fish. A fish was placed on paper towels to remove surface water, the fresh weight recorded, and the fish cut into small pieces with a single edge razor blade. Fish pieces were then extracted by blending three times with about 3ml acetone/gm fish and centrifuged at 1000 g for 15 min. The precipitates were resuspended in 6ml acetone per 1 gm fish and centrifuged as described before. The combined supernatant was concentrated to 10ml under a stream of N_2 then partitioned with 10ml of 2% sodium chloride solution and 30ml CH_2Cl_2 in a 250ml separatory funnel. The remaining acetone aqueous salt solution was extracted with 30ml CH_2Cl_2. The extracts were combined, dried over 25gm of anhydrous sodium sulfate, and concentrate to 1ml under N_2. A 1.5 x 30 cm glass column was slurry packed with 15gm of florisil (previously deactivated with 2% water) in methylene chloride and topped with 1gm of anhydrous sodium sulfate. The concentrated extract was added to the column in a small volume of methylene chloride, and then eluted with 10ml of methylene chloride (which was discarded) followed by 50ml 2 : 8 ethyl ether - hexane. The later eluate was concentrated just to dryness by rotary evaporator. The carbofuran residue was re-dissolved in 5ml methanol for HPLC analysis .

Determination of carbofuran remaining in the formulations

At the end of the experiment, granular and prepared formulations (CRF) were filtered off and added to methanol : water (1 : 1 V/V) in a flask and sonicated for 1 hour. The solution was filtered and the formulations were extracted three more times. The combined filtrate was analyzed by HPLC (*6*).

RESULTS AND DISSCUSSION

Distribution and Bioaccumulation in a Model Rice Fish Ecosystem

In many rice growing zones, fish are often combined into the field to form a more productive rice/fish ecosystem while ponds and rivers are usually connected to paddy fields as irrigation water sources. In either case, the application of pesticides might cause acute and long-term effect to the fish and other aquatic organisms. Hence, it is necessary to evaluate the safety of the alginate controlled release formulations versus the commercial formulation of carbofuran under these conditions.

The residues of carbofuran from water, plant, soil, and fish samples were extracted and subject to chromatographic analysis by HPLC. The efficiency of these extraction procedures were evaluated by determining the percent recovery of known amount of carbofuran from fortified samples. The recovery percentages were 94.24, 96.19, 95.46 and 92.16% for water, soil plant and fish samples, respectively.

Residual concentration in paddy water

The dynamic changes of carbofuran residues in paddy water released from CF7, CF8, and the granular formulation (10G) in the rice/fish ecosystem were determined at 2hrs and 1, 4, 7, 14, 21 and 28 days after treatments. The data presented in Table III indicated that the maximum level of carbofuran residue (0.593 ppm) in water was recorded on the fourth day following the addition of the granular formulation to the ecosystem. This concentration gradually decreased with time and reached the lowest concentration (0.00339 ppm) at the end of the experiment. Generally, the granular formulation of carbofuran quickly released the active ingredient into water due to carbofuran's high water solubility (approximately ≥ 320 ppm) which increases its loss to the environment (*9*). This loss of carbofuran and its residues from water may be affected by several environmental factors such as, pH, evaporation, photo-decomposition, uptake by plants and fish, and microbial degradation of carbofuran (*3,10,11*). On the other hand, the controlled release formulations (CF7 and CF8) released the carbofuran at slower rates during the first 4 days after treatment. After this period, the rate of release started to increase with time until the end of the experiment . Thus, the maximum levels were recorded in both case at 14 and 21 days for CF8 (0.5422 ppm) and CF7 (0.414 ppm) respectively. It appears that the release of carbofuran from alginate controlled-release formulations was more controlled and persisted significantly longer in the paddy water than the granular formulation (10G) under the experimental conditions. Moreover, this experiment also indicated that controlling the rate of release of carbofuran resulted in reduction of rate of loss due to photo decomposition, hydrolysis and evaporation from water.

Table III Water Residues of Carbofuran Released From Different Formulations

Time after Application (day)	*Formulation types*		
	CF_7	CF_8	G_{10}
	concentration (ppm) ± SE		
0.083	0.00829± 0.00	0.0155± 0.007	0.0521± 0.020
1	0.024± .006	0.0398± 0.012	0.1100± 0.032
4	0.0412± 0.013	0.0556± 0.006	0.5930± 0.033
7	0.299± 0.081	0.320± 0.050	0.2110± 0.020
14	0.405± 0.095	0.5422± 0.082	0.1462± 0.0128
21	0.414±0.099	0.418± 0.093	0.1300± 0..011
28	0.414±0.087	0.370±0.073	0.00339± 0.00

This reduction in the loss of carbofuran by the use of controlled release formulations gave higher concentration of carbofuran available in water from one week for CF8 and two week for CF7 after the application onward. This would in turn be expected to improve the performance of the carbofuran formulations in controlling pests.

Residual concentration in paddy soil

The residue level in paddy soil was formulation type and time dependent. Carbofuran residue levels in the soil of ecosystem treated with the granular formulation were the highest between the 1st and 4th days. However, after the fourth day there was a rapid decrease in the concentration of carbofuran. These results were consistent with the rate of release of carbofuran from the commercial formulation and its concentration trend in paddy water.

The maximum soil concentration of carbofuran released from the granular formulation peaked 1 day (1.714 ppm) after treatment and the lowest concentration (0.02 ppm) was found at the end of 28 days after application (Table IV). Singh and Kalra (*12*) reported that the half life of carbofuran residues in the flooded soil was 88 days when carbofuran applied at a rate of 1 kg a.i./ha. Several studies reported that the degradation of carbofuran was greater in soils with higher organic carbon content (*13*), alkaline pH and flooded conditions (*14,15*). On the other hand, soil concentrations of carbofuran released from CR formulations did not reach maximum value until a number of days after application. The maximum concentrations were 1.92 and 1.34 ppm at 21 and 28 days for CF8 and CF7, respectively. In this respect, it can be concluded that carbofuran formulated as CR formulations will persist longer in the soil than the commercial formulation. This means that the carbofuran enclosed alginate controlled-release formulations might provide more protection against the critical factors affecting the persistence of carbofuran under flooded soil condition such as soil pH, soil temperature and microflora. However, the longer persistence achieved in CR formulations would not be an environmental disadvantage as with some of the persistent pesticides because the persistence could be adjusted as necessary.

Table IV. Soil Residues of Carbofuran Released from Different Formulations

Time after Application (day)	*Formulation types* CF_7	CF_8	G_{10}
	concentration (ppm) ± SE		
0.083	0.0607± 0.009	0.105± 0.047	0.2368± 0.031
1	0.143± 0.036	0.250± 0.024	1.714± 0.056
4	0.4348± 0.035	0.680± 0.090	1.254± 0.066
7	0.620± 0.044	0.711± 0.068	0.171± 0.0042
14	0.720± 0.018	1.057± 0.042	0.056±0.001
21	1.106± 0.056	1.920± 0.090	0.0551±0.00
28	1.340± 0.072	0.955±0.044	0.020±0.00

Uptake and accumulation of carbofuran residues in rice plant

The carbofuran released from the different formulations into water was adsorbed by soil particles and then absorbed by the plants (Table V). Carbofuran residues in water could then be taken up by the root system and translocated to all parts of rice plants in the ecosystem treated with the granular formulation more readily than in ecosystems treated with CR formulations. The concentration of carbofuran increased with time and the highest concentration (1.012 ppm) was recorded on the seventh day following the transplanting rice seedlings in the ecosystem treated with the granular formulation. Thereafter, the concentration of carbofuran declined rapidly over time and this might be due to the degradation by plants and/or the dilution caused by new growth. The lowest concentration in plant (0.101 ppm) was associated with the granular formulation, 28 days after application. It has been reported by many investigators (*16,17*) that plants can uptake significant quantities of carbofuran residues from the soil in which they grow. Carbofuran is metabolized in plants mainly via hydroxylation to produce 3-hydroxy carbofuran and its conjugated glycosides (*18,19,20*). It is also obvious from the results in Table V that the concentration of carbofuran accumulated in rice increased gradually in the ecosystem treated with CR formulations. Residue levels detected in whole rice plant treated with CF8 formulation indicated that the concentration increased from 0.068 to 0.616 ppm over the first week after transplanting the rice seedlings, while the concentration in rice plant treated with the granular formulation was almost 2 fold higher at the same period. The highest concentration (1.27 ppm) was recorded at 28 days after application of CF8 formulation. These results support our findings in water and paddy soil at 28 days after treatment with CF8. The concentration of carbofuran residues in plant treated with the CF7 formulation increased gradually with time and reached the maximum value (0.809 ppm) 28 day after transplanting the rice. The results show that the carbofuran residues in rice were absorbed at a lower rate for the CRF than

Table V. Plant Residues of Carbofuran Released from Different Formulations

Time after transplanting rice (day)	*Formulation types* CF_7	CF_8	G_{10}
	Concentration (ppm) ± SE		
0.083	0.055± 0.003	0.068± 0.00	0.056± 0.003
1	0.275± 0.0062	0.373± 0.052	0.530±0.090
4	0.283± 0.091	0.462± 0.084	0.890± 0.091
7	0.403± 0.018	0.616± 0.081	1.012± 0.044
14	0.553± 0.056	0.671± 0.059	0.364± 0.088
21	0.707± 0.034	0.807± 0.061	0.289± 0.005
28	0.809± 0.080	1.270± 0.101	0.101± 0.006

the granular formulation as well as the rice plants treated with CF7 than plants treated with CF8 throughout the entire 28 day test period. This is in agreement with the previously reported release rate (*6*) and is likely related to the portion of carbofuran available in water and soil which could be absorbed by plants. Quantities of the parent compound retained in plants treated with CR formulations were higher than that treated with the granular formulation after 14 days which may provide more protection to the rice plant against pests.

Uptake and accumulations in fish

The total residual concentrations in fish were determined for 28 days after treatments as shown in Table VI. It is obvious from the results that the concentration of carbofuran residues in fish after a few hours of exposure was the lowest for all the ecosystems treated either with CRF than with the granular formulation. Greater residues were absorbed 4 days after fish exposure (0.947

Table VI. Fish Residues of Carbofuran Released from Different Formulations

Time after introducing fish (day)	*Formulation types* CF_7	CF_8	G_{10}
	concentration (ppm) ± SE		
0.083	0.047± 0.011	0.082± 0.019	0.1135± 0.004
1	0.08± 0.015	0.212± 0.033	0.585± 0.079
4	0.162± 0.02	0.410± 0.041	0.947± 0.064
7	0.45± 0.065	0.953± 0.102	1.2989± 0.110
14	0.719± 0.054	0.519± 0.082	0.517± 0.024
21	0.59± 0.021	0.214± 0.036	0.151± 0.065
28	0.214± 0.032	0.168± 0.0055	0.1388± 0.038

ppm) to the granular formulation and the residues reached the maximum levels (1.295 ppm), 7 days after exposure. Thereafter the residue levels decreased with the time and were reduced to 0.1388 ppm at 28 days. The granular formulation released the carbofuran into paddy water very quickly, reached the maximum levels in water 4 days after treatment, and varied more thereafter. This resulted in a large uptake and accumulation of residue in fish in the first week and the consequent death of fish within the first 16 days of exposure. No mortality was recorded after the first 16 days. This can be considered as an indication that the fish was able to bioaccumulate the carbofuran residues at levels below the lethal dose required to kill fish. These results support the findings of Tejada (*21*) who reported that the carbofuran remained in fish up to 30 days at low levels but the bioconcentration factor was much lower compared to the organochlorine pesticides. Also, Sun *et al.* (*22*) and Tajada *et al.* (*23*) reported that most of the carbofuran residues in fish were concentrated in entrails, some in fillet and least in the head. The fish was able to metabolize and excrete carbofuran easily and the residues declined to an acceptable limit near harvest. However, the release of carbofuran from CR formulation was much slower and the water concentrations increased gradually to 0.542 ppm for CF8 at 14 days while CF7 was 0.414 ppm at the termination of this study. Thus, carbofuran accumulated less in CR exposed fish than those exposed to the granular formulation. The highest carbofuran concentration found in fish treated with CF7 was 0.719 ppm, while that treated with the granular formulation was 1.2989 ppm. The ratio of carbofuran accumulated in fish from G10 relative to that from CF7 was almost 1.81. Only 20% of the tested fish were killed with CF7. This might have been due to abnormal physiological and metabolic conditions of the dying fish. It was also found that fish in the ecosystem treated with CF8 did not accumulate residues comparable to the highest concentrations in the granular formulation treated ecosystem. The highest residue detected in fish (day 7) was 0.953 ppm. Two weeks later the observed concentration was 0.214 ppm. This could be attributed to the ability of the fish to metabolize and excrete carbofuran. In conclusion, from the point of view of acute toxicity and bioaccumulation of residues, CF7 followed by CF8 proved to be significantly safer than the granular formulation to the common carp.

Behaviour and Mortality of common Carp exposed for 28 day to different formulations of carbofuran insecticide

Behaviour and mortality of fish after the application of carbofuran formulations were recorded as shown in Table VII. Fish treated with the granular formulation (10G) began to show abnormal symptoms such as loss of balance and fast swimming 4 days after the fish were released into compartment II. During this time, 2 fish of the test fish died . Thereafter, similar abnormal behaviour was observed accompanied by lack of eating during the second period of exposure to carbofuran, i.e. days 5-10. During this time, another 2 fish died. During the third period (11-16 days), most fish showed normal behaviour though some exhibited depression of their activity levels. A single fish died while the remaining fish resumed normal activity and survived until the end of the experiment. Fish treated with CF8 started to have less activity 7 days after exposure to carbofuran. Only 2 fish died during the first 2 weeks after exposure. Those treated with CF7 started to behave abnormally during the third period of exposure (11-16 days) and only one fish died over 28 days exposure to the carbofuran released from this formulation.

Table VII. Behaviour and Mortality (in parentheses) Observed After Application

Time after introducing fish (day)		*formulation types*		*Control*
	CF_7	CF_8	G_{10}	
0.083-4	normal (0)	normal (0)	abnormal behaviour,2 fish died (20%)	normal (0)
5-10	normal (0)	abnormal behaviour , loss balance,1 fish died (10%)	not eating , abnormal behaviour, 3 fish died (30%)	normal (0)
11-16	not eating, 1 fish died (10%)	not eating , less active,1 fish died (10%)	less active, I fish died (10%)	normal (0)
17-22	normal (0)	less active (0%)	normal (0)	normal (0)
23-28	normal (0)	normal (0)	normal (0)	normal (0)

Distribution of carbofuran residues in components of the model ecosystem

Distribution of carbofuran residue concentrations over 28 days in paddy water, paddy soil, rice plants and fish are summarized in Table VIII. The results show that carbofuran released from the different formulations and transferred to each component of the model rice/fish ecosystem were very different. The carbofuran residues in the ecosystem treated with (10G) were highest in fish (3.75 ppm), followed by soil (3.51 ppm), then plant (3.242 ppm) and at least in water (1.25 ppm). Most of the residues in the ecosystem treated with CR formulation were found in the soil (5.68 and 4.42 ppm), followed by plant (4.277 and 3.09 ppm) and fish (2.56 and 2.26 ppm), then water (1.9 and 1.605 ppm) for CF8 and CF7 respectively.

In comparing the two types of formulations (CRF and 10G), there were significant differences in the behaviour of the residues in the soil, plants, water, and fish. Residues in soil of the ecosystem treated with CRF were much higher than in the treated with the granular formulations. This suggests that the residues of carbofuran in soil were negatively correlated with the release rate of carbofuran from the formulation types. The carbofuran adsorbed to soil more when the rate of carbofuran released by CRF decreased. In contrast, there was a positive correlation between the release rate of carbofuran into water, the residues recovered from fish and the percent mortality observed. Also the results indicate that the concentration of the carbofuran residues distributed in plant treated with CF8 was a little higher than observed in plants treated with the granular formulations. The carbofuran remaining in the beads and granules at the end of the experiment was analyzed by HPLC. The data in Table VIII show that CF7 still contained 12.7% of the added carbofuran, while CF8 contained 7.38%. The granular formulation contained smaller proportions of unreleased carbofuran (0.0354%).

Table VIII. Distribution of Carbofuran Residues in a Model Rice/Fish Ecosystem Treated with Different Formulations

Formulation types	*Total residual concentration (ppm) recovered over 28 days*				*% carbofuran unreleased from beads (% of the originally applied)*
	water	*soil*	*rice*	*fish*	
CF_8	1.90	5.68	4.27	2.56	7.38
CF_7	1.605	4.42	3.09	2.26	12.7
G_{10}	1.25	3.51	3.242	3.75	0.0356

Generally, the results suggest that the release of carbofuran from formulation can be controlled by using alginate matrix with kaolin as filling. By changing the amount of alginate and kaolin during the preparation of the alginate beads it is possible to alter the release rate into water. Results also indicated that carbofuran in CF8 and CF7 formulations persisted significantly longer than the granular formulation in the paddy soil under the experimental conditions. This might reduce the loss via leaching and provide more protection to rice plant against nematodes and insects for a longer time than would the granular formulation. The concentration in fish was lower in fish treated with CR formulations. Therefore, when applied in a paddy field, the CR formulations are able to extend the persistence of carbofuran and at the same time reduce the environmental risks. Further field evaluations of the environmental behaviour of CR formulations are recommended. Also biological evaluations are needed to test the relative performance of these formulations in the field.

References

1. Jinhe, S.; G. Jianying and C. Ziyuan. *Pestic. Sci.*, **1989,** *26(2)*:147-157.
2. Yu, C.C.; G. M.Booth; D.J.Hansen; and J.R. Larsen *J. Agric. Food Chem.*, **1974**, *22(3)*:431-434.
3. El-Shoura, M.Y.; M.I. Aly and A. Abdel-All. *Proceed. Int. Conf. on Environ. Hazards of Agrochemicals in developing countries*, **1983**, 2:890-896.
4. Schuh, J. and Blakely, B.Can. Vet. J. 1988, 29: 58-59.
5. Rolsi, B.M.; J.H. Arshad and I. Abdul-Rahman. *J. Plant Protec. Tropics.* **1990**, *7(1)*:27-37.
6. Soltan, H. R. *Alex. J. Agric. Res.*, **1996**, *41(1)*:241-252.
7. Nicosia, S.; N. Carr; D.A. Gonzales; and M.K. Orr. *J. Environ. Qual.* **1991**, *20(3)*:532-539.
8. Ferreira, G.A.L.; and J.N. Seiber. *J. Agric. Food Chem.*, **1981**, *29(1)*:93-99.
9. Trotter, D.M.; R.A. Kent and M.P. Wong. *Critical Reviews in Environmental Control.* **1991**, *21(2)*:137-176.
10. Racke, K.D.; and J.R. Coats. *J. Agric. Food Chem.* **1988**, *36(4)*:1067-1072.
11. Scheunert, I.; M.Masour; U.Dorfler and R.Schroll. *Sci. of the Total Environ.* **1993**, *132 (2-3)*: 361-369.
12. Singh, B.; and R.L. Kalra. *India J. Ecol.* **1990** *17 (1)*: 51-53.
13. Sukop, M.; and C.G. Cogger. *Gov. Rep. Announce. Index.* **1992**, *15*:130.
14. Talekar, N.S.; L.T.Sun; E.M. Lee; and J. S. Chen. *J. Agric. Food Chem.* **1977**, *25(2)*: 348-352.
15. Venkateswarlu, K.; T.K. S. Gowda; and N. Sethunathan. J. Agric. Food Chem. **1977** *25(3)*: 533-536.
16. Zayed, S.M. A. D.; A.E.EL- Arab; and I.Y. Mostafa. *J. Environ. Sci E. Health B.* **1988** *23(4)*: 401-407.
17. Radder, B.M.; and R. Siddaramappa. *Karnataka J. Agric. Sci.* **1989**, *2(3)*:226-228.
18. Knaak, J.B.; D.M. Munger; J.F. McCarthy; and L.D. Satter. J. Agric. Food Chem., **1970**, 18(4):832-837.

19. Metcalf, R.L.; T.R. Fukuto; C. Collins; K. Brock; S. Abdel-Aziz; R. Munoz; and C.C. Cassil. *J. Agric. Food Chem.* **1968**, *16(2)*:300-311.
20. Mostafa, I.Y.; S.M. Zayed; M. Farghaly; and F. Mahdy. *J. Environ. Sci. Health B.* **1992**, *27(4)*:399-405.
21. Tejada, A. W. Monthly Seminar at the National Crop Protection Center. College, Laguna (Philippines). **1985**, 7pp.
22. Sun, J. H; J. Y. Gan; and Y. X. Zhang. *Proceed. Int. Symp. Pestic. Food Environ. Implic.* **1988**, 235-245.
23. Tajada, A. W.; S. M.F. Calumpang; and E. D. Magallona. *Tropical Pest Management.* **1990**, *36(3)*:237-243.

Reducing Wildlife Exposure to Pesticides

Chapter 23

Comparison of Primary and Secondary Repellents for Aversive Conditioning of European Starlings

Roger W. Sayre and Larry Clark

APHIS/WS/National Wildlife Research Center, U.S. Department of Agriculture, 4101 Laporte Avenue, Fort Collins, CO 80521–2154

Secondary repellents have been demonstrated to effectively reduce avian crop depredation. However, these compounds frequently are toxins and there are concerns about environmental safety. Many primary repellents are toxicologically and environmentally safer, but these chemicals are not as effective when traditional delivery methods are used. We set out to determine whether the difference in efficacy of primary vs. secondary repellents was due to differences in potency of the chemical classes or site of action within the animal. We hypothesized that the efficacy of primary repellents could be enhanced if they could be delivered directly to the gastrointestinal system, thus by-passing the peripheral senses. Our experiments were conducted on captive European starlings (*Sturnus vulgaris*). We showed that by-pass of the peripheral senses via oral gavage enhanced the efficacy of 2 hydroxyacetophenone (2HAP), a primary repellent. However, we found that starlings were sensitive to changes in their visual environment, and that learned avoidance to a colored food cup could be mitigated by changing the cage door of the test system. Topical application of repellents to foods allowed starlings to self-mediate their exposure to the repellents. We found that methiocarb and methyl anthranilate (MA), a primary repellent, both induced food aversion learning, but 2HAP did not. Detailed analyses of feeding behavior elucidated the mechanistic differences between Methiocarb, MA, and 2HAP. Methiocarb induced gastrointestinal illness in starlings, which resulted in post-ingestive suppression of food intake, and subsequent food aversion learning. Some starlings with MA in their food also developed gastrointestinal illness, and food aversion learning. Starlings with 2HAP in their feed did exhibit reduced feeding efficiency, but these starlings did not develop gastrointestinal illness or food aversion learning. These experiments demonstrate that the location and intensity of illness are of paramount importance to induce food aversion learning.

Introduction

Chemical repellents are important management tools for reducing avian crop damage. Chemical repellents are categorized as either primary or secondary, and these categories operate by different mechanisms[1]. With secondary repellents, animals learn to avoid foods because they associate an aversive experience caused by the chemical (the unconditional stimulus), with paired ancillary sensory cues such as visual, olfactory, taste, or auditory stimuli (the conditional stimulus). Secondary repellents have been demonstrated to be effective at reducing crop depredation, and they have the advantage of promoting long-term learned avoidance. However, there are concerns about environmental and toxicological safety, which has made use of these chemicals less desirable[2,3].

In contrast, primary repellents are immediately noxious to an animal and avoidance behavior is reflexive; i.e., learning is not required. In most cases primary repellents appear to cause irritation[4]. Chemically sensitive nociceptors of the trigeminal nerve detect chemicals and the animal perceives this neuronal activity as a sensation of pain. The reflexive avoidance response results in a limited exposure or dosage. Thus, the animal self-limits its exposure to the chemical and thereby limits the magnitude of the unpleasant experience. Because the strength of aversive conditioning is directly related to the intensity of the punishment[5], primary repellents are less likely to promote strong conditioned avoidance. That is to say, primary repellents by their nature are infrequently consumed in sufficient quantity, in a short period of time, to be useful as effective agents in promoting strong learned avoidance. Rather, birds frequently sample foods containing primary repellents. Insufficient quantities of the repellents are consumed to promote learned avoidance. Consequently birds will frequently return to feed once these chemicals are removed[6,7,8].

In spite of the limitations, there has been increased interest in developing primary repellents, especially those that are considered to be less toxic and environmentally harmful than secondary repellents[9,10]. The primary repellents of interest are derived from natural food additives, which are safe for human consumption. In contrast, most effective secondary repellents (e.g., methiocarb) are toxicants derived from pesticides. The primary repellent methyl anthranilate (MA) is a grape flavored food additive that it is an irritant to birds. The current body of knowledge indicates that MA should be toxicologically and environmentally safe when used as a bird repellent. MA is toxic only at very high concentrations, and it biodegrades rapidly. MA is approved by the U.S. Food and Drug Administration as a food flavor additive and is classified as Generally Recognized as Safe (GRAS).

Although primary repellents are safe, they are not considered to be as effective as secondary repellents. A possible method to circumvent the limitation that primary repellents do not promote long-term learned avoidance would be to by-pass the peripheral senses so the primary repellent is delivered into the gastrointestinal system. This would result in several advantages. The animals would not self-mediate the dosage, and could be exposed to more repellent. Second, gastrointestinal illness due to a repellent would increase the saliency of the unconditional stimulus in relation to feeding behavior, resulting in stronger aversive conditioning.

Our objective was to by-pass the peripheral nociceptors and directly compare the illness promoting potential of primary and secondary repellents when delivered enterically. We tested 2 hydroxyacetophenone (2HAP) and MA, which were identified as primary repellents from structure-activity models and drinking trials[11,12,13]. The secondary repellent, methiocarb, has been demonstrated to be an effective aversive conditioner[6,14,15,16,17,18].

Methods

Study Subjects

During September 1998 we trapped European starlings ($n = 40$) in Fort Collins, Colorado. We conducted the 3 experiments from November 1998 to June 1999. Starlings were housed in visually isolated metal cages (36 cm × 23 cm × 28 cm) (Animal Care Products, Bryan Texas). Temperature remained constant at 22° C during the study, and lighting was maintained on a seasonally shifting photo period. Each cage was provisioned with a food cup and a water bottle holder attached to the cage door, 5 cm above the wire mesh cage floor. We provided food (Purina Layena Checker) *ad libitum* in metal food cups (8.3 cm diameter), in addition to unlimited access to water. To establish baseline food intake, we measured food consumption of each starling for a period of 2 hours (commencing 2 hours after the onset of light) for 2 days prior to testing.

Experiment 1

The objective of this experiment was to by-pass the peripheral senses to compare the effect that primary and secondary repellents have on food avoidance learning in starlings. The primary repellent was 2 hydroxyacetophenone (2HAP), and the secondary repellent was methiocarb.

Test Protocol

We ranked starlings according to food intake and assigned them to 1 of 3 groups (n = 8/group) so that each group had individuals with high, moderate, and low consumption rates. Treatments were randomly assigned to the groups. Treatments included control (no handling); gavage with methiocarb (4 mg/kg) diluted into propylene glycol (PG) at 2 ml/kg; and 2HAP (250 mg/kg), also diluted into PG (2 ml/kg). Dosages of methiocarb and PG followed Mason and coworkers[15,16,19]. We conducted preliminary range finding determine dosages of 2HAP and found that 250 mg/kg was sufficient to cause temporary irritation and delay of food consumption (Sayre and Clark unpublished data).

The experiment was conducted during a period of 2 days for each bird. During the first day, i.e., training, we provided food in a colored cup (the conditional stimulus) for 2 hours. The cup was colored with orange and black vertical stripes, because previous reports have indicated that this was the most effective color pattern for a visual stimulus in conditioning trials[15,16]. We removed the food cup and immediately gavaged (orally intubated) the starlings with repellent (the unconditional stimulus). We measured food intake after gavage to evaluate the immediate effect of the repellent (the unconditional response). After 2 hours starlings were left undisturbed and were provided food *ad libitum* in uncolored cups. On the second day, 24 hours after exposure to the conditional stimulus, we conducted a two-choice feeding test, with food from an uncolored and a colored food cup, to determine if learned avoidance occurred (the conditional response).

Training

Approximately 16 hours prior to training we replaced the standard wire cage doors with plexiglass doors (0.64 cm thick). We installed these doors to enhance video taping and analysis for concurrent behavioral analyses, and we believed that 16 hours would be sufficient for the starlings to habituate. The placement of cup and water bottles on these doors was identical to the wire cage doors. On the training day, 2 hours after the onset of light, we replaced the standard metal food cup with a cup with orange and black vertical stripes (the conditional stimulus). Each cup contained 30 g of standard chow. After 2 hours of exposure to the conditional stimulus we removed the cup and immediately took the test starling from its cage and gavaged it with repellent solution. Control starlings were not handled. We held each starling for 1 min after gavage to prevent immediate regurgitation, and then returned the starling to its cage and presented it with 30 g of food in an uncolored food cup. We returned to measure food intake at 2, 4, and 6 hours post-gavage. Starlings were then left undisturbed until the following day.

2-Choice Learning Test

Two hours after the onset of light on the second day we conducted a 2-choice learning test, with an uncolored metal cup and a colored cup with orange and black vertical stripes. Each cup contained 30 g of chow. The position of the cup (left or right) was determined randomly. After 2 hours, we measured food intake from each cup. We determined preference ratios by dividing consumption from the colored cup by consumption from both cups combined.

Analysis

We used a 2-factor repeated measures ANOVA to evaluate food intake on the day of training. The model included the independent variable for treatment with 3 levels (2HAP, methiocarb, and control). The dependent variable was the repeated measure of food intake at 2, 4, and 6 hours post-gavage. We used planned orthogonal contrasts to compare the average food intake by controls vs. 2HAP and methicarb (averaged during the 6 hour feeding period). In a separate analysis, to look for learning effects of treatment, we evaluated the data from the 2-choice learning test with a 1-way ANOVA. The preference ratios of controls vs. the treatment groups were compared with planned contrasts.

Experiment 2

The objective of Experiment 2 was to evaluate the effect of replacing the standard cage door with plexiglass. We tested starlings in 3 treatment groups (n = 6/group). Treated starlings were selected randomly among the starlings used in Experiment 1. To minimize any effect of carryover learning, we waited $\geq$31 days before re-testing any starling. Mason[16,19] reported that learned avoidance lasted 10-12 days following treatment with methiocarb, and we believe that the time between experiments was sufficient to minimize possible bias.

Treatments included oral gavage of PG (2 ml/kg) as control, 2HAP (250mg/kg), and methiocarb (4 mg/kg). The repellents were diluted into PG (2 ml/kg). We used the same training and testing procedures as Experiment 1 with the following exceptions: (1) the standard metal cage doors remained on the cages and these were not replaced by the clear plexiglass doors; (2) we measured food consumption only during the first 2 hours post-gavage.

Analysis

We used a 1-way ANOVA and planned contrasts to compare food consumption of controls (PG) vs. 2HAP and methiocarb during the 2 hours immediately after gavage. The 2-choice preference ratios of starlings gavaged with PG, 2HAP, and methiocarb, measured on the following day, were also compared with a 1-way ANOVA and contrasts.

Experiment 3

The first objective of Experiment 3 was to evaluate the effect of increasing the duration of habituation to the plexiglass cage door from 16 hours to 60 hours. A second objective was to compare the behavioral responses of starlings to primary and secondary repellents when the peripheral senses are not bypassed.

Adaptation

To provide starlings a longer habituation period to the plexiglass cage door, we replaced the standard cage door with the clear plexiglass door 60 hours prior to training. In addition, we adapted the starlings to a schedule of mild food deprivation during the pre-training period. On the days prior to training (i.e., 48 and 24 hours before), we removed the food cup for 2 hours. After 2 hours we returned and provided the starlings with 30 g of standard chow for another 2 hours. We then measured food consumption to obtain a baseline food intake after mild food deprivation.

Test Protocol

We randomly assigned starlings to 1 of 4 groups, and randomly assigned treatments to the groups. Treatments included methiocarb (0.2% w/w), a secondary repellent; and primary repellents, 2HAP (1% w/w) and methyl anthranilate (MA) (0.8% w/w). All repellents were diluted into ethyl alcohol, and the solution was spread topically onto the food 48 hours prior to feeding. The control was ethyl alcohol added to the food at 1% w/w, which was the concentration we used with the repellent dilutions. After applying the repellent/ethyl alcohol mixture, we air dried the foods for 48 hours prior to feeding.

Like Experiments 1 and 2, Experiment 3 was conducted over 2 days with training and 2-choice learning phases. Two hours after sunrise on the training day, we removed each starlings's food cup for 2 hours. After 2 hours we provided food, according to preassigned treatment (the unconditional stimulus), in a colored cup (the conditional stimulus) for 2 hours. Each test starling was video-taped during the 2 hour training phase. We then removed the colored cup and weighed food consumption. Starlings were provided food *ad libitum* from a standard metal cup until the next day.

On the second day, 24 hours after exposure to treated food, we conducted a two-choice feeding test to determine if learning in response to the unconditional and conditional stimuli occurred. Two hours after a light onset we removed the food cup for two hours. We then provided standard chow in two food cups, an uncolored cup and a colored cup with orange and black vertical stripes, with each cup containing 30 g of standard chow. After 2 hours we measured food consumption from each cup and determined preference ratios by dividing consumption from the colored cup by consumption from both cups combined.

Behavioral Analysis

We used an integrated video hardware and computer software package (Noldus Observer, Wanningen, The Netherlands) to quantify feeding behavior. During each 2 h observation period we measured the duration and frequency of feeding bouts; the frequency of eaten pellets, dropped pellets, failed attempts to eat, and drinks (Table 1). We calculated feeding efficiency as the number of pellets consumed per minute of feeding. To further evaluate the behavioral mechanisms in response to repellent-treated foods, we analyzed the frequency of feeding bouts, number of pellets eaten, and frequency of drinks at 30 minute intervals (i.e., 0 - 30 min, 30 - 60 min, 60 - 90 min and 90 - 120 min). To evaluate the relationship between type of repellent and gastrointestinal illness we quantified the frequency of regurgitations during each 30 min interval.

Table I. Categories and definitions for feeding and drinking behavior of European starlings.

Category of Behavior	*Measure of Behavior*	*Definition*
Feeding bout	Duration & Frequency	Bird's bill is in food cup or bird has food pellet in bill.
Eat	Frequency	Bird consumes food pellet.
Drop	Frequency	Bird drops food pellet before eating.
Attempt	Frequency	Bird searches in food cup but does not hold food in bill.
Drink	Frequency	Bird drinks from water tube.

Analysis

We used a 1-way ANOVA to evaluate the effect of repellent on food consumption during training. Food consumption of controls were compared against the repellent treatment groups with planned contrasts. We also used a 1-way ANOVA and planned contrasts to evaluate preference ratios of treated vs. controls. The feeding behavior data were analyzed by combining the primary repellents (2HAP and MA), which we compared with controls and methiocarb using planned contrasts.

Results

Experiment 1

The objective of Experiment 1 was to compare the effectiveness of gastrointestinal delivery of 2HAP, a primary repellent, and methiocarb, a secondary repellent.

Food Intake During Training

Food intake on the day of training did not vary as a function of time and the type of treatment (Table II). However, starlings varied their food intake in response to the main effects of repellent treatment and time (Table II). Contrasts showed that relative to controls, starlings did not reduce their food intake after 2HAP gavage (P = 0.279) (Figure 1a). However, starlings treated with methiocarb reduced food consumption during the 2 hours following methiocarb gavage (P = 0.003).

Table II. ANOVA for Food Consumption by starlings Following Gavage.

Experiment	*df*	*F*	*P*
Experiment 1			
Repellent	2, 21	5.62	0.011
Time	2, 21	69.86	0.001
Repellent × Time	4, 42	2.26	0.078
Experiment 2			
Repellent	2,15	5.67	0.015
Experiment 3			
Repellent	3, 20	2.01	0.14

2-Choice Learning Test

Results from the 2-choice learning experiment indicated that the starlings did not develop a learned aversion to the visual cue ($F_{2,21}$ = 0.76 P = 0.48), indicating that they failed to make the association between the physiological effects of the chemical used in the gavage and the visual stimulus (i.e., the vertically striped black and orange food cup). The contrasts showed that preference ratios of starlings treated with 2HAP

($P = 0.53$) and methiocarb ($P = 0.56$) were similar to those observed in the control group (Figure 1b).

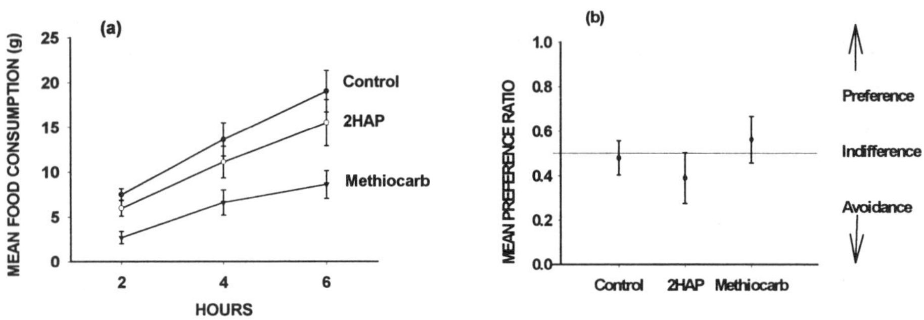

Figure 1a. Mean consumption (g) of food by starlings 2, 4, and 6 hours after gavage treatment (n = 8 birds/group). Figure 1b. Mean preference ratios (± SE) of starlings during 2 choice test in Experiment 1.

Experiment 2

Experiment 2 also was conducted to compare the effect primary and secondary repellents had when the peripheral senses are bypassed. However, we first wanted to test the hypothesis that the starlings failed to develop an aversion to the conditional stimulus in Experiment 1 because they were distracted by the clear plexiglass door, and learning failed because of concurrent interference of the sign stimulus. That is to say the door was a more salient visual cue than the vertically striped color cup and, as a consequence, starlings failed to associate illness with feeding. We treated starlings with PG, 2HAP, and methiocarb but did not replace the standard wire cage door with plexiglass.

Food Intake During Training

The repellents affected food consumption (Table II). Relative to controls, food consumption by starlings was suppressed for starlings treated with 2HAP ($P = 0.010$) and methiocarb ($P = 0.011$) (Figure 2a).

2-Choice Learning Test

Under the conditions of the test we found that the starlings could be conditioned to avoid the colored food cup. Compared to controls, starlings treated with 2HAP ($P = 0.076$) and methiocarb ($P = 0.033$) formed a conditioned avoidance to the visual cue (Figure 2b).

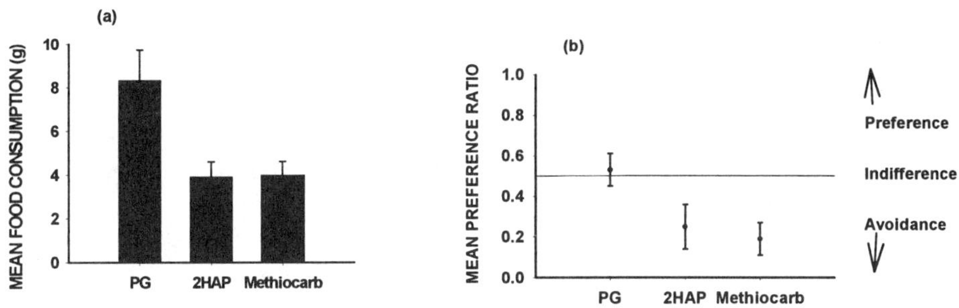

Figure 2a. Mean (± SE) consumption (g) of food during training. Figure 2b. Mean preference ratios (± SE) of starlings during 2 choice test with standard cage door.

Experiment 3

The objectives of Experiment 3 were to (1) evaluate the effect of lengthening the duration of habituation to the plexiglass cage door, and (2) compare the behavioral mechanisms of starlings to primary and secondary repellents when the peripheral senses are not bypassed.

Food Intake During Training

Food consumption was not affected by the combined effects of repellents (Table II). Starlings that fed on chow dosed with 2HAP ($P = 0.18$) and MA ($P = 0.11$) slightly reduced food consumption compared to controls (Figure 3a). However, starlings with methiocarb in their feed exhibited more of a reduced food intake ($P = 0.026$).

2-Choice Learning Test

Application of repellents on food affected the subsequent 2-choice preference ratios of the starlings ($F_{3,20} = 6.09$, $P = 0.004$). Starlings that were

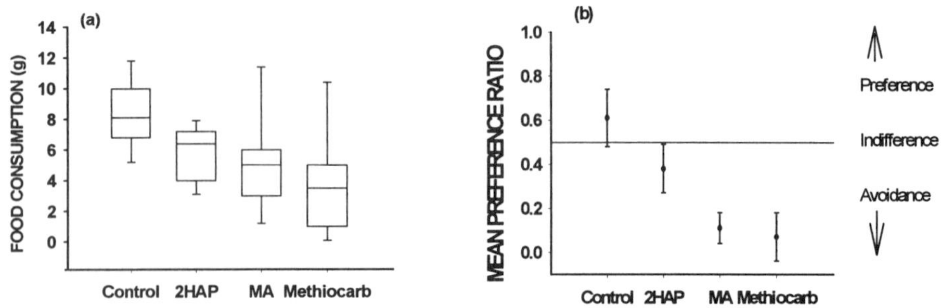

Figure 3a. Mean (± SE) consumption (g) of food by starlings during 2 hour exposure to topically applied repellents. The upper and lower capped vertical bars represent 95th and 5th percentiles and the upper and lower ends of the boxes represent the 75th and 25th percentiles, respectively. The horizontal mark through the box represents the median. Figure 3b. Mean preference ratios (± SE) of starlings during 2-choice test after extended adaptation to plexiglass cage door (60 hours).

provided food treated with 2HAP did not develop an aversion to the colored food cup ($P = 0.123$) (Figure 3b). However, starlings fed chow treated with MA ($P = 0.002$) and methiocarb ($P = 0.001$) ate less from the colored food cup, indicating strong food aversion learning.

Behavioral Analyses

Compared to controls, starlings fed pellets treated with methiocarb tended to reduce the proportion of time feeding ($P = 0.023$) and decrease the duration of feeding bouts ($P = 0.101$) (Figure 4a, 4b). Starlings that ate pellets treated with the primary repellents, 2HAP and MA, did not reduce their total feeding time ($P = 0.97$) or duration of feeding bouts ($P = 0.709$) in comparison to controls. Comparison of primary repellents and the secondary repellent, methiocarb, indicates that the methiocarb starlings devoted less of their activity budget to feeding ($P = 0.014$), and demonstrated a slight reduction in the duration of feeding bouts ($P = 0.132$).

We measured the feeding efficiency of starlings by quantifying the number of pellets consumed per minute of feeding (Figure 5a). Starlings that were provided food with the primary repellents showed a tendency to reduce efficiency compared to both controls ($P = 0.129$) and starlings with methiocarb in their feed ($P = 0.140$).

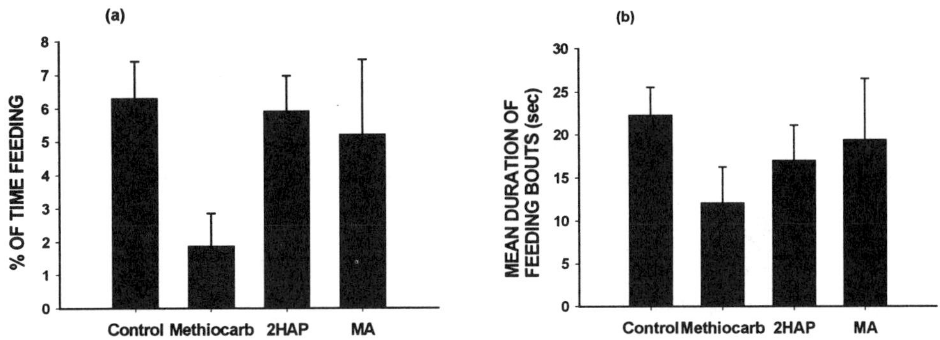

Figure 4a. Mean percent time (± SE) starlings were engaged in feeding bouts during 2 hour training period. Figure 4b. Mean duration (± SE) of feeding bouts of starlings during 2 hour training period.

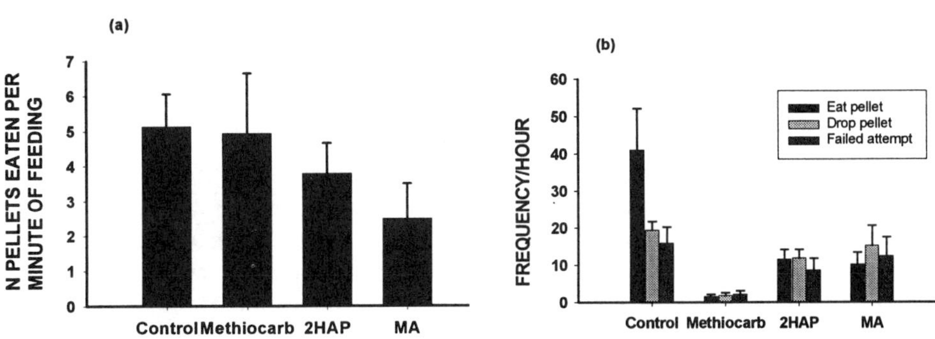

Figure 5a. Mean feeding efficiency (# of pellets consumed/minute of feeding) (± SE) of starlings during 2 hour training period. Figure 5b. Mean frequency (± SE) of food pellets eaten, food pellets dropped, and failed attempts (bird searches in food cup but does not pick up any pellets) of starlings during 2 hour training period.

Feeding efficiency of starlings treated with methiocarb was not different from controls ($P = 0.679$) (Figure 5a), although the latter group spent less total time feeding (Figure 4a). Controls ate more frequently than those fed primary repellents ($P = 0.018$) or methiocarb ($P = 0.004$). Starlings fed primary repellents ate slightly more frequently than birds with methiocarb ($P = 0.18$). The frequency of dropped pellets and failed attempts among controls was not different from the primary repellents ($P = 0.56$), but controls demonstrated these behaviors more frequently than those fed methiocarb ($P = 0.048$). Starlings fed primary repellents dropped and failed to secure food more frequently than with methiocarb ($P = 0.037$) (Figure 5b).

To evaluate behavioral mechanisms we measured feeding frequency and number of pellets eaten at 30 minute intervals. The control and primary repellent groups ate most frequently at the 0 - 30 minute and 30 - 60 minute time periods, followed by a slight decrease at the 60 - 90 minute and 90 - 120 minute intervals (Figure 6). The methiocarb group exhibited several differences: these starlings decreased food consumption rates and frequency of feeding (Figure 6).

Evaluation of the number of pellets eaten per 30 minute interval demonstrates the operation of different mechanisms in response to the different treatments (Figure 7). Control starlings increased the number of pellets eaten during the 60 - 90 minute and 90 - 120 minute intervals, indicating a possible compensation for the reduced frequency of feeding bouts as observed in Figure 6. Starlings with MA in their feed ate about the same number of repellents as controls during the first 30 minutes, but then exhibited a sharp decline in consumption rates. Finally, starlings with 2HAP in their food exhibited consistent consumption rates throughout each 30 minute interval. Among the starlings treated with methiocarb, pellet consumption was minimal: 94% ($n = 17$) of pellets eaten were consumed within the first 30 minutes; and no pellets were eaten after 60 minutes.

Compared to controls, starlings fed repellent-treated foods reduced the frequency of drinks ($P < 0.01$) (Figure 8). The frequency of drinks for the 30 minute time intervals was consistent among controls (Figure 8). Starlings with methiocarb reduced the frequency of drinks after the first 30 minutes (Figure 8). Although starlings fed foods treated with the primary repellents also drank less frequently than the controls, these starlings appeared to exhibit a different pattern than methiocarb (Fig 8). The MA and 2HAP starlings drank less frequently during the first 30 minutes, then tended to increase drinking in the 30 - 60 minute interval.

In addition to measuring feeding behavior in response to the repellents, we evaluated evidence of gastrointestinal illness (i.e., the unconditional response) by quantifying the frequency of regurgitation. Control starlings (2 of 6) visibly regurgitated their food on 3 occasions during the 2 hours of observation, but no pattern was apparent (Figure 9). A majority of starlings with MA (4 of 6) in their feed regurgitated during the 2 hour observation period. Regurgitation occurred most frequently during the first 30 minutes ($x = 1.5 \pm 0.9$ SE). Only 2 of 6 starlings with 2HAP in their feed regurgitated during the 2 hours, and like controls no pattern was

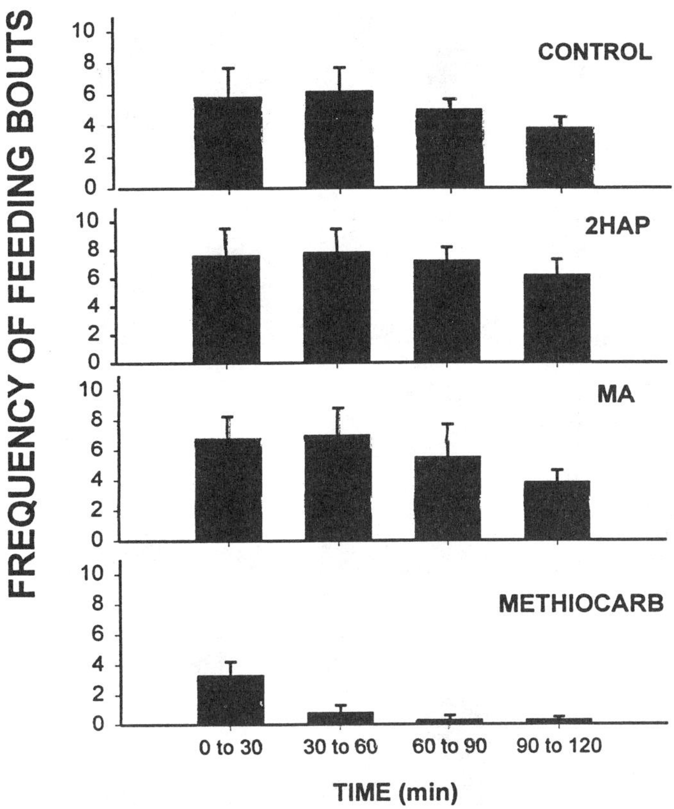

Figure 6. Mean frequency (± SE) of feeding bouts during 30 min intervals.

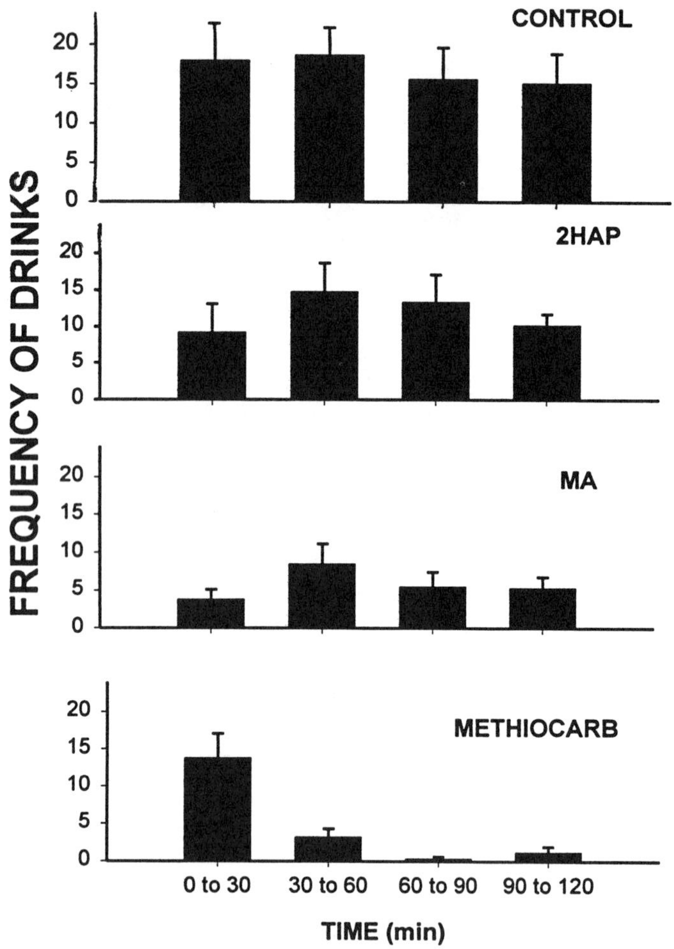

Figure 7. Mean frequency (± SE) of food pellets eaten during 30 min intervals.

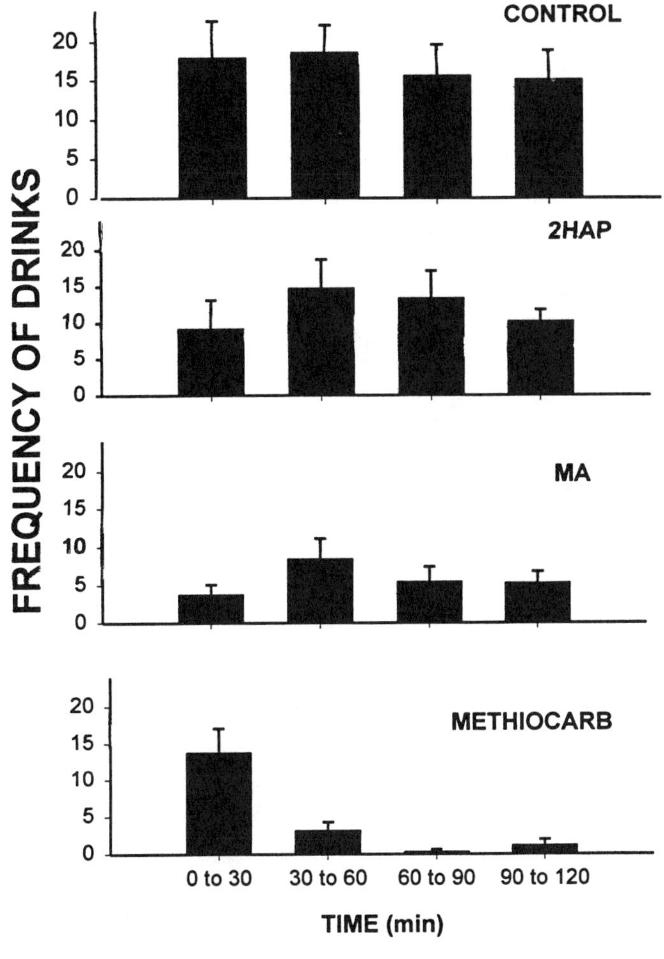

Figure 8. Mean frequency (± SE) of drinks during 30 min intervals.

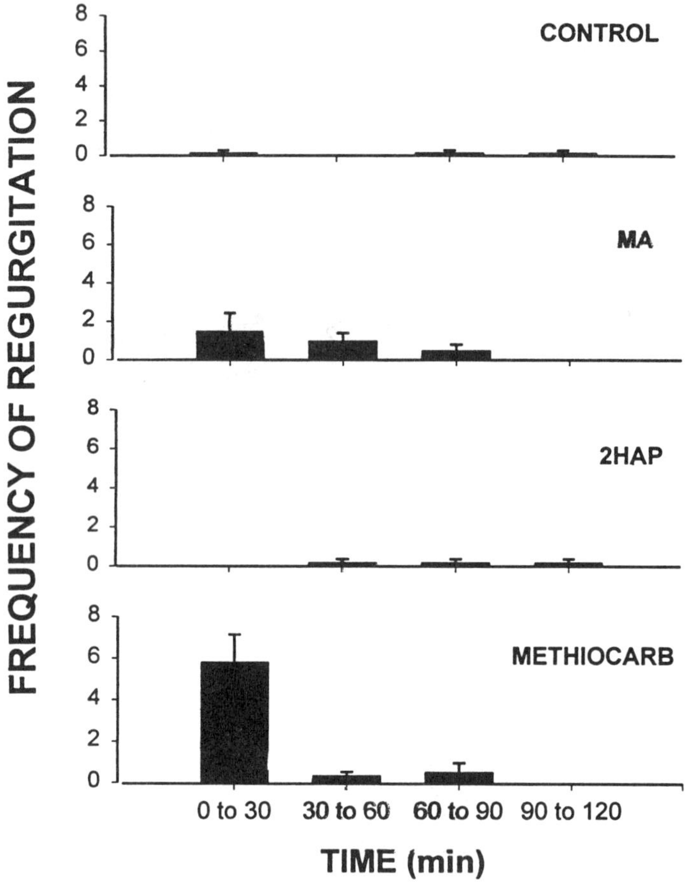

Figure 9. Mean frequency (± SE) of regurgitation during 30 min intervals of 2 hour training period.

apparent. As expected, evidence of gastrointestinal illness was observed most frequently among the starlings treated with methiocarb, where 6 of 6 regurgitated. Mean regurgitation during the first 30 minutes was 5.8 $\pm$ 1.9 SE (Figure 9).

Discussion

Results from these experiments, and from a similar set of studies (Sayre and Clark in review), indicate that the efficacy of primary repellents can be enhanced via gastrointestinal delivery. Topical application of 2HAP in Experiment 3 showed that this repellent was completely ineffective as an aversive conditioner. However, when the peripheral senses were by-passed in Experiment 2, the starlings displayed a tendency to avoid food they associated with the visual cue. When MA was compared with methiocarb in a parallel experiment (Sayre and Clark in review), this primary repellent was as effective as methiocarb.

Starlings in Experiment 1 did not develop food aversion learning from either primary or secondary repellents. This was unexpected because methiocarb has been demonstrated to effectively condition birds to avoid foods[13,14,17]. We hypothesized that the anomalous results were due to concurrent interference[20,21]. Specifically, that replacement of the standard cage door with a clear plexiglass door 16 hours prior to training distracted the starlings from learning to associate the colored food cup with the illness caused by the repellent. Results from Experiments 2 and 3 supported our hypothesis that alteration of the visual environment mitigated aversive conditioning. Starlings exposed to methiocarb developed aversions to the colored food cup when their cage door was not replaced (Experiment 2), and when they habituated to the plexiglass cage door (Experiment 3).

Experiment 3 compared the effect of topical (i.e., unmasked) application of primary repellents 2HAP and methyl anthranilate, and methiocarb a secondary repellent. The amount of food consumed during training did not decrease with the primary repellents, but the starlings did reduce consumption of foods treated with methiocarb.

The behavioral data demonstrate different behavioral mechanisms in response to the 3 repellents. Methiocarb, the secondary repellent, is an emetic and a cholinesterase inhibitor. The starlings in the methiocarb group initially ate a few pellets, but quickly became ill. Thereafter these starlings exhibited a marked reduction in feeding and drinking during the 2 hour training period. Although starlings that were fed methiocarb demonstrated a relatively high feeding efficiency per minute of feeding, the overall proportion of time feeding was reduced.

As expected, the behavioral response to primary repellents, 2HAP and MA, were different than controls or to methiocarb. In comparison to controls, starlings that ate food treated with 2HAP and MA showed only a very slight reduction in proportion of time feeding or duration of feeding bouts. However, the foraging efficiency decreased compared to controls. Compared to controls, starlings fed primary repellents also had a greater ratio of drops and failed attempts in relation to pellets consumed. These data demonstrate that 2HAP and MA act as irritants.

There were also some interesting behavioral differences between responses to the 2 primary repellents. Comparison between 30 minute time blocks indicate that starlings with MA in their diet reduced the frequency of feeding bouts and the number of pellets eaten from the 0 - 30 minute to 90 - 120 minute time blocks. These data suggest that MA also might act as a secondary repellent. On the other hand, starlings with 2HAP in their food did not reduce the frequency of feeding bouts or number of pellets eaten as time progressed in the training period. These data demonstrate that 2HAP irritates the peripheral senses but birds continue to sample the food.

These subtle behavioral differences help explain the results from the 2-choice learning trials. As expected, adding the secondary repellent, methiocarb, into food resulted in markedly reduced food consumption and gastrointestinal illness during training. The starlings demonstrated food aversion learning during the 2-choice test. MA appears to have acted as both a primary and a secondary repellent. Compared to starlings with methiocarb in their food, those provided foods treated with MA continued feeding, although their efficiency decreased. Several of these starlings (4 of 6) developed gastrointestinal illness, although the illness did not appear to be as intense as with methiocarb. Starlings with MA in their feed subsequently demonstrated a strong aversion to the colored food cup. Trigeminal irritants such a methyl anthranilate do not result in conditioned odor avoidance in birds[21]. Evidence suggests that the although odor is a good potentiator of avoidance learning, it is not as strong as taste at inducing conditioned avoidance of foods[22,23,24,25,26]. In contrast, gustatory and chemesthetic cues (which irritate peripheral as well as gastrointestinal nerves) will result in food avoidance learning.

Treatment of food with the primary repellent 2HAP resulted in reduced feeding efficiency, but the test starlings did not develop gastrointestinal illness. Moreover, as time progressed the frequency of feeding bouts, and the number of pellets eaten per bout did not change. As expected with a primary repellent, 2HAP irritated the peripheral senses, but did not induce illness. Thus, the starlings did not develop learned aversion.

The reduction in drinking among all groups with repellent in their food was unexpected. The avoidance of drinking could be explained by a location effect[17], whereby starlings avoided a proximally located food cup and drink tube. It is also possible that application of repellent from food simply reduces all consumptive behavior, perhaps due to irritation of nerves in the oral and nasal cavity.

Management Implications

The challenge ahead is to mask the irritating qualities of primary repellents so birds will ingest sufficient quantities to incur gastrointestinal illness and subsequent food aversion learning. This can be achieved by micro-encapsulating repellents so that the aversive properties are masked to a bird's peripheral senses, but are released enterically. Conversion of primary repellents to secondary repellents holds promise, both for a new way of conceptualizing the formulations of primary repellents but also for paving the way to develop new repellent formulation that can rely on compounds that are less toxic or environmentally harmful. The development of primary repellents will focus on naturally occurring products that are considered toxicologically and environmentally safe. Some candidates, such as MA, are food additives that have been approved for human consumption. MA has been demonstrated to be safe, even when consumed in high doses. The formulations used for bird repellents will not be at toxic concentrations. Using chemicals such as these would provide wildlife managers and agriculturists with a tool that is effective, environmentally and safe.

The sensitivity of birds in these experiments to visual stimuli (i.e., interference due to the plexiglass cage door) demonstrates the critical importance involved with the presentation and timing of cues. Signals from the sender (the wildlife manager) may not be interpreted by the receivers (the animals) in the way that was intended. A successful operation to develop conditioned food avoidance learning requires careful planning to ensure that birds learn to associate a visual cue with exposure to the repellents. Otherwise, the effects of concurrent interference can counteract even the most effective repellents.

References

1. Rogers, J. G., Jr. *Journal of Wildlife Management* 1974, 38, 418-423.
2. Dolbeer, R. A.; Avery, M. L; Tobin, M. E. *Pesticide Science* 1994, 40, 147-161.
3. Fagerstone, K. A.; Shafer, E. W. Jr. *Proceedings of the Vertebrate Pest Conference* 1998, 18, 319-324.
4. Clark, L. *Current Ornithology* 1998, 14, 1-37.
5. *The principles of learning and behavior*; Domjan, M.; Brooks/Cole: Pacific Grove, CA, USA, 1998.
6. Conover, M. R. *Journal of Applied Ecology* 1984, 21, 437-443.
7. Glahn, J. F.; Mason, J. R.; Woods, D. R. *Wildlife Society Bulletin* 1989, 17, 313-320.
8. Mason, J. R.; Glahn, J. F.; Reidinger, R. F.; Dolbeer, R. A. *Journal of Wildlife Management* 1985, 49, 636-642.
9. Mason, J. R.; Clark, L. *Proceedings of the Vertebrate Pest Conference* 1992, 15, 115-129.

10. Fenaroli's Handbook of Flavor Ingredients. 1971. Furia, T. E. and N. Bellanca, Editors. The Chemical Rubber CO, Cleveland, OH.
11. Clark, L.; Shah, P. S. *Journal of Chemical Ecology* 1994, 20, 321-339.
12. Shah, P. S.; Mason, J. R.; Clark, L. *Chemical Signals in Vertebrates, VI*, Doty, R. L.; Muller-Schwarze, D., editors, Plenum Press. New York, USA, 1992, 291-296.
13. Mason, J. R.; Shah, P. S.; Clark, L. *Journal of Wildlife Management* 1991, 55, 334-340.
14. Avery, M. L. *Journal of Wildlife Management* 1985, 49, 1116-1121.
15. Mason, J. R.; Reidinger, R. F. *Journal of Wildlife Management* 1983, 47, 383-393.
16. Mason J. R. *Social Learning: physiological and biological perspectives*, Zentall, T.R.; Galef, B. G., editors, Lawrence Erlbaum Associates, Publishers, Hilldale, New Jersey, USA, 1987, 99-115.
17. Mason J. R; Reidinger, R.F. *The Auk* 1982, 99, 548-554.
18. Tobin, M. E. *Journal of Wildlife Management* 1985, 49, 1102-1108.
19. Mason, J. R.; Arzt, A. H.; Reidinger, R. F. *The Auk* 1984, 101, 796-803.
20. Kamin, L. J. *Punishment and Aversive Behavior*, Campbell, B. A.; Church, R.M., editors. Appleton-Century-Crofts. New York, USA, 1969.
21. Revusky, S. *Food Aversion Learning*, Milgram, N. W.; Krames, L.; Alloway, T.M., editors, Plenum Press. New York, USA, 1977, 1-51.
22. Clark, L. Wilson Bulletin 1996, 108, 36-52.
23. Hargrave, G. E.; Bolles R. C. *Psyconomic Science* 1971, 23, 91-92.
24. Launchbaugh, K. L.; Provenza, F. D.; Burritt, E. A. *Journal of Chemical Ecology* 1993, 19, 1047-1056.
25. Mason J. R.; Adams, M. A.; Clark, L. *Journal of Wildlife Management* 1989, 53, 55-64.
26. Nachman, M.; Rauschenberger, J.; Ashe, J. H. *Food Aversion Learning*, Milgram, N. W.; Krames, L.; Alloway, T. M., editors, Plenum Press. New York, USA, 1977, 105-131.

Chapter 24

Avian Use of Various Bait Mixtures Offered in Harvested Cornfields during Spring Migration in South Dakota

George M. Linz, H. Jeffrey Homan, and Ryan L. Wimberly

National Wildlife Research Center, U.S. Department of Agriculture, 2110 Miriam Circle, Bismark, ND 58501

The avicide, DRC-1339, is used to cull populations of spring-migrating blackbirds in eastern South Dakota to reduce damage to ripening sunflower in late summer. We investigated nontarget bird hazards associated with using various grain mixtures to attract blackbirds (Icteridae), especially red-winged blackbirds (*Agelaius phoeniceus*), to avicide-treated bait broadcast in harvested cornfields. During spring 1997 and 1998, we recorded the species and numbers of birds attracted to 0.2-ha plots baited with cracked corn and brown rice, fine-chipped sunflower meats and brown rice (1997 only), and unsupplemented brown rice. Unbaited plots were used as reference sites. In 1997 and 1998, nontarget birds showed no preference among bait mixtures (P's > 0.16). In 1997, blackbird preference did not differ among bait mixtures ($P > 0.12$); whereas, in 1998, blackbirds preferred the corn/rice mixture over rice ($P = 0.02$). Our data show that augmenting brown rice with cracked corn or sunflower meats poses little additional risk to nontarget birds and may help attract blackbirds to the baitsite.

In the northern Great Plains, millions of blackbirds (Icteridae) begin feeding on maturing sunflower in mid-August (*1*), causing annual losses of 5 –$10 million before the crop is harvested in October (*2,3*). Sunflower producers began calling for the reduction of blackbird populations in the northern Great Plains shortly after the crop became economically viable in the early 1970s (*4*). DRC-1339 (3-Chloro-*p*-toluidine), an avicide developed by the U.S. Fish and Wildlife Service for controlling European starlings (see tables for scientific names) at feedlots (*5*), was touted as an environmentally safe avicide for reducing spring-migrating blackbird populations responsible for damaging sunflower (*4*). Designated by the U. S.

Environmental Protection Agency (EPA) as a restricted-use pesticide, DRC-1339 can only be applied by trained personnel in the U.S. Department of Agriculture, Animal Plant Health Inspection Service, Wildlife Services program.

In March 1993, we initiated a long-term study to evaluate the use of DRC-1339-treated rice for culling spring-migrating blackbird populations congregating in eastern South Dakota (*6*). The management goal is to cull the breeding population thereby reducing late-summer populations of blackbirds responsible for damaging sunflower (*7*). The treated baits are spread in harvested grain fields raising concerns among wildlife agencies about acute and chronic poisoning of nontarget species (*8*). Currently, DRC-1339 is applied to brown rice and broadcast in harvested cornfields within 50 m of a road (*9*). Although blackbirds eat rice in the southern United States, rice may not be a favored food of spring-migrating blackbirds in South Dakota, and by providing an additional food item along with the treated rice may entice more blackbirds to feed in our baited plots.

Field data comparing preferences of blackbirds and nontarget birds for various bait combinations are not available. Our objective was to assess the preferences of blackbirds and nontarget birds for different bait mixtures.

Study Area and Methods

Study Area

Our study was conducted in Brookings, Miner, and Lake counties in east-central South Dakota. This region of low, rolling hills has been developed for agriculture but still has an abundance of undrained lakes and potholes. Of about 370,000 ha of cropland in these counties, 46% was soybean, 43% corn, and 8% wheat (10). The remaining 156,100 ha of land consisted of hayland (39%), Conservation Reserve Program lands (14%), and wetlands (47%). The long-term average temperature and precipitation during our March-April study period were 2.2° C and 4.2 cm, respectively. In 1997, temperatures averaged 0° C and precipitation totaled 4.1 cm; while in 1998, the average temperature was 2.2°C and precipitation totaled 2.9 cm (10).

Baiting Procedure

In 1997, we placed a pair of 0.8-ha (2-acre) plots in each of three harvested corn fields in attractive locations for blackbirds and nontarget birds (e.g., near woodlots and grasslands). The nearest edge of each plot was about 25 m from the road. Each 0.8-ha plot was subdivided into four 0.2 ha (0.5 acre) subplots and randomly assigned one of the following treatments: 2.8 kg rice (6.25 lb) and 2.8 kg fine-chipped sunflower meats, 2.8 kg brown rice and 2.8 kg cracked corn, 2.8 kg rice, and no bait. An all-terrain vehicle, equipped with a seed spreader, was used to apply baits. An unbaited buffer zone of about 3-m was left between subplots.

In 1998, we used a similar design except the rice/sunflower treatment was canceled and sample size was increased to four harvested fields. Plots were subdivided into three 33 x 33 m (0.2 ha) subplots with about a 3-m buffer between them. Each 0.2-ha subplot randomly received either 11.3 kg brown rice and 11.3 kg cracked corn, 11.3 kg rice, or no bait.

In both years, we established four bait stations (30.5 x 30.5 cm) in the reference and baited subplots to monitor bait consumption. The stations contained 10 rice kernels and were checked at least every three days, weather permitting. Subplots were rebaited after 75% of the bait at the stations was eaten or when precipitation exceeded 10 mm (0.4 inches).

Food Characteristics

Percent analyses of dry matter, ash, crude protein, acid detergent fiber, and gross energy (kcal/g) of each food were reported previously (11). Briefly, percent dry matter was essentially equal among the three foods, averaging 92%. Sunflower contained 29% crude protein whereas, rice and corn had about 10% protein. Brown rice had 4% ash and 2% fiber, which was three times more ash and three times less fiber than corn and sunflower. Finally, sunflower contained 7.3 kcal/g; whereas, cracked corn and brown rice contained about 4.5 kcal/g.

Plot Observations

Between 1 April and 25 April 1997, we observed each plot at least every third day beginning at sunrise to three hours after sunrise and from three hours before sunset to sunset. The observer arrived about ½ hr before the start of each one hour observation period and erected a blind on a 3-m high platform affixed to a truck (*5*). The nearest edge of each plot was about 25 m from the truck. After waiting 15 min, the observer began recording all birds that landed in the first randomly selected subplot for 30 sec. The observer then paused 30 sec before recording data in the next plot. After all four subplots had been censused, the observer rested for one minute and repeated the procedure.

In 1998, plots were observed from 27 March to 22 April. Procedures for observing plots were altered so that each 0.2-ha subplot was observed for one minute with a one minute data-recording period between observation periods. Observations were not conducted during steady precipitation or if the wind exceeded 32 km/hr because of poor visibility.

Statistical Analyses

Target birds, as defined under the FIFRA (Federal Insecticide Fungicide and Rodenticide Act) Section 3 Label 'Compound DRC-1339 Concentrate-Staging Areas'

(EPA registration number 56228-30), included red-winged blackbirds, yellow-headed blackbirds, common grackles, Brewer's blackbirds, European starlings, and brown-headed cowbirds (12). All other birds were considered nontargets.

We used Kruskal-Wallis tests to examine the null hypothesis that the number of blackbirds and nontarget birds recorded per hour were similar among the bait treatments and the reference plots and the three bait treatments (13). For the 1998 data, we used the Wilcoxon 2-sample test because we only had two bait treatments (13).

The significance level was set at 0.05 for all statistical tests. Means and standard errors are reported as $\bar{x} \pm$ SE. To illustrate the temporal dynamics of bird migration, we graphed 7-day moving averages of the mean numbers of birds/hour in the plots.

Results

1997

Blackbirds were observed in greater numbers in baited subplots than unbaited reference subplots ($\chi^2 < 9.63$, df = 3, $P > 0.02$); however, we detected no difference in the number of birds using the three bait mixtures ($\chi^2 = 4.22$, df = 2, $P = 0.12$). In comparison, nontarget use of baited and unbaited subplots was similar ($\chi^2 = 1.96$, df = 3, $P = 0.58$). During 86 hours of observations, numbers of blackbirds averaged 84.0 (± 16.6) and nontarget birds 1.5 (± 0.3). Blackbird migration peaked during mid-April (Figure 1), one week later than the peak for nontargets (Figure 1).

1998

More blackbirds ($\chi^2 = 21.8$, df = 2, $P < 0.01$) and nontarget birds ($\chi^2 = 8.64$, df = 2, $P = 0.01$) were recorded using subplots baited with corn/rice or unsupplemented rice than unbaited plots. We counted more blackbirds in subplots baited with corn/rice than brown rice ($Z = 2.26$, df = 1, $P = 0.02$). However, abundance of nontargets did not differ between these two baits ($Z = 1.40$, df = 1, $P = 0.16$). During 66 hours of observation, mean numbers of blackbirds and nontargets observed per hour in all subplots were 29.4 ± 8.1 and 3.2 ± 0.6, respectively. Migration peaks for blackbirds and nontarget birds were similar, occurring in early April. (Figure 1).

Years Combined

We observed 21 nontarget birds species during 152 hours of observations. After pooling the data across years, we found that the numbers of blackbirds differed ($\chi^2 = 6.49$, df = 2, $P = 0.04$) among the baited subplots (Table I), with the blackbirds preferring sunflower/rice over corn/rice and rice alone. In comparison, over-all

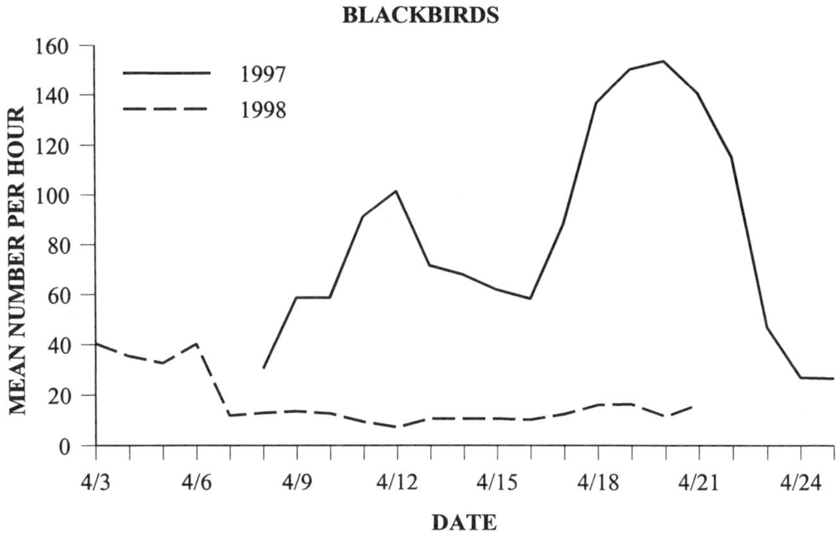

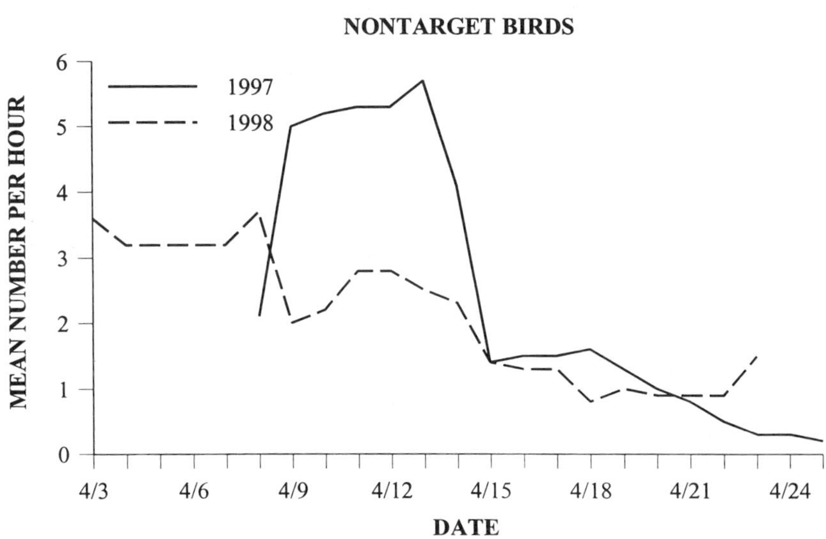

Figure 1. Seven-day moving average of the mean number of blackbirds and nontarget birds using 6-0.2 ha subplots in harvested cornfields in east-central South Dakota during April 1997 and 1998.

Table I. Target birds observed in east-central South Dakota cornfields containing 0.2 ha areas of bait mixture combinations of fine-chipped sunflower and brown rice, cracked corn and brown rice, and brown rice during March and April 1997 and 1998.

		Bait Type			
		Sunflower[a]	*Corn*[b]	*Rice*[b]	*Reference*[b]
Common Name	*Scientific Name*	Mean Number Per Hour (Percent Frequency of Occurrence)			
European starling	*Sturnus vulgaris*	0.00(0)	0.09(2)	0.01(<1)	0.00(0)
Yellow-headed blackbird	*Xanthocephalus xanthocephalus*	4.58(31)	4.35(24)	3.01(16)	1.64(15)
Brewer's blackbird	*Euphagus cyanocephalus*	0.80(5)	0.20(1)	0.05(1)	0.20(<1)
Brown-headed cowbird	*Molothrus ater*	0.15(8)	0.31(11)	0.13(8)	0.03(1)
Red-winged blackbird	*Agelaius phoeniceus*	70.01(64)	48.65(53)	42.65(47)	27.06(35)
Common grackle	*Quiscalus quiscula*	22.15(24)	13.66(20)	13.95(14)	17.06(10)
Mean		97.8(65)	67.3(55)	59.8(49)	46.0(36)

[a]*n = 86 observation hours*
[b]*n = 152 observation hours*

numbers of nontarget birds recorded were similar among bait mixtures ($\chi^2 = 2.13$, df = 2, $P = 0.34$) (Table II).

Discussion

Bait Selection

In both years, blackbirds consistently selected against unbaited subplots and tended to select the highest energy food mixture available, suggesting that blackbirds choose feeding sites based on food quantity and probably food quality. An assessment of food preference is confounded by several factors, including food characteristics such as color, size, taste, and caloric content (13, 14, 15), and the morphology of the bird. For example, in a cage study, female red-winged blackbirds, which are 30% smaller than males, clearly chose brown rice over cracked-corn (*11*). On the other hand, males showed no consistent food preference. It probably was easier for the smaller-billed females to handle rice than cracked corn.

Migration and feeding ecology

In 1997 and 1998, we observed 21 nontarget bird species in our study plots. In comparison, Kenyon (*16*) recorded 12 species in 1995, and Knutsen (*9*) detected 13 species in 1996 and 1997 (Table III). Across all four years, 24 nontarget species were recorded in cornfields in east-central South Dakota. We believe that nearly all nontarget birds that commonly use harvested corn in east-central South Dakota from mid-March to late April were recorded during these three studies.

In 1997, Lapland longspurs, which normally migrate north from east-central South Dakota by mid- to late March, were abundant in the study area until mid-April because of harsh weather. In comparison, large numbers of blackbirds and other early migrant species such as western meadowlarks, mallards, snow geese, and American robins were not observed in appreciable numbers in South Dakota until late March 1997. Differences in migrational timing between study years were probably related to the mild, late-winter weather that created open agricultural land two weeks earlier in 1998 than in 1997, allowing for an earlier arrival of migrant birds (17, 18).

From 1995 to 1998, six members of the sparrow family *(Emberizidae)* were observed in our study plots; none, however, have been tested for susceptibility to DRC-1339. We recommend LD_{50s} for some or all of these birds because their granivorous food habits, small size, and flock feeding behavior may make them vulnerable to DRC-1339 poisoning.

Other birds frequently observed in the plots included American robins, killdeers, and American coots. The foraging habits and diets of these species make it unlikely that they will be adversely affected by DRC-1339-treated rice baits (*8, 19*). On the other hand, western meadowlarks, as members of the highly vulnerable blackbird

Table II. Nontarget birds observed in east-central South Dakota cornfields containing 0.2 ha areas of bait mixture combinations of fine-chipped sunflower and brown rice, cracked corn and brown rice, and brown rice during March and April 1997 and 1998.

		Bait Type			
		Sunflower[a]	*Corn*[b]	*Rice*[b]	*Reference*[b]
Common Name	*Scientific Name*	Mean Number Per Hour (Percent Frequency of Occurrence)			
Canada goose	*Branta canadensis*	0.01(1)	0.00(0)	0.00(0)	0.00(0)
Wood duck	*Aix sponsa*	0.12(1)	0.06(1)	0.00(0)	0.00(0)
Northern pintail	*Anas acuta*	0.00(0)	0.01(<1)	0.00(0)	0.00(0)
Blue-winged teal	*Anas discors*	0.00(0)	0.01(<1)	0.00(0)	0.00(0)
Northern shoveler	*Anas clypeata*	0.06(1)	0.00(0)	0.00(0)	0.00(0)
Ring-necked pheasant	*Phasianus colchicus*	0.00(0)	0.01(<1)	0.00 (0)	0.03(1)
Killdeer	*Charadrius vociferus*	0.04(4)	0.07(5)	0.11(6)	0.11(5)
Common snipe	*Gallinago gallinago*	0.00(0)	0.03(<1)	0.01(1)	0.01(1)
Mourning dove	*Zenaida macroura*	0.02(1)	0.09(4)	0.10(3)	0.01(<1)
Downy woodpecker	*Picoides pubescens*	0.00(0)	0.01(1)	0.01(<1)	0.01(<1)
Northern flicker	*Colaptes auratus*	0.01(1)	0.20(<1)	0.02(1)	0.01(<1)
Horned lark	*Eremophila alpestris*	0.05(1)	0.16(4)	0.12(4)	0.05(2)
American robin	*Turdus migratorius*	0.00(0)	0.07(2)	0.10(4)	0.17(4)
Water pipit	*Anthus spinoletta*	0.00(0)	0.00(0)	0.03(2)	0.00(0)

Table II. Continued

		Bait Type			
		Sunflower[a]	*Corn*[b]	*Rice*[b]	*Reference*[b]
Common Name	*Scientific Name*	Mean Number Per Hour (Percent Frequency of Occurrence)			
American tree sparrow	*Spizella arborea*	2.23(14)	1.61(9)	0.70(7)	0.62(10)
Vesper sparrow	*Pooecetes gramineus*	0.01(1)	0.01(<1)	0.01(<1)	0.00(0)
Savannah sparrow	*Passerculus sandwichensis*	0.00(0)	0.24(7)	0.26(7)	0.01(<1)
Song sparrow	*Melospiza melodia*	0.00(0)	0.25(3)	0.33(2)	0.03(2)
Dark-eyed junco	*Junco hyemalis*	0.07(2)	0.13(5)	0.05(3)	0.03(2)
Lapland longspur	*Calcarius lapponicus*	0.05(2)	0.07(<1)	0.01(1)	0.01(<1)
Western meadowlark	*Sturnella neglecta*	0.12(8)	0.15(9)	0.09(5)	1.70(6)
Unknown species		0.00(0)	0.01(<1)	0.20(3)	0.00(0)
Mean		2.80(29)	2.98(38)	1.85(34)	1.85(26)

[a]*n = 86 observation hours*
[b]*n = 152 observation hours*

Table III. Nontarget birds observed in cornfields in east-central South Dakota from 1995 through 1998.

Common Name	Year						
	1995[a]	*1996*[b]	*1997*[c]	*1997*[d]	*1998*[e]	*Mean*	*Rank*
	(Percent Frequency of Occurrence)						
American tree sparrow	1	28	5	36	10	16	1
Western meadowlark	3	23	9	13	17	13	2
Canada goose	11	22	12	1	0	9	3
Ring-necked pheasant	35	7	3	0	2	9	3
American robin	10	1	12	6	6	7	4
Killdeer	4	4	8	8	12	7	4
American coot	0	0	35	0	0	7	4
Dark-eyed junco	5	3	0	8	6	4	5
Song sparrow	0	9	3	0	7	4	5
Unknown species	12	1	1	1	3	4	5
Mourning dove	1	0	5	4	5	3	6
Downy woodpecker	6	0	1	4	2	3	6
Horned lark	0	0	0	7	6	3	6
Savannah sparrow	0	0	0	0	15	3	6
Northern flicker	7	1	0	1	3	2	7
Water pipit	0	0	0	0	2	2	7
Mallard	5	2	3	0	0	2	7
Lapland longspur	0	0	0	6	0	1	8
Wood duck	0	0	0	2	1	<1	9
Green-winged teal	1	0	0	0	0	<1	9

Table III. Continued

Common Name	*1995*[a]	*1996*[b]	*1997*[c]	*1997*[d]	*1998*[e]	*Mean*	*Rank*
	Year (Percent Frequency of Occurrence)						
Northern shoveler	0	0	0	1	0	<1	9
Common snipe	0	0	0	0	3	<1	9
Vesper sparrow	0	0	1	2	1	<1	9
Number of Identified Species	12	10	12	16	16		
Total Individuals	177	111	147	101	150		

[a]*n = 72 observation hours, 4 fields with 2-0.8 ha plots in each field; Kenyon 1996*
[b]*n = 58 observation hours, 3 fields with 2-0.8 ha plots in each field; Knutsen 1998*
[c]*n = 77 observation hours, 4 fields with 2-0.8 ha plots in each field; Knutsen 1998*
[d]*n = 86 observation hours, 3 fields with 2-0.8 ha plots in each field; This study*
[e]*n = 66 observation hours, 4 fields with 2-0.8 ha plots in each field; This study*

family, may be susceptible to DRC-1339 because they feed in open fields and eat rice grains (20).

The number of ring-necked pheasants, mourning doves, and waterfowl recorded in the plots were too small for valid statistical analyses. Ring-necked pheasants and waterfowl are both wary, particularly during the early spring, and thus probably avoided our study plots because of their proximity to traveled roads. Mourning doves are not susceptible to spring baiting with DRC-1339 because they do not arrive in east-central South Dakota in significant numbers until late April, after the baiting has been completed.

Conclusions

The purpose of our study was to evaluate bird use of three bait offerings placed within harvested cornfields. Of the 10 most commonly observed nontarget birds, the western meadowlark was probably the most susceptible to poisoning due to its abundance, feeding habits, size, and possible chemical susceptibility to DRC-1339. Even so, this bird is very common (*21*), and DRC-1339-treated brown rice is unlikely to significantly affect the population. The ring-necked pheasant may also be vulnerable to acute and chronic DRC-1339 poisoning (22). The other eight species of nontargets are not susceptible to a spring-baiting program because of their size, food habits, behavior, or high tolerance to DRC-1339.

Future Research

We recommend two studies to further clarify the hazards of DRC-1339 to spring-migrating nontarget birds: 1. A food habits study of bird species that frequent rice-baited plots to determine which species eat rice and how much they eat. 2. Determine the LD_{50} values for rice-eating species observed in the baited plots.

Acknowledgments

We thank B. Bonesteel and T. Slowik for their help with the field observations. The Animal Nutrition Laboratory of the Department of Animal and Range Sciences, North Dakota State University, conducted the chemical analyses of the food. P. Pochop and R. Sayre provided helpful suggestions on an earlier draft. This study was conducted under the auspices of National Wildlife Research Center study protocol QA-437. Procedures involving use and care of birds were reviewed and approved by the Animal Use and Care Committee, National Wildlife Research Center.

References

1. Cummings, J. L., J. L. Guarino, and C. E. Knittle. Chronology of blackbird damage to sunflowers. *Wildl. Soc. Bull.* 1989. 17:50-52.
2. Hothem, R. L., R. W. DeHaven, and S. D. Fairazl. Bird damage to sunflower in North Dakota, South Dakota and Minnesota, 1979-1981. *U.S. Fish and Wildlife Tech. Rep. 15.* 1988. 11 pp.
3. Linz, G. M., and J. J. Hanzel. Birds and sunflower. *In* A. Schneider, ed. *Sunflower Science and Technology*. Am. Soc. of Agron., Madison, Wis. 1997. Pages 381-394.
4. Linz, G. M., and H. J. Homan. Tracing the history of blackbird research through an industry's looking glass: *The sunflower magazine.* Vertebr. Pest Conf. 1998. 18:35-42.
5. DeCino, T. J., D. J. Cunningham, and E. W. Schafer. Toxicity of DRC-1339 to starlings. *J. Wildl. Manage.* 1966. 30:249-253.
6. Barras, A. M.S. thesis, North Dakota State University, Fargo, North Dakota, 1996.
7. Knittle, C. E., G. M. Linz, J. L. Cummings, J. E. Davis, Jr., B. E. Johns, and J. F. Besser. Spring migration patterns of male red-winged blackbirds (*Agelaius phoeniceus*) from two migratory roosts in South Dakota and Minnesota. *Am. Midl. Nat.* 1996. 136: 134-142.
8. Gober, P. *Section 7 Biological Opinion.* September 29, 1998. U.S. Department of Interior, Fish and Wildlife Service, Pierre, South Dakota. Letter - 9 pp.
9. Knutsen, G. A. M.S. thesis, North Dakota State University, Fargo, North Dakota, 1998.
10. South Dakota Agricultural Statistics Service. *South Dakota Agricultural Statistics*. S.D. Dep. Agric., Sioux Falls, SD, 1999.
11. Linz, G. M, L. A. Mendoza, D. L. Bergman, and W. J. Bleier. Preference of three blackbird species for sunflower meats, cracked corn, and brown rice. *Crop Protection.* 1995. 14:375-378.
12. U.S. Department of Agriculture. *Animal Damage Control Program Final Environmental Impact Statement.* Appendix P. Risk assessment of wildlife damage control methods used by the USDA Animal Damage Control Program. Washington, D.C. 1994. pp. 194-210.
13. Mah, J. Ph.D. dissertation, North Dakota State University, Fargo, North Dakota, 1988.
13. Conover, W. J. *Practical Nonparametric Statistics*, 2nd ed. John Wiley & Sons, New York. 1980.
14. Mason, J. R., R. A. Dolbeer, A. H. Arzt, R. F. Reidinger and P.P. Woronecki. Taste preferences of male red-winged blackbirds among dried samples of ten corn hybrids. *J. Wildl. Manage.* 1984. 48:611-616.
15. Mason, J. R., G. Nuechterlein, G. Linz, R. A. Dolbeer, and D. L. Otis. Oil concentration differences among sunflower achenes and feeding preference of red-winged blackbirds. *Crop Protection.* 1991. 10:299-304.

16. Kenyon, M. J. M.S. thesis. North Dakota State University, Fargo, North Dakota, 1996.
17. DeGraaf, R. M., N. G. Tilghman, and S. H. Anderson. Foraging guilds of North American birds. *Environ. Manage.* 1985. 9:493-536.
18. Hirst, S. M., and C. A. Easthope. Use of agricultural lands by waterfowl in southwestern British Columbia. *J. Wildl. Manage.* 1981. 45:454-462.
19. Schafer, Jr., E. W., R. B. Brunton, D. J. Cunningham, and N. Lockyer. The chronic toxicity of 3-chloro-4-methyl benzamine HCL to birds. *Arch. Environ. Contam. Toxicol.* 1977. 6:241-248.
20. Flickinger, E. L., and K. A. King. Some effects of aldrin-treated rice on Gulf Coast wildlife. *J. Wildl. Manage.* 1972. 36:706-727.
21. Igl, L. D., and D. H. Johnson. Changes in breeding bird populations in North Dakota: 1967 to 1992-93. *Auk.* 1997. 114:74-92.
22. Avery, M. L., M. J. Kenyon, G. M. Linz, D. L. Bergman, D. G. Decker, and J. S. Humphrey. Potential risk to ring-necked pheasants from application of toxic bait for blackbird control in South Dakota. *J. Wildl. Manage.* 1998. 62:388-394.

Chapter 25

Hurdles in Compiling Pesticide Incidents

Patrick J. Hannan

NOWCC/Senior Environmental Employee Program, Mail Code 7507C, U.S. Environmental Protection Agency, Ariel Rios Building, Washington, DC 20460

Pesticides must be registered by the United States Environmental Protection Agency (EPA) before they can be offered for sale. They also must be re-registered periodically. A factor to be considered in the re-registration is the number of recorded incidents in which they have been shown to be hazardous to humans, to wildlife, or to the environment. This paper discusses the Ecological Incident Information System that is concerned primarily with the damaging effects of pesticides on plants and wildlife, primarily fish and birds. Included is an account of how EPA is made aware of pesticide incidents, and how the information obtained is used to classify the degree of blame assigned to a pesticide. This particular database does not include human exposure incidents which are within the scope of another reporting system.

Being Made Aware

The principal hurdle in compiling incidents is learning that they exist. EPA must depend on the voluntary contributions of State environmental or agriculture offices, its own regional offices, and interested citizens who report bird kills, fish kills, and crop damage that have been caused by pesticides. Many incidents are reported by pesticide manufacturers (under 6(a)2 requirements) who inform EPA of any complaints they have received regarding their products. Recent changes in the reporting requirements have lowered the threshold of such reportings; whereas one or two initial incidents may have been needed to justify disclosure previously, now even allegations are a sufficient justification. On the other hand, under the new guidelines a registrant might provide less information on a given incident than had been required before.

Prominent among the 6(a)2 reportings are instances in which lawns are damaged, or field crops diminished, by herbicides that have either run off of treated areas, or have drifted during crop spraying. Of course, there have been substantial fish kills resulting from the contamination of creeks or ponds by the runoff of pesticides that have been applied in a manner consistent with their registration limits but, afterwards, are washed away by substantial rains.

Even if an interested citizen becomes aware of an environmental problem caused by pesticide usage, there often is a substantial delay before an incident reaches the database. The uninformed public faces information barriers in reporting any incident. What phone number should be called? Whose task is it to investigate? Even if one calls the correct number and receives an automated impersonal response, will he have the patience necessary to complete the call? It is certain that only a fraction of the incidents caused by pesticides reach EPA's attention, simply because people do not know how to report them.

Once the News Has Reached EPA

Certainty Discussion

For every incident reported to the EPA, a judgment must be made concerning its cause. Toward this end the database includes options ranging from "Unlikely" to "Highly Probable." Of primary importance in entering an incident is the matter of certainty. For a pesticide to be judged the "Highly Probable" cause of an incident there must be convincing evidence that it was, indeed, the cause. Generally this must include an analysis of the carcass showing the presence of the suspected pesticide at a substantial concentration. Additional evidence might be an effect on an organ that is known to be affected by the pesticide. For example, an organophosphate may have been detected in the esophagus contents of a bird, and the bird's cholinesterase activity may have been far below normal; in such a case, the "Highly Probable" classification would be appropriate.

For the term " Probable" to be applied, the evidence would be less convincing. Perhaps the cholinesterase activity was measured and found to be below normal, and it was known that an adjacent field was sprayed with a pesticide having anti-cholinesterase activity. Such a combination of factors would warrant the judgment that this pesticide was probably responsible for the incident.

In many cases the term "Possible" is the only alternative in judging whether a given pesticide is responsible for an incident. Suppose multiple analyses are performed on the carcass of a bird, showing a variety of DDT metabolites and several highly toxic compounds at very low concentrations; it could be that there was a combined influence of all of those components and the only reasonable approach would be to classify each as a possible cause of death. Also, as mentioned above, another large source in the "Possible" category is the plethora of incidents reported by pesticide manufacturers under the 6(a)2 requirements imposed by EPA. Typically, a citizen complains that a product was used to kill the weeds in his lawn and the whole lawn was affected. The manufacturer reports that such a complaint was made but supplies no other information -- and might not have any. In the absence of any analytical information, it can only be concluded that the pesticide in question was a possible cause.

The term "Unrelated" is another potential tag to apply to the relationship between a pesticide and an environmental incident. No further explanation for that should be required for this forum.

It is pertinent here to add a note about the significance of cholinesterase activity levels. Carbamates, for example, will severely inhibit cholinesterase activity but a partial recovery from their effect is possible with the passage of time. Organophosphates, on the other hand, have an irreversible effect on cholinesterase activity. An intermediate cholinesterase activity might result from a situation in which the brain has recovered partially from the presence of a carbamate. Another possibility is that the sample being analyzed included tissue that was extraneous to the brain.

Use/Misuse Variations

It would be unjust to blame a pesticide for an incident if it has been used in a way that is not permitted according to the label. Suppose someone is sufficiently exasperated with bird droppings on his car that he laces a slice of bread with Pesticide X and leaves it in an area where it would attract birds. That is clearly a misuse of the pesticide and the resultant bird kill should be in the misuse category.

Generally, misuses of this type are easily identified. In recent years there have been many such instances in New York City and, while the perpetrator may not have been discovered, there is no doubt about the source of the lethal dose. There are many other instances in which the aerial application of a pesticide has caused severe contamination of ponds, the result being many dead fish. Often such an occurrence is judged to be a misuse of the offending pesticide.

In all of these cases, however, the person making the entry into the database must depend on the accuracy of the account forming the basis of the case.

Dissolved Oxygen vs Pesticide

Fish present a unique case in that they must depend on sufficient oxygen in their waters to sustain life. If a substantial number of fish have died in a pond, was the cause a lack of dissolved oxygen (D.O.) or was it the presence of a pesticide? In many instances the cause is clear cut, such as a heavy rain storm immediately following the application of chlorpyrifos to the foundation of a new home nearby. But there are many instances in which there has been a trace concentration of a pesticide in a water body having a marginal dissolved oxygen concentration. If the investigator has determined the D.O. content, that is an obvious help. Still, it is probable that the D.O. sample was taken in daylight hours when the reading may be considerably higher than if it had been taken at 2:00 AM. That is one complication. Another is that some fish cope with a low D.O. concentration far better than others.

Time of sampling

In conjunction with the D.O. problem is the matter of how and when a sample was taken. Let us suppose that there has been a fish kill and a subsequent analysis of the water, showing less than a part-per-billion of a pesticide that might be suspected as the source of the problem. Was the pesticide responsible? That might be hard to determine unless information related to the sampling is available. If the sample is taken within hours of the apparent fish kill, perhaps the concentration found is representative of the conditions pertaining to the kill. But suppose the sample is taken a week after the incident was noticed - - what does it mean? It might not mean anything, and the judgment of the one entering the incident into the database must depend on an evaluation of what is known. A problem often encountered, however, is that the time the sample was taken is not known.

Each reporting agency has its own form - - if, in fact, it uses a form rather than a simple narrative account. It is çommon to encounter forms which mention the date of the report being made, and the date of the incident, but with no mention of the time a sample was taken. Then, too, there is the problem of the adsorption of trace amounts of a pesticide to the walls of a sampling vessel, causing the analysis to reflect a concentration considerably lower than it had been. That is a matter too complicated for a complete discussion in this short report.

Talents of the Investigator

With the difficulties cited above in the collection of data surrounding a pesticide incident, it is clear that a weak link in the chain can be the background of the one making the investigation. In the case of a fish or bird kill, the most important evidence would be the carcass of a victim; if a substantial concentration of a pesticide is found, then the decision becomes automatic. It is obvious, however, that reports are often submitted by personnel having insufficient training. They might be knowledgeable regarding fish classifications, for example, but have no knowledge of chemistry. Rather than collect a carcass, freeze it, and make arrangements to have it analyzed, they might be content to take a water sample even though it would be days after the incident has taken place.

It is a fact, too, that a number of investigations are carried out by people who are not trained. In today's world, where spelling seems to be a lost art, it would be risky to assume that several misspelled words in a report would reduce its credibility. But when phrases such as " - - the bream was intoxication," or " - - he had did everything possible" are present, there is some doubt about the capability of the investigator.

An Interesting Disconnect

EPA maintains a large Pesticide Ecotoxicity Database in which summaries of pesticide toxicities toward plants, fish, and birds can be found. This is valuable in making judgments about which pesticide, among several identified in a carcass, may have been responsible for the death of the subject.

Upon reflection, however, one is aware of an inadequacy in the process of assigning a cause for an environmental incident. The Pesticide Ecotoxicity Database provides one measure, namely the concentration required to kill something. But that does not match up with what the compiler of the incident database has to work with, viz the concentration of pesticide X in a fish or bird. If there were tabular data relating the concentration of a compound in a fish, to that in the water in which it was swimming, it would be useful. Of course, the situation in the real world is much more complex. There may have been a large pulse of pesticide input to the water followed by rapid dilution, which would produce results that are radically different from a chronic situation in which a considerably lower concentration in the water would have prevailed for an extended period.

I discussed the matter with Dr. Emmett Braselton, Chief of the Toxicology Unit at Michigan State University and he told of two examples he often relates to his students:

Case #1: A liver specimen from an animal was received for study. A previous study showed that the brain cholinesterase activity was normal, so one would tend to rule out an organophosphate as the cause of death. However, they found a high concentration of malathion (an organophosphate) in the liver. This was a case of acute poisoning in which the animal had died before the organophosphate reached the brain.

Case #2: A liver specimen was analyzed by GCMS and no organophosphates were found. But the brain cholinesterase activity was reduced. So then they ground up the brain and extracted it, and found that it contained malathion. This was considered a case of chronic poisoning that had occurred over time and, even though there was no malathion detectable in the liver, it had been exerting its effect on the brain.

OVERALL SUMMARY

In writing this report it has been necessary to point out the limitations of an incident database related to pesticides. There are many problems associated with making judgments on an environmental situation when one is far removed from it. On the other hand, such an instrument is necessary and useful despite its inherent limitations. Pesticides by their nature are toxic to a certain segment of the earth's biota. If the dangers associated with them outweigh their benefits, they should be removed from the market. A pesticide incident database can shed light on those compounds that pose more risks than are warranted.

It is in everyone's interest that the incident database be as complete and accurate as possible. Toward this end the active support of local and State agencies is required, and a greater emphasis on selecting personnel with technical backgrounds to investigate the causes is necessary.

For those who have knowledge of an incident related to pesticides, they are asked to send the information to :

Document Processing Desk
Office of Pesticide Programs - - 7504C
U.S. Environmental Protection Agency
401 M St. S.W.
Washington D.C. 20460

Another option would be to FAX it to EIIS (Ecological Incident Information System) at 703-305-6309.

Chapter 26

Regulatory Aspects of Protecting Threatened and Endangered Species with County Pesticide Bulletins

J. Allen White[1], Richard A. Marovich[2], and Gerald H. Miller[3]

[1]Fish and Wildlife Service, U.S. Department of the Interior, 10711 Burnet Road, #200, Austin, TX 78758
[2]Department of Pesticide Regulation, California Environmental Protection Agency, 830 K Street, Sacramento, CA 95814
[3]California Department of Food and Agriculture, Room A357, 1220 N Street, Sacramento, CA 95814

County pesticide bulletins serve as extensions of FIFRA pesticide labels by informing pesticide applicators of necessary protective measures for Federally-listed threatened and endangered species. Examples of protective measures involving pesticides include buffer zones, reduced application rates, modified application methods, restricted periods of application, and other limitations on pesticide use. County pesticide bulletins are developed by the U.S. Environmental Protection Agency and various state agencies that together are responsible for assembling information on species, habitat, pesticides, and protective measures for individual bulletins. The regulatory development and application of county pesticide bulletins are discussed.

County pesticide bulletins are the current means by which the U.S. Environmental Protection Agency and various state agencies inform pesticide users in most states on how to appropriately apply pesticides in proximity to Federally-protected threatened and endangered species. Bulletins for a particular county typically contain four elements: 1) a list of threatened and endangered species that are located near areas where pesticides are used, 2) habitat maps showing spatial distribution of these species or areas requiring limitations on pesticide use, 3) pesticides that are normally used in the county, and 4) measures considered necessary for protecting species from pesticides. The list of threatened and endangered species in the county bulletin is derived from the U.S. Fish and Wildlife Service which has primary responsibility for enforcement of the Endangered Species Act of 1973. Locations of these species in a county may be represented by a map and/or description of habitat specific to individual species. Tables of pesticide active ingredients are based on local patterns of pesticide use and are developed by entities such as county extension services. Protective measures in a county bulletin are preventive concepts that are intended to minimize impacts on species by individual pesticides according to characteristics of toxicology, physicochemical parameters, exposure potential, transport, and environmental fate.

The county pesticide bulletin is a programmatic attempt by regulatory agencies to prevent pesticide impacts on threatened and endangered species in areas harboring these species. The purpose of this paper is to discuss the regulatory process of developing county pesticide bulletins and to assess bulletin formulation from the perspective of actual field application.

Discussion

Legal Authorities

The need for county pesticide bulletins arises from the interaction of two Federal laws: the Federal Insecticide, Fungicide, and Rodenticide Act (FIFRA) of 1947, as amended (7 U.S.C. 135 *et seq.*) and the Endangered Species Act of 1973, as amended (16 U.S.C. 153 *et seq.*). The specific intent of FIFRA is to regulate pesticides sold, distributed, or used in the United States. The U.S. Environmental Protection Agency (EPA) regulates pesticides under the act by requiring either full registration under FIFRA's Section 3 for use in all states or exemption status under Section 18 for use during emergency conditions involving individual counties or states. FIFRA also has provisions for special local needs (i.e., individual state registrations) under Section 24(c) and experimental use under Section 5. Before registration can occur, EPA must make a determination for each pesticide under FIFRA Section 3(c)(5) such that "when used in accordance with widespread and commonly recognized practice it will not generally cause unreasonable adverse effects on the environment." EPA may suspend a previously registered pesticide under FIFRA Section 6(c)(1) if the pesticide is determined by EPA to be an imminent hazard jeopardizing a threatened or endangered species.

During the registration process, EPA determines whether a particular pesticide has adverse environmental effects in part by evaluating the pesticide's potential to harm threatened and endangered species or the habitat considered critical to survival of these species. Such species and their critical habitat are given Federal statutory protection under the Endangered Species Act (ESA). An endangered species is any species that is likely to become extinct within the foreseeable future throughout its entire range or at least a significant portion. In turn, a threatened species is a species likely to become endangered. The term species itself involves any distinct population segment of a species and can mean a species, subspecies, or a distinct vertebrate population. The ESA is jointly administered by the Fish and Wildlife Service (FWS) in the U.S. Department of Interior and the National Marine Fisheries Service (NMFS) in the U.S. Department of Commerce. However, FWS has the primary role in administering the act since NMFS is responsible for only a few marine and anadromous fish species. In September 1999, there were 1,197 species (481 animals,

716 plants) listed as threatened or endangered in the United States with 1,754 species listed worldwide (*1*).

The ESA is composed of 18 sections of which four most directly relate to pesticide use and county pesticide bulletins:

Section 4	Authorizes the determination and listing of species as threatened or endangered. Such species are called "listed" species and are given Federal protection under the act at such time as designated in the Federal Register. Areas of critical habitat necessary for a species survival may be designated at the time of listing or in a later action. Critical habitat designation requires Federal agencies to provide greater protection for these areas to avoid actions that could cause "adverse modification."
Section 7	Requires Federal agencies to ensure that any action that they authorize, fund, or carry out is not likely to jeopardize the continued existence of listed species or adversely modify critical habitat of those species. Agencies are required to consult with FWS or NMFS when an agency action is likely to affect a listed species. Such actions include registration of pesticides.
Section 9	Prohibits the following: 1) "take" of any animal species that is Federally-listed, 2) the removal, damage, or destruction of any listed plant species on areas under Federal jurisdiction, and 3) the removal, damage, or destruction of listed plant species on non-Federal areas in knowing violation of state law or regulations. Taking of a listed animal species can involve harassing, harming, pursuing, hunting, shooting, wounding, killing, trapping, capturing, collecting or attempting any such conduct. This language is essentially derived from the Migratory Bird Treaty Act of 1918, as amended (16 U.S.C. 703 *et seq.*), and the Bald and Golden Eagle Act of 1940, as amended (16 U.S.C. 668 *et seq.*). Use of pesticides must not involve unauthorized take of a listed animal species or cause damage to listed plant species with respect to Federal or non-Federal areas of jurisdiction. Take of listed animal species can only be done under limited circumstances by either an exception (e.g., a permit issued by either FWS or NMFS as defined in Section 10 of the ESA) or in accordance with incidental take provisions of biological opinions as specified in Section 7.
Section 11	Assesses civil penalties under 16 U.S.C. 1540(a) for up to $25,000 for each count of violation in situations where provisions of the ESA are knowingly violated. In cases where unknowing violations occur, a civil penalty may be assessed for up to $500 per count. For criminal violations, imprisonment for up to one year can occur with a fine of up to $50,000 per count (16 U.S.C. 1540(b)). In addition to penalties associated with FIFRA violations, pesticide applicators

may incur penalties under this section when provisions of the ESA are violated.

Further information on the ESA may be found on the Internet at the FWS website (*2*).

Biological Opinions

As part of their responsibility under Section 7 of the ESA, all Federal agencies must consult with the FWS or NMFS whenever they issue a license or permit that may affect listed species. In evaluating pesticides for registration, EPA makes risk assessments for pesticides which includes an analysis of their potential for harming listed species or their critical habitat. If application of a pesticide has no discernible impact on species and protection measures are not required, EPA may request concurrence of FWS on a "no effect" determination. When a "may affect" determination is reached for individual pesticides, EPA is obligated to undertake consultation with FWS (or NMFS in limited cases). The result of the formal consultation is a Biological Opinion issued by FWS on the pesticides in question that finds: 1) no jeopardy and no adverse modification of critical habitat, or 2) jeopardy and/or adverse modification of critical habitat. In making a determination as to whether use of a particular pesticide involves jeopardy for a specific species, FWS evaluates factors such as species exposure, toxicity characteristics, and possible secondary impacts (e.g., indirect toxicity). In cases where pesticide application does not involve jeopardy for a listed animal species or adverse modification of critical habitat, the Biological Opinion will typically have a statement that allows incidental take of the species given normal, expected usage of a particular pesticide. Non-discretionary protective measures called Reasonable and Prudent Measures (RPMs) are stipulated to minimize incidental take during use of the pesticide. RPMs are not stipulated for pesticides affecting plants in a Biological Opinion; however, the Biological Opinion may suggest "Conservation Measures" to minimize impacts. In the event that potential use of a particular pesticide has been found by FWS to jeopardize the continued existence of a listed species (either animal or plant) or have an adverse modification on its critical habitat, a Biological Opinion may contain non-discretionary protective measures called Reasonable and Prudent Alternatives (RPAs). RPAs are considered to be actions that: 1) can be implemented within a Federal agency's legal authority and jurisdiction, 2) are economically and technologically feasible, and 3) can avoid the likelihood of jeopardy to listed species or avoid adverse modification of critical habitat (50 Code of Federal Regulations 402.02). RPAs can be used by EPA to reduce impacts of pesticides to acceptable levels for listed species and thereby allow use of these pesticides in the environment. After the Biological Opinion has been issued, formal consultation is concluded between the two agencies. EPA then translates information obtained from the Biological Opinion into use limitations for individual pesticides.

Biological opinions can be issued on a state, regional, or national level by FWS in response to requests by EPA for consultation on particular pesticides. RPAs and

RPMs in biological opinions are essentially protective measures such as pesticide buffer zones, specific pesticide formulations, daily and seasonal timing of applications, pesticide user education requirements, etc., that are intended to conserve listed species. RPAs are usually more restrictive (e.g., larger buffer zones) than RPMs to preclude jeopardy for a listed species or adverse modification of critical habitat. RPAs or RPMs in a Biological Opinion may be amended by FWS to address pesticide operations in special circumstances. For example, a 1989 Biological Opinion involving the endangered northern aplomado falcon (*Falco femoralis septentrionalis)* was subsequently amended to accommodate pesticide use in the area of the falcon's initial release in Cameron County, Texas after RPAs in the opinion were found to conflict with local agricultural practices.

Historical Background of County Pesticide Bulletins

EPA originally assumed responsibility for registration of pesticides from the U.S. Department of Agriculture (USDA) in the early 1970s and retroactively accepted all existing registrations at that time. After the ESA was implemented in 1973, EPA approached pesticide consultation with FWS on a case-by-case basis at the time of registration (*3*). However, the process was cumbersome and tended to emphasize new pesticide registrations rather than older and potentially more toxic pesticides for which there was no agency action (*3*). As a consequence, EPA devised the "cluster" consultation with FWS where pesticides would be grouped together based on threat levels to listed species and patterns of use. EPA initially consulted with FWS on the five largest use patterns--corn, cotton, sorghum, soybeans and small grains (wheat, oats, barley and rye)--to address approximately 67% of all pesticide usage in the U.S. on a quantity basis. EPA also consulted with FWS on a cluster basis for mosquito larvicides and for pesticides involving forests and rangeland/pastureland. Although EPA switched to a species-based approach to pesticide consultation in 1989 because of public concerns with the cluster approach, biological opinions resulting from the earlier cluster consultations are still used since they provide protective measures for many pesticides.

In 1982, EPA began development of an Endangered Species Protection Program that would implement RPAs in biological opinions by using county-specific pesticide bulletins to communicate information on pesticide use limitations near listed species habitat (*3*). In contrast to providing information for listed species on pesticide labels, the proposed county bulletin format would allow more detailed information on listed species and use limitations that could be updated on a more timely basis. Under this approach, FWS would provide information to EPA for developing habitat maps of listed species and then would review the bulletin maps for accuracy. Use limitations were to be developed by EPA for those pesticides that had been identified in FWS biological opinions as having the potential to harm listed species. The compiled information would then be used in county-specific bulletins which were to have detailed county maps identifying listed species habitat and any pesticide use limitations for areas considered necessary to protect the species. A standard format for county pesticide bulletins would be provided by EPA for counties in the U.S. (see Figure 1-a,b,c).

With its Endangered Species Protection Program, EPA originally proposed that pesticide labels should carry a statement requiring compliance by pesticide applicators with use limitations in county bulletins (*3*). Negative public reaction to this approach resulted in a 1988 amendment to the ESA (Public Law 100-478; 102 U.S.C. 2306) whereby EPA was directed to work in cooperation with FWS and USDA in studying ways of implementing a pesticide labeling program that would minimize impacts on pesticide users while remaining protective of listed species. A joint EPA-FWS-USDA task force modified the pesticide consultation process to emphasize the ranking of listed species according to their vulnerability to pesticides. EPA then proposed a revised Endangered Species Protection Program in 1989 (*3*). The revised program addressed the new consultation process, "may affect" determinations, public notification for upcoming pesticide consultations, and regulatory requirements for pesticide labels and county bulletins; however, final action for implementing the revised program is still pending.

Until the revised Endangered Species Protection Program is finally approved, EPA has implemented an interim protection program. With the interim program, EPA and the states work together to distribute county bulletins to pesticide users. EPA is still required to determine whether incidental take limits set by biological opinions are being exceeded by pesticide users while following bulletin use limitations and must modify the use limitations if warranted. Compliance with use limitations in county bulletins is voluntary. However, pesticide users have considerable incentive to comply with the interim program due to the substantial penalties associated with the ESA. Applicators are protected from the liability of Section 9 of the act for take of a listed animal species or damage to a listed plant species only if the applicator is following bulletin use limitations. Pesticide application not in compliance with a bulletin could be prosecuted pursuant to Section 9.

Both the interim and proposed final protection programs allow states and tribes to develop alternative approaches for protection of listed species. Alternative plans can be used by individual states with concurrence by EPA. In lieu of county bulletins, Wisconsin has taken the approach of developing species protection agreements with landowners because of the limited number of listed species in the state. As part of its species protection plan, Texas has organized regional "coexistence" teams to provide local input to EPA as a basis for county pesticide bulletins. Representatives of agricultural communities, state and Federal agencies, conservation groups, and chemical companies serve on the regional teams.

The greatest limitation with the interim protection program since its inception in 1989 has been the inability of Federal and state agencies to develop and update bulletins for all counties with listed species. Only 300 county bulletins have been developed and distributed by EPA to date (*5*). To facilitate bulletin development, new bulletins are being done in a GIS (geographic information systems) format which provides greater flexibility for revision than earlier bulletin versions.

United States
Environmental Protection
Agency

20T-3063
September 1990

Pesticides And Toxic Substances (H-7506C)

Protecting Endangered Species

Interim Measures

Colorado County, Texas

The information in this pamphlet is similar to what the U.S. Environmental Protection Agency (EPA) expects to distribute once our Endangered Species Protection Program is in effect. The limitations on pesticide use are not law at this time, but are being provided now for your use in voluntarily protecting endangered and threatened species from harm due to pesticide use. We encourage you to use this information. We also welcome your comments.

The Endangered Species Act is intended to protect and promote recovery of animals and plants that are in danger of becoming extinct due to the activities of people. Under the Act, EPA must ensure that use of pesticides it registers will not result in harm to the species listed as endangered or threatened by the U.S. Fish and Wildlife Service, or to habitat critical to those species' survival. To accomplish this, the EPA expects to implement program requirements beginning in 1991. This program will protect endangered and threatened species from harm due to pesticide use.

EPA requests your comments regarding the information presented in this publication. Please drop us a line to let us know whether the information is clear and correct. Also tell us to what extent following the recommended measures would affect your typical pesticide use or productivity. This information will be considered by EPA during the final stages of program development.

Please submit comments to:
Interim Endangered Species Protection Program (H7506C)
Public Docket and Information Section
U.S. EPA
401 M Street, SW
Washington, DC 20460

About This Publication

This publication contains a County Map showing the area within the county where pesticide use should be limited to protect listed species. These areas are identified on the map by a shaded pattern. Each shaded pattern corresponds to a species in need of protection.

The Shading Key shows the name of the species that each shaded pattern represents and describes the shaded area. The area may be described in terms of Township, Range, and Section or by giving details about the habitat of the species.

The first column of the "Table of Pesticide Active Ingredients" lists the active ingredients for which there should be limitations on use to protect certain species. The next columns are headed by the shaded pattern of the species with Codes listed underneath them.

The Code indicates the specific limitation that is necessary to protect the species. The section titled Limitations on Pesticide Use explains the code.

Does This Information Apply To You?

To determine whether this information applies to your use of a pesticide, review the questions below. The information applies only if you answer "yes" to both questions:

- Do you intend to use pesticides within the shaded area on the county map?
- Are any of the ingredients listed on the front panel of your pesticide product label named in the "Table of Pesticide Active Ingredients"?

If you answer "yes" to both questions, you should follow the instructions on "How to Use This Information" to determine if you should limit use of the pesticide to help protect listed species.

If you answer "no" to either question, you should follow the usage directions on the pesticide product label.

Figure 1-a. EPA Pesticide Bulletin for Colorado County, Texas. Face page of bulletin with information on EPA's Endangered Species Protection Program. (Reproduced from reference 4.)

How To Use This Information

1) On the county map, find the specific shading patterns that cover the area where you will apply pesticides.

2) Read the shading key for those patterns to identify the specific area involved.

3) In the "Table of Pesticide Active Ingredients," locate the active ingredients in the pesticide you intend to apply.

4) Locate the code to the right of the active ingredient name and under the shading patterns that apply to you.

5) When using the pesticide, you should follow the limitations indicated for those codes described under "Limitations on Pesticide Use."

6) If you are applying more than one listed active ingredient or applying a listed active ingredient in an area with more than one shaded pattern (species), multiple codes may apply. If so, you should follow the most restrictive limitation.

Table Of Pesticide Active Ingredients

Active Ingredient	Shading Pattern/Code
4-AMINOPYRIDINE (4-AP)	31
ALDICARB	31
AZINPHOS-METHYL	31
CARBOFURAN	31
CHLORPYRIFOS	31
DIAZINON	31
DICROTOPHOS	31
DIMETHOATE	31
ENDOSULFAN	31
ENDRIN	31
EPN	31
ETHOPROP (GRANULAR)	31
FENAMIPHOS	31
FENSULFOTHION	31
FONOFOS	31
ISOFENPHOS	31
METHYL PARATHION	31
MEVINPHOS	31
OXAMYL	31
OXYFLUORFEN	31
PARAQUAT	31
PARATHION (ETHYL PARATHION)	31
PHORATE (GRANULAR)	31
TOXAPHENE (CAMPHECHLOR)	31
ZINC PHOSPHIDE	28

Limitations On Pesticide Use

Code	Limitation
28	Do not apply within 100 yards of species habitat for aerial applications or within 20 yards of species habitat for ground applications.
31	Do not apply within one-quarter mile of species habitat on any crop, except as approved in the following seasonal chart:

CROP	JAN	FEB	MAR	APR	MAY	JUN	JUL	AUG	SEP	OCT	NOV	DEC
SOYBEANS	A	A	A	A	A	A	-	-	-	A	A	A
SORGHUM	A	-	-	-	-	-	-	-	A	A	A	A
WHEAT, BARLEY	-	-	-	A	A	A	A	A	A	-	-	-

A = Use is approved during the month indicated.

Figure 1-b. EPA Pesticide Bulletin for Colorado County, Texas. Use limitations for pesticide active ingredients that are commonly used in the county. (Reproduced from reference 4.)

Colorado County, Texas

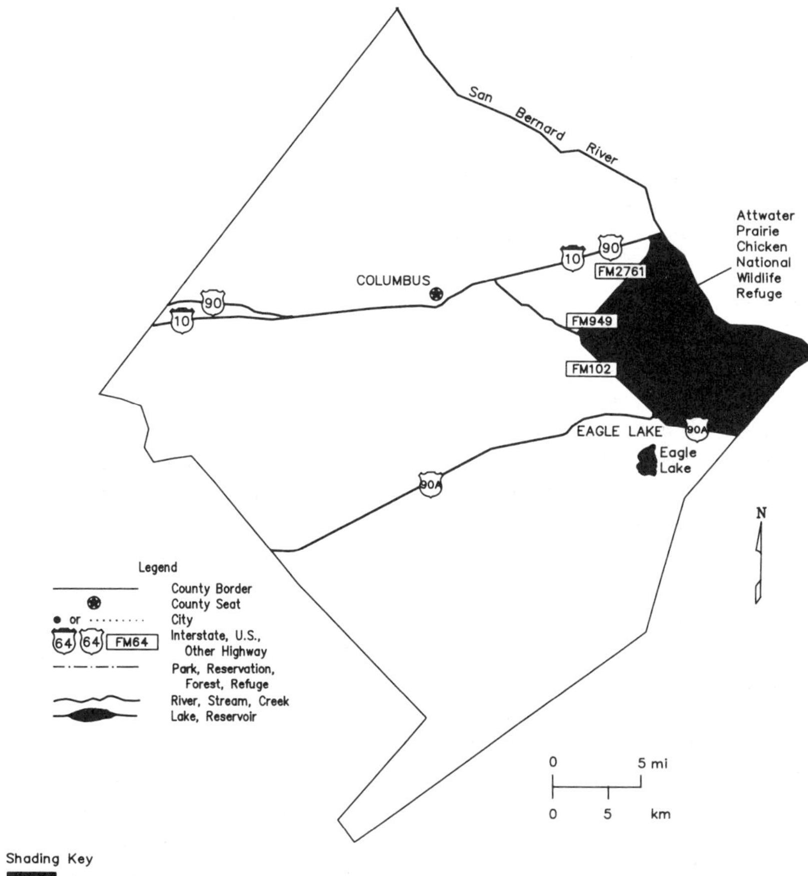

Shading Key

Attwater's greater prairie-chicken, *Tympanuchus cupido attwateri*. The shaded area shown on the map is bound on the southwest by FM 102; on the northwest by FM 949, FM 2761 and Interstate 10/U.S. Highway 90; on the northeast by the San Bernard River; on the southeast by the Wharton County line; and on the south by U.S. Highway 90A. Use limitations only apply within this area.

Figure 1-c. EPA Pesticide Bulletin for Colorado County, Texas. Map of the county detailing area where use limitations on pesticides are to be implemented with respect to listed species- Attwater's greater prairie-chicken, Tympanachus cupido attwateri. (Reproduced from reference 4.)

Considerations in Developing County Pesticide Bulletins

Biological opinions to date have not been able to completely address all listed species or pesticides since the consultation process between EPA and FWS lags both listing of new species and registration of new pesticides. EPA has approached this problem in part by developing use limitations for county bulletins that are essentially modifications of protective measures in RPAs and RPMs. Pesticides with comparable toxicological characteristics are given the same use limitations in a county bulletin when applied in proximity to species that share similar features of taxonomy, habitat, etc.

Habitat mapping for listed species has been an area of difficulty in developing county bulletins. Biological opinions can provide other elements of county bulletins (i.e., pesticides, species, and protective measures), but they generally do not contain habitat maps showing the distribution of species within a given county. Although FWS may provide habitat maps as an addendum to biological opinions or as part of informal consultations, development of habitat maps has generally been left to EPA and state agencies. Bulletin habitat maps are usually based on local knowledge of a species by area biologists rather than derived from detailed surveys. However, locations for different populations of a species and their habitat are not always known by biologists with FWS, state wildlife agencies, or conservation groups. Migratory birds or other highly mobile species can be particularly difficult to locate with respect to their current habitat. In some cases, biologists may be reluctant to reveal precise details of habitat areas for fear that species may be harmed by illegal collection or other forms of trespass.

A major consideration encountered with habitat mapping for county bulletins is the delineation of areas requiring pesticide use limitations. To protect the endangered salt marsh bird's-beak plant species (*Cordylanthus maritimus ssp. maritimus*) from runoff of herbicides in upland agricultural areas, an entire watershed ranging from salt marshes on California's coast to the Pacific Coastal Mountain Range was initially considered but was eventually deemed to be impractical. EPA's present policy on county bulletin maps is to depict and protect only habitat that is known to be currently occupied, but this approach may be too restrictive to provide adequate protection for some species. Maps showing only occupied habitat cannot account for unsurveyed populations of a listed species nor do they reflect potential habitat (e.g., border areas) that is needed for habitat expansion and recovery. Listed plant species in particular may be negatively impacted by the occupied habitat approach by not allowing for reestablishment through dispersal of seed or other types of propagules.

Another concern encountered with habitat mapping is the possible lack of topographical features that can be used to spatially locate boundaries of species habitat. Even with small scale maps, landscapes involving listed species may offer few landmarks to readily locate habitat areas. In the past, EPA has attempted to address this problem by generalizing habitat maps to the nearest road, river, or other easily recognizable landmark boundary. However, this approach may potentially lead to gross overestimates of actual habitat in remote, featureless areas. For the threatened Kern primrose sphinx moth (*Euproserpinus euterpe)* in California's Kern County, generalizing to the nearest landmark roads on a habitat map would encompass an area exceeding 100,000 acres (247,000 ha) although the species actually inhabits a single acre (2.5 ha) within the county. Lack of identifiable landmarks has required the use of a township-range-section approach in some states to approximate areas with pesticide use limitations, but the general public may not always understand this concept.

Technical depiction of habitat areas in county bulletin maps can be difficult if the ranges of different species overlap. Various shading patterns are typically used in habitat maps to depict overlap of species ranges when the habitat maps are overlaid on the base map of a particular county. Although the shading pattern approach is workable in county bulletin maps if a small number of species are involved, it becomes cumbersome when more than three species occupy the same area.

Finally, updating habitat maps is itself problematical given the necessary mapping revisions for new listings, newly found habitat, etc. EPA originally proposed to update county pesticide bulletins on an annual basis, but this has been unworkable given the complexities of bulletin formation. In California, habitat data evaluations are typically required every five years; however, resources are seldom adequate to support comprehensive reviews of available information much less perform species surveys during that interval. Habitat data that is routinely compiled into natural heritage databases or other similar databases can provide useful information to facilitate bulletin updates. The GIS approach in bulletin formation should also make updating bulletins more timely as new information on species and pesticides becomes available.

Distribution and Public Use of County Pesticide Bulletins

County bulletins are intended to be widely available to pesticide users through a variety of channels. Distribution of bulletins at the point of sale (i.e., pesticide retailers) is possibly the most effective means; however, bulletin distribution for individual states can vary. The lead agency for bulletin development in each state (usually an agricultural or environmental agency) is responsible for determining the best methods of distribution. Most states choose to involve pesticide retailers and county extension agents as sources for bulletins. Other approaches include direct mail and use of state field personnel. In California, almost all field-use rodenticides are sold directly to a pesticide user by a local county agricultural commissioner who is able to provide county bulletins as well as discuss proper application techniques for protection of non-target species.

Use of county pesticide bulletins may vary according to the public's perception for the necessity of protecting listed species. The public has responded to protecting well-publicized listed species such as the bald eagle (*Haliaeetus leucocephalus*) but has not been as responsive in protecting unfamiliar or reclusive species such as reptiles, amphibians, insects, etc. Pesticide users in particular may be reluctant to implement protective measures in county bulletins such as buffer zones that they perceive as uneconomical or impractical. In California, a surcharge collected on rodenticide sales is used in part to fund research studies on alternative protective measures for use in county bulletins while allowing such pesticides to remain available for targeted pest species.

One misconception with county bulletins is that habitat maps indicate the physical presence of listed species on private property. Regulatory agencies have spent considerable effort in explaining to the public that habitat maps are only subjective guides for purposes of management rather than an actual indication of species presence. Another difficulty associated with county bulletins has been a confusion by the public as to precisely what species are protected with a bulletin. Individual states may have their own lists of endangered species and provide protection for these species under state law. The term "endangered" itself is used loosely in reference to both endangered and threatened species. As a matter of practicality, county bulletins promulgated by EPA are restricted to Federally-listed species. The state wildlife agency in California provides supplemental pages of protective measures for state-listed species for distribution with EPA county bulletins.

County bulletins can provide benefits beyond protection of listed species. Bulletins may serve as a focal point for sensitizing the public to pesticide use in regard to non-target, unlisted species in addition to listed species. In some states, use limitations in county bulletins reinforce local or state restrictions on pesticides. County bulletins may also benefit the pesticide applicator by specifying the most useful techniques of application and preventing uneconomical misapplications of certain pesticides.

County Pesticide Bulletins on the Internet

County pesticide bulletins can be shown over the Internet, and EPA currently displays county bulletins for a number of states at its website (*6*). Internet county bulletins have the advantages of not requiring a paper copy inventory that quickly becomes obsolete and of being more readily disseminated with updated information in regard to pesticide registrations, use limitations, etc. However, Internet county bulletins may not always be readily accessible to all potential users of pesticides and may require supplementation with paper copy versions at points of sale for pesticides.

As a refinement of the Internet county bulletin, the state of California has developed an interactive database application that produces a customized set of use limitations for specific pesticides in a given township-range-section as selected by the pesticide applicator (*7*). Regulatory advantages for this approach include: 1) avoidance of the problem of deriving specific use limitations from habitat maps with various shading patterns, 2) less ambiguity in explaining use limitations for a particular pesticide application, and 3) reduction in unnecessary reading for the pesticide applicator. The process of communicating use limitations during actual field operations is also simplified since the website's output can be used as a work order attachment for pesticide applicators or as a written recommendation from pest control advisors.

Conclusions

Although county pesticide bulletins were designed to be extensions of FIFRA pesticide labels, the labels themselves ordinarily do not refer pesticide users to county bulletins. Pesticide labels sometimes direct users to consult with FWS or state wildlife agencies for approval of individual pesticide applications, but such referrals are generally unrealistic and have the potential for negating previous pesticide consultations under the ESA. In some cases, a registrant or EPA has amended a pesticide label to include protective measures for listed species in addition to the general protective measures found in the label's Environmental Hazards section. This approach may be required when use limitations have not been developed from biological opinions or else are not listed in a county bulletin. Inclusion of protective measures for listed species is particularly necessary for pesticide labels that involve FIFRA Section 18 emergency exemptions or Section 24(c) special local needs on a state-specific basis. However, placement of protective measures for listed species on pesticide labels has proved to be impractical in general. Such information tends to be out of date and usually does not provide sufficient geographical context for determination of species habitat. Because of limitations with the pesticide label format, the county pesticide bulletin is the most effective means currently available for informing pesticide users of protective measures for listed species. Given the gradual progress in bulletin availability, pesticide labels should contain language that refers users to county bulletins for more explicit information on species and their habitat. Development of bulletins should also be given a higher priority by regulatory agencies due to the potential liability for pesticide users pursuant to Section 9 of the ESA.

A major problem in developing county pesticide bulletins has been the lack of a comprehensive regulatory approach in providing required elements for the bulletins, i.e., species lists, habitat maps, commonly used pesticides, and protective measures. The current process of consultation and development of biological opinions by Federal agencies has allowed only partial coverage of these necessary elements. As an alternative, a joint EPA-FWS task force could be formed to establish specific criteria for bulletin elements. In particular, such a task force could facilitate development of use limitations by defining criteria by which species and pesticides can be grouped according to similarities in required RPAs and RPMs.

We advocate the Internet approach to county bulletins because of its flexibility and potential for providing pesticide applicators with timely information on currently listed species, habitat, and necessary protective measures. To inform pesticide users who are not trained on computers or else lack computer access, output from a bulletin website could be distributed as paper copy bulletins at points of sale for pesticides and other alternative locations such as pesticide training facilities. Regulatory agencies may need to adopt a multi-jurisdictional strategy to facilitate Internet county bulletins. For a given state, an interagency agreement could be implemented whereby individual Federal and state agencies would be responsible for maintaining and/or updating different components of the state's Internet bulletins.

Disclaimer: This paper does not represent official policy of the regulatory agencies discussed herein.

Acknowledgments

The authors wish to thank Larry Turner of EPA and Kathryn Kennedy of FWS for reviewing this paper.

References

1. *Box Score-Listings and Recovery Plans as of September 30, 1999,* URL http://endangered.fws.gov/boxscore.html.
2. *Endangered Species Act of 1973*, URL http://endangered.fws.gov/esa.html.
3. US Environmental Protection Agency. Endangered species protection program: notice of proposed program; Federal Reg., Monday, July 3, 1989, Vol. 54, No. 126, pp. 27984-28008.
4. US Environmental Protection Agency (H7506C). *Protecting Endangered Species: Interim Measures: Colorado County, Texas;* Washington, D.C., Sept. 1990.
5. Turner, L. Personal communication; US Environmental Protection Agency, 1999.
6. *Endangered Species Protection Program (ESPP)*, URL http:www.epa.gov/espp/.
7. *Endangered Species Project: Species-Section Database and Index to Pesticide Use Limitations*, URL http://www.cdpr.ca.gov/docs/es/db_desc.htm.

Author Index

Subject Index

B

C

D

G

H

I

M

P

Q

R

S

T

U

V

W